SCHAUM'S OUTLINE OF

THEORY AND PROBLEMS

of

HEAT TRANSFER

•

DONALD R. PITTS, Ph.D.

Professor of Mechanical Engineering
Tennessee Technological University

and

LEIGHTON E. SISSOM, Ph.D., P.E.

Professor and Chairman
Department of Mechanical Engineering
Tennessee Technological University

SCHAUM'S OUTLINE SERIES

McGRAW-HILL, INC.

New York St. Louis San Francisco Auckland Bogotá Caracas Lisbon
London Madrid Mexico City Milan Montreal New Delhi
San Juan Singapore Sydney Tokyo Toronto

0-07-050203-X

19 20 21 22 23 24 BAW BAW 9 8 7 6 5 4

 This book is printed on recycled paper containing a minimum of 50% total recycled fiber with 10% postconsumer de-inked fiber. Soybean based inks are used on the cover and text.

Library of Congress Cataloging in Publication Data

Pitts, Donald R.
  Schaum's outline of theory and problems of heat transfer.

  (Schaum's outline series)
  Includes index.
    1. Heat—Transmission.   2. Heat—Transmission
—Problems, exercises, etc.   I. Sissom, Leighton E.,
joint author.   II. Title.   III. Title: Theory
and problems of heat transfer.

QC320.P55   621.4'022   77-20255
ISBN 0-07-050203-X

Cover design by Amy E. Becker.

Preface

The topic "Heat Transfer" covers a wide range of subtopics. Besides treatments of conductive, convective and radiative transport of thermal energy, the subject is often assumed to include heat transfer involving phase change and applications to heat exchanger designs. To some people, the approach to heat transfer centers on a mathematical derivation of the thermal transport rate equations; to others, the treatment focuses upon the ability to apply a collection of equations which are in the main empirical. In this Schaum's Outline we have tried to achieve a compromise between these extreme viewpoints, and it is believed that the result will serve a variety of individuals.

In general, each chapter begins with a clear definition of the class of heat transfer problems introduced. In formulating the specific governing equation, particular attention is given to a correct application of the first law of thermodynamics to an appropriate control volume and specification of a complete set of boundary and initial conditions. The treatment generally proceeds to development of useful working relationships, whether analytically or experimentally obtained; these expressions are highlighted with a bullet ■.

Each chapter contains a large number of solved problems, approximately 60% of them in the British Engineering system of units with the remainder being in SI units. Numerous derivations are also included in the solved problems. The additional Supplementary Problems with answers serve to provide a review of the chapter.

The book has been designed either as a supplement to current textbooks or as a textbook for the junior- or senior-year college course in heat transfer. It should be a useful aid to all engineering and physics students studying this subject, as well as to practicing engineers and scientists whose work includes heat transfer calculations. Considerably more material for each major topic has been included than can be covered in most single courses.

We wish to express our appreciation to the staff of the Schaum Paperback Division of McGraw-Hill Book Company, and to editor David Beckwith in particular, for their splendid efforts.

D. R. Pitts
L. E. Sissom

Contents

CONTENTS

CONTENTS

Chapter 1

Introduction

The engineering area frequently referred to as *thermal science* includes *thermodynamics* and *heat transfer*. The role of heat transfer is to supplement thermodynamic analyses, which consider only systems in equilibrium, with additional laws that allow prediction of time rates of energy transfer.

These supplemental laws are based upon the three fundamental modes of heat transfer, namely *conduction*, *convection*, and *radiation*.

1.1 CONDUCTION

A temperature gradient within a homogeneous substance results in an energy transfer rate within the medium which can be calculated by

$$q = -kA\frac{\partial T}{\partial n} \tag{1.1}$$

where $\partial T/\partial n$ is the temperature gradient in the direction normal to the area A. The *thermal conductivity* k is an experimental constant for the medium involved, and it may depend upon other properties such as temperature and pressure, as discussed in Section 1.4. The units of k are Btu/hr-ft-°F or W/m-K. (For units systems, see Section 1.5.)

The minus sign in *Fourier's law*, (1.1), is required by the second law of thermodynamics: thermal energy transfer resulting from a thermal gradient must be from a warmer to a colder region.

If the temperature profile within the medium is linear (Fig. 1-1), it is permissible to replace the temperature gradient (partial derivative) with

$$\frac{\Delta T}{\Delta x} = \frac{T_2 - T_1}{x_2 - x_1} \tag{1.2}$$

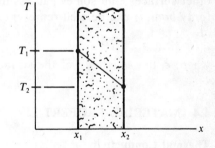

Fig. 1-1

Such linearity always exists in a homogeneous medium of fixed k during *steady-state* heat transfer.

Steady-state transfer occurs whenever the temperature at every point within the body, including the surfaces, is independent of time. If the temperature changes with time, energy is either being stored in or removed from the body. This storage rate is

$$q_{\text{stored}} = mc_p\frac{\partial T}{\partial x} \tag{1.3}$$

where the mass m is the product of volume V and density ρ.

1.2 CONVECTION

Whenever a solid body is exposed to a moving fluid having a temperature different from that of the body, energy is carried or *convected* away by the fluid.

1

If the upstream temperature of the fluid is T_∞ and the surface temperature of the solid is T_s, the heat transfer per unit time is given by

$$q = hA(T_s - T_\infty) \qquad (1.4)$$

which is known as *Newton's law of cooling*. This equation defines the *convective heat-transfer coefficient h* as the constant of proportionality relating the heat transfer per unit time and unit area to the overall temperature difference. The units of h are Btu/hr-ft^2-°F or W/m^2-K. It is important to keep in mind that the fundamental energy exchange at a solid-fluid boundary is by conduction, and that this energy is then convected away by the fluid flow. By comparison of (1.1) and (1.4), we obtain, for $y = n$,

$$hA(T_s - T_\infty) = -kA\left(\frac{\partial T}{\partial y}\right)_s \qquad (1.5)$$

where the subscript on the temperature gradient indicates evaluation in the fluid at the surface.

1.3 RADIATION

The third mode of heat transmission is due to electromagnetic wave propagation, which can occur in a total vacuum as well as in a medium. Experimental evidence indicates that radiant heat transfer is proportional to the fourth power of the absolute temperature, whereas conduction and convection are proportional to a linear temperature difference. The fundamental *Stefan-Boltzmann law* is

$$q = \sigma A T^4 \qquad (1.6)$$

where T is the absolute temperature. The constant σ is independent of surface, medium, and temperature; its value is 0.1714×10^{-8} Btu/hr-ft^2-°R^4 or 5.6697×10^{-8} W/m^2-K^4.

The ideal emitter, or *blackbody*, is one which gives off radiant energy according to (1.6). All other surfaces emit somewhat less than this amount, and the thermal emission from many surfaces (*gray bodies*) can be well represented by

$$q = \epsilon \sigma A T^4 \qquad (1.7)$$

where ϵ, the *emissivity* of the surface, ranges from zero to one.

1.4 MATERIAL PROPERTIES

Thermal Conductivity of Solids

Thermal conductivities of numerous pure metals and alloys are given in Table B-1, Appendix B. The thermal conductivity of the solid phase of a metal of known composition is primarily dependent only upon temperature. In general, k for a pure metal decreases with temperature; alloying elements tend to reverse this trend.

The thermal conductivity of a metal can usually be represented over a wide range of temperature by

$$k = k_0(1 + b\theta + c\theta^2) \qquad (1.8)$$

where $\theta = T - T_{ref}$ and k_0 is the conductivity at the reference temperature T_{ref}. For many engineering applications the range of temperature is relatively small, say a few hundred degrees, and

$$k = k_0(1 + b\theta) \qquad (1.9)$$

The thermal conductivity of a nonhomogeneous material is usually markedly dependent upon the *apparent bulk density*, which is the mass of the substance divided by the total volume occupied. This total volume includes the void volume, such as air pockets within the overall boundaries of the piece of material. The conductivity also varies with temperature. As a general rule, k for a nonhomogeneous material increases both with increasing temperature and increasing apparent bulk density. Table B-2, Appendix B, contains thermal conductivity data for non-homogeneous materials.

Thermal Conductivity of Liquids

Table B-3, Appendix B, lists thermal conductivity data for some liquids of engineering importance. For these, k is usually temperature dependent but insensitive to pressure. The data of this table are for saturation conditions, i.e. the pressure for a given fluid and given temperature is the corresponding saturation value. Thermal conductivities of most liquids decrease with increasing temperature. The exception is water, which exhibits increasing k up to about 300 °F and decreasing k thereafter. Water has the highest thermal conductivity of all liquids except the so-called liquid metals.

Thermal Conductivity of Gases

The thermal conductivity of a gas increases with increasing temperature but is essentially independent of pressure for pressures close to atmospheric. Table B-4, Appendix B, presents k-data for several gases at atmospheric pressure. For high pressure (i.e. pressure of the order of the critical pressure or greater), the effect of pressure is significant. The generalized chart of Figure B-3, Appendix B, allows approximation of the thermal conductivity at elevated pressure. This chart should be used only in the absence of specific conductivity data at high pressure.

Two of the most important gases are air and steam. (No distinction is made between a gas and a vapor in this chapter.) For air the atmospheric values listed in Table B-4 are suitable for most engineering purposes over the ranges: (i) $32\,°F \leq T \leq 3000\,°F$ and $1\,atm \leq p \leq 100\,atm$; (ii) $-100\,°F \leq T \leq 32\,°F$ and $1\,atm \leq p \leq 10\,atm$. Figure B-3 should not be used for air, as it results in significant error (overcorrection).

Thermal conductivity data for steam exhibit a strong pressure dependence. For approximate calculations the atmospheric data of Table B-4 may be used together with Figure B-3.

Density

Density is defined as the mass per unit volume. All systems considered in this book will be sufficiently large for statistical averages to be meaningful; that is, we will consider only a *continuum*, which is a region with a continuous distribution of matter. For systems with variable density we define *density at a point* (a specific location) as

$$\rho \equiv \lim_{\delta V \to \delta V_c} \frac{\delta m}{\delta V} \tag{1.10}$$

where δV_c is the smallest volume for which a continuum has meaning.

Density data for most solids and liquids are only slightly temperature dependent and are negligibly influenced by pressure up to 100 atm. Density data for solids and liquids are presented in Tables B-1, B-2, and B-3. The density of a gas, however, *is strongly dependent upon the pressure* as well as upon the temperature. In the absence of specific gas data the atmospheric density of Table B-4 may be modified by application of the ideal gas law:

$$\rho = \rho_1 \left(\frac{p}{p_1}\right) \tag{1.11}$$

The *specific volume* is the reciprocal of the density,

$$v = \frac{1}{\rho} \tag{1.12}$$

and the *specific gravity* is the ratio of the density to that of pure water at a temperature of 4 °C and a pressure of one atmosphere (760 mmHg). Thus

$$S = \frac{\rho}{\rho_w} \tag{1.13}$$

where S is the specific gravity.

Specific Heat

The *specific heat* of a substance is a measure of the variation of its stored energy with temperature. From thermodynamics the two important specific heats are:

$$\text{specific heat at constant volume:} \quad c_v \equiv \frac{\partial u}{\partial T}\bigg|_v \tag{1.14}$$

$$\text{specific heat at constant pressure:} \quad c_p \equiv \frac{\partial h}{\partial T}\bigg|_p \tag{1.15}$$

Here u is the internal energy per unit mass and h is the enthalpy per unit mass. In general, u and h are functions of two variables: temperature and specific volume, and temperature and pressure, respectively. For substances which are incompressible, i.e. solids and liquids, c_p and c_v are numerically equal. For gases, however, the two specific heats are considerably different. The units of c_v and c_p are Btu/lbm-°F or J/kg-K.

For solids, specific-heat data are only weakly dependent upon temperature and even less affected by pressure. It is usually acceptable to use the limited c_p-data of Tables B-1 and B-2 over a fairly wide range of temperatures and pressures.

Specific heats of liquids are even less pressure dependent than those of solids, but they are somewhat temperature influenced. Data for some liquids are presented in Table B-3.

Gas specific-heat data exhibit a strong temperature dependence. The pressure effect is slight except near the critical state, and the pressure dependence diminishes with increasing temperature. For most engineering calculations the data of Table B-4 can be used for pressures up to 200 psia.

Thermal Diffusivity

A useful combination of terms already considered is the *thermal diffusivity* α, defined by

$$\alpha \equiv \frac{k}{\rho c_p} \tag{1.16}$$

It is seen that α is the ratio of the thermal conductivity to the thermal capacity of the material. Its units are ft²/hr or m²/s. Thermal energy diffuses rapidly through substances with high α and slowly through those with low α.

Some of the tables of Appendix B list thermal diffusivity data. Note the strong dependence of α for gases upon both pressure and temperature; these data for gases are only for atmospheric pressure, and they are only valid for the specified temperature.

Viscosity

The simplest flow situation involving a real fluid, i.e. one that has a nonzero viscosity, is laminar flow along a flat wall (Fig. 1-2). In this model, fluid layers slide parallel to one another, the molecular layer adjacent to the wall being stationary. The next layer out from the wall slides along this stationary layer, and its motion is impeded or slowed because of the frictional shear between these layers. Continuing outward, a distance is reached where the retardation of the fluid due to the presence of the wall is no longer evident.

Consider plane P–P. The fluid layer immediately below this plane has velocity $u - \delta u$, and the fluid layer immediately above has velocity $u + \delta u$. Here u is the value of the velocity in the x-direction at the y-location of the P–P plane. The difference in velocity between these two adjacent fluid layers produces a shear stress τ. Newton postulated that this stress is directly proportional to the velocity gradient normal to the plane:

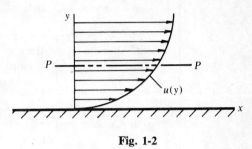

Fig. 1-2

$$\tau = \mu_f \frac{du}{dy} \tag{1.17}$$

The coefficient of proportionality is called the *coefficient of dynamic viscosity*, or more simply the *dynamic* or *absolute viscosity*.

Viscosity units. As shown by (1.17), the units of μ_f are lbf-sec/ft^2 or N-s/m^2. In many applications it is convenient to have the dynamic viscosity expressed in terms of a mass, rather than a force, unit. In this book, μ_m will denote the mass-based viscosity coefficient. In the SI system the units of μ_m are kg/m-s, and μ_m is numerically equal to μ_f. In the British Engineering system μ_m has units lbm/ft-sec, and, numerically, $\mu_m = (32.17)\mu_f$. In those contexts in which the units are irrelevant we shall simply write μ for the dynamic viscosity.

For gases and liquids dynamic viscosity is markedly dependent upon the temperature but rather insensitive to pressure; data are presented in Tables B-3 and B-4.

As in the case of gas thermal conductivity, gas dynamic viscosity is pressure dependent at pressures approaching the critical value, or greater. The generalized chart of Figure B-4 may be used in the absence of specific gas viscosity data at high pressure. For air, however, the variation of μ with pressure is negligible for most engineering problems; in particular, use of the generalized chart will seriously overcorrect the viscosity.

The ratio of dynamic viscosity to density is called the *kinematic viscosity* ν:

$$\nu \equiv \frac{\mu_m}{\rho} \tag{1.18}$$

The units of ν are ft^2/sec or m^2/s.

Warning: Unlike the dynamic viscosity, the kinematic viscosity is *strongly pressure dependent* (because the density is). The data of Table B-4 are for 1 atm only; they must be modified for use at higher pressure (if used at all).

1.5 UNITS

Table 1-1 summarizes the units systems in common use. The proportionality constant g_c in Newton's second law of motion,

$$F = \frac{1}{g_c} ma \tag{1.19}$$

is given in the last column.

In this book the SI and British Engineering systems will be employed. For convenience, conversion factors from non-SI into SI units are given in Appendix A.

Table 1-1

Units System	Defined Units	Derived Units	Proportionality Constant, g_c
Metric Absolute	Mass, g Length, cm Time, sec Temp., °K	Force: $dyne = \dfrac{g\text{-}cm}{sec^2}$	$1\,\dfrac{g\text{-}cm}{dyne\text{-}sec^2}$
English Absolute	Mass, lb Length, ft Time, sec Temp., °R	Force: $poundal = \dfrac{lbm\text{-}ft}{sec^2}$	$1\,\dfrac{lbm\text{-}ft}{poundal\text{-}sec^2}$
British Technical	Force, lbf Length, ft Time, sec Temp., °R	Mass: $slug = \dfrac{lbf\text{-}sec^2}{ft}$	$1\,\dfrac{slug\text{-}ft}{lbf\text{-}sec^2}$
British Engineering	Force, lbf Mass, lbm Length, ft Time, sec Temp., °R	(None)	$32.17\,\dfrac{lbm\text{-}ft}{lbf\text{-}sec^2}$
International System (SI)	Length, m Mass, kg Time, s Temp., K	Force: $newton\ (N) = \dfrac{kg\text{-}m}{s^2}$	$1\,\dfrac{kg\text{-}m}{N\text{-}s^2}$

Solved Problems

1.1. Determine the steady-state heat transfer per unit area through a 1.5-in.-thick homogeneous slab with its two faces maintained at uniform temperatures of 100 °F and 70 °F. The thermal conductivity of the material is 0.11 Btu/hr-ft-°F.

The physical problem is shown in Fig. 1-3. For the steady state, (*1.1*) and (*1.2*) combine to yield

$$\frac{q}{A} = -k\left(\frac{T_2 - T_1}{x_2 - x_1}\right)$$

$$= -\frac{0.11\ \text{Btu}}{\text{hr-ft-}°\text{F}}\left(\frac{70 - 100}{1.5/12}\right)\frac{°\text{F}}{\text{ft}}$$

$$= +26.40\ \frac{\text{Btu}}{\text{hr-ft}^2}$$

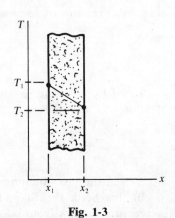

Fig. 1-3

1.2. The forced convective heat-transfer coefficient for a hot fluid flowing over a cool surface is 40 Btu/hr-ft²-°F for a particular problem. The fluid temperature upstream of the cool surface is 250 °F, and the surface is held at 50 °F. Determine the heat transfer per unit surface area from the fluid to the surface.

Equation (*1.4*) applies to this problem, but the sign must be reversed to represent heat flux *from* the surface *to* the fluid. Thus

$$\frac{q}{A} = h(T_\infty - T_s) = \frac{40 \text{ Btu}}{\text{hr-ft}^2\text{-}°\text{F}} [(250 - 50) °\text{F}] = 8000 \frac{\text{Btu}}{\text{hr-ft}^2}$$

1.3. After sunset, radiant energy can be sensed by a person standing near a brick wall. Such walls frequently have surface temperatures around 110 °F, and typical brick emissivity values are on the order of 0.92. What would be the radiant thermal flux per square foot from a brick wall at this temperature?

Equation (*1.7*) applies.

$$\frac{q}{A} = \epsilon\sigma T^4 = (0.92)(0.1714 \times 10^{-8} \text{ Btu/hr-ft}^2\text{-}°\text{R}^4)[(110 + 459.7)^4 °\text{R}^4]$$

$$= (0.92)(0.1714 \times 10^{-8})(5.697)^4(10)^8 \text{ Btu/hr-ft}^2 = 166.11 \text{ Btu/hr-ft}^2$$

Note that absolute temperature must be used in all radiant energy calculations.

1.4. Determine the thermal conductivity of gaseous hydrogen at 500 °F and a pressure of 15 atm.

For this pressure one would suspect that the value at 500 °F and 1 atm given in Table B-4 would require pressure modification. From Table B-4, by linear interpolation,

$$k_1 \approx 0.165 \text{ Btu/hr-ft-}°\text{F}$$

where the subscript 1 denotes the pressure is 1 atm.

Table B-5, Appendix B, gives critical constants for various gases. From the table, $p_c = 12.8$ atm and $T_c = 59.9$ °F. Thus, the reduced pressure and temperature are

$$P_r = \frac{15}{12.8} = 1.172 \qquad T_r = \frac{500 + 459.7}{59.9} = 16$$

From Figure B-3 at $T_r \approx 16$, we see that $k/k_1 \approx 1.0$ and so $k \approx 0.165$ Btu/hr-ft-°F. The absence of pressure effect in this case is due to the relatively high temperature.

1.5. Estimate the maximum value of thermal conductivity in W/m-K for air at 273.15 K and 13.789×10^6 N/m².

In order to use Table B-5, it will be convenient to convert from SI temperature and pressure units to British Engineering units. Using Table A-1,

$$T \text{ °R} = \frac{9}{5}(273.15 \text{ K}) = 491.7 °\text{R}$$

$$p = (13.789 \times 10^6 \text{ N/m}^2)\left(\frac{\text{atm}}{1.01325 \times 10^5 \text{ N/m}^2}\right) = 136.1 \text{ atm}$$

From Table B-5, the critical state is $p_c = 37.2$ atm; $T_c = 238.4$ °R. Thus,

$$P_r = \frac{136.1}{37.2} \approx 3.66 \qquad T_r = \frac{491.7}{238.4} \approx 2.06$$

From Figure B-3 (which probably overcorrects for air), $k/k_1 \approx 1.37$. From Table B-4, $k_1 \approx$ 0.0139 Btu/hr-ft-°F. With the aid of Table A-2,

$$k \approx 0.0139 \times 1.37 \frac{\text{Btu}}{\text{hr-ft-°F}} \times \frac{1.72957 \text{ J/m-s-K}}{\text{Btu/hr-ft-°F}} = 0.0330 \text{ J/m-s-K}$$

or 0.0330 W/m-K, since 1 watt = 1 joule/second.

1.6. Determine k in British Engineering units (Btu/hr-ft-°F) for nitrogen gas at 80 °F and 2000 psia. (This is a common bottle pressure for commercial sales.)

From Table B-5, the critical state is $p_c = 33.5$ atm; $T_c = 226.9$ °R. Thus,

$$P_r = \frac{(2000/14.7) \text{ atm}}{33.5 \text{ atm}} = 4.06 \qquad T_r = \frac{539.7 \text{ °R}}{226.9 \text{ °R}} = 2.379$$

From Figure B-3, $k/k_1 \approx 1.25$. From Table B-4, $k_1 = 0.01514$ Btu/hr-ft-°F, so

$$k = 0.01514 \times 1.25 = 0.0189 \text{ Btu/hr-ft-°F}$$

1.7. What is the approximate density of air at 60 psia and 350 °F?

From Table B-4 the density of air at 1 atm and 300 °F is $\rho_1 = 0.0489$ lbm/ft. By (*1.11*),

$$\rho = \rho_1\left(\frac{p}{p_1}\right) = \left(0.0489 \frac{\text{lbm}}{\text{ft}^3}\right)\frac{(60/14.7) \text{ atm}}{1 \text{ atm}} = 0.1996 \text{ lbm/ft}^3$$

1.8. A manometer fluid has a specific gravity $S = 2.95$. Determine the density of this fluid in (*a*) lbm/ft^3, (*b*) kg/m^3.

(*a*) From Table B-3, $\rho_w \approx 62.4$ lbm/ft^3 at ambient temperature. By (*1.13*),

$$\rho = S\rho_w = (2.95)(62.4) \text{ lbm/ft}^3 = 184.08 \text{ lbm/ft}^3$$

(*b*) Using the density conversion factor from Table A-4,

$$\rho = \left(184.08 \frac{\text{lbm}}{\text{ft}^3}\right)\left(\frac{16.0184 \text{ kg/m}^3}{\text{lbm/ft}^3}\right) = 2948.67 \text{ kg/m}^3$$

1.9. Determine the thermal diffusivity of helium gas at 500 °F and 120 psia.

From Table B-4, by interpolation,

$$k_1 = 0.122 \text{ Btu/hr-ft-°F} \qquad \rho_1 = 0.0058 \text{ lbm/ft}^3 \qquad c_p = 1.24 \text{ Btu/lbm-°F}$$

From Table B-5 the critical constants are $p_c = 2.26$ atm; $T_c = 9.47$ °R. Thus,

$$P_r = \frac{120/14.7}{2.26} = 3.61 \qquad T_r = \frac{500 + 459.7}{9.47} = 101$$

At this state, $k/k_1 \approx 1.0$ (see Figure B-3). By (*1.11*),

$$\rho = \rho_1\left(\frac{p}{p_1}\right) = (0.0058 \text{ lbm/ft}^3)\left(\frac{120 \text{ psia}}{14.7 \text{ psia}}\right) = 0.0473 \text{ lbm/ft}^3$$

By (*1.16*),

$$\alpha \approx \frac{0.122 \text{ Btu/hr-ft-°F}}{(0.0473 \text{ lbm/ft}^3)(1.24 \text{ Btu/lbm-°F})} = 2.08 \text{ ft}^2/\text{hr}$$

1.10. A hollow cylinder having a weight of 4 lbf slides along a vertical rod which is coated with a light lubricating oil [see Fig. 1-4(*a*)]. The steady velocity (terminal) of the slide is 3.0 ft/sec. The hole diameter in the slide is 1.000 in. and the radial clearance between the slide and the rod is 0.001 in. The slide length is 5 in. Neglecting all drag forces except the viscous shear at the inner surface of the cylinder and assuming a linear velocity profile for the lubricant, estimate the dynamic viscosity of the oil. Compare this with tabulated values for engine oil at 158 °F.

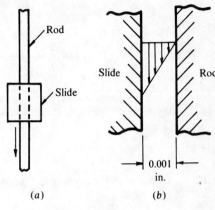

Fig. 1-4

Since the radial clearance is small, the suggested linear velocity profile, shown in Fig. 1-4(*b*), will be a suitable first-order approximation. The velocity in the fluid ranges from 0 at the rod surface to 3 ft/sec at the slide surface, and

$$\frac{du}{dy} \approx \frac{\Delta u}{\Delta r} = \frac{(3-0)\ \text{ft/sec}}{(0.001/12)\ \text{ft}} = 3.6 \times 10^4\ \text{sec}^{-1}$$

From a free-body diagram of the slide, the total downward force is the weight, and the total upward force is the product of the shear stress and the inner surface area. So $\tau A = 4$ lbf, whence

$$\tau = \frac{4\ \text{lbf}}{A} = \frac{4\ \text{lbf}}{\pi(\frac{1}{12}\ \text{ft})(\frac{5}{12}\ \text{ft})} = 36.7\ \frac{\text{lbf}}{\text{ft}^2}$$

By (*1.17*),

$$\tau = \mu_f \frac{du}{dy} = 36.7\ \frac{\text{lbf}}{\text{ft}^2}$$

$$\mu_f = \frac{36.7\ \text{lbf/ft}^2}{\dfrac{du}{dy}} \approx \frac{(36.7\ \text{lbf/ft}^2)\ \text{sec}}{3.6 \times 10^4} = 1.019 \times 10^{-3}\ \frac{\text{lbf-sec}}{\text{ft}^2}$$

From Table B-3, the dynamic viscosity of engine oil at approximately 158 °F (70 °C) is

$$\mu_f = \rho\nu\frac{1}{g_c} = \left(53.57\ \frac{\text{lbm}}{\text{ft}^3}\right)\left(0.6535 \times 10^{-3}\ \frac{\text{ft}^2}{\text{sec}}\right)\left(\frac{\text{lbf-sec}^2}{32.17\ \text{ft-lbm}}\right)$$

$$= 1.08 \times 10^{-3}\ \frac{\text{lbf-sec}}{\text{ft}^2}$$

which compares well with the preceding estimate. Note that the conversion from lbm to lbf was accomplished by use of the factor

$$32.17\ \frac{\text{ft-lbm}}{\text{lbf-sec}^2} = 1$$

obtained from Table 1-1. This factor can always be introduced when using British Engineering units if needed for dimensional conversion.

1.11. What is the kinematic viscosity ν of air at 350 °F and 150 psia?

From Table B-4 at 300 °F,

$$\mu_m = 1.669 \times 10^{-5}\ \text{lbm/ft-sec} \qquad \rho_1 = 0.0489\ \text{lbm/ft}^3$$

By (1.13),

$$\rho = \rho_1 \left(\frac{p}{p_1}\right) = \left(0.0489 \frac{\text{lbm}}{\text{ft}^3}\right)\left(\frac{150}{14.7}\right) = 0.499 \text{ lbm/ft}^3$$

Thus, by (1.18),

$$\nu = \frac{1.669 \times 10^{-5} \text{ lbm/ft-sec}}{0.499 \text{ lbm/ft}^3} = 3.34 \times 10^{-5} \text{ ft}^2/\text{sec}$$

which is approximately one-tenth of the tabulated value at atmospheric pressure. A calculation of the reduced pressure yields a value of 0.27, and there is no pressure effect upon the dynamic viscosity (Figure B-4).

1.12. Determine the dynamic viscosity μ_m and the kinematic viscosity ν of hydrogen gas at 90 °R and 38.4 atm.

From Table B-4 at 90 °R = −370 °F,

$$\mu_{m1} = 1.691 \times 10^{-6} \text{ lbm/ft-sec} \qquad \rho_1 = 0.03181 \text{ lbm/ft}^3$$

From Table B-5 the critical state is $p_c = 12.8$ atm; $T_c = 59.9$ °R. So,

$$P_r = \frac{38.4}{12.8} = 3.0 \qquad T_r = \frac{90}{59.9} = 1.50$$

From Figure B-4, $\mu_m/\mu_{m1} \approx 1.55$, so that

$$\mu_m \approx 1.55 \times 1.691 \times 10^{-6} = 2.62 \times 10^{-6} \text{ lbm/ft-sec}$$

At high pressure ($P_r = 3.0$) the ideal gas law must be modified in order to calculate ρ and thence ν. For a real gas we write the equation of state as $pv = ZRT$, where Z is the *compressibility factor* and $R = \mathcal{R}/M$ is the universal gas constant divided by the molecular weight. It follows that

$$\frac{\rho}{\rho_1} = \frac{v_1}{v} = \frac{p/Z}{p_1/Z_1}$$

For the present problem, standard tables give $Z \approx 0.81$, $Z_1 \approx 1$. Thus,

$$\frac{\rho}{0.03181 \text{ lbm/ft}^3} \approx \frac{38.4/0.81}{1/1}$$

$$\rho \approx 1.51 \text{ lbm/ft}^3$$

and

$$\nu = \frac{\mu_m}{\rho} = \frac{2.62 \times 10^{-6} \text{ lbm/ft-sec}}{1.51 \text{ lbm/ft}^3} = 1.74 \times 10^{-6} \text{ ft}^2/\text{sec}$$

In general, whenever the pressure effect upon viscosity μ or thermal conductivity k is significant, the deviation from ideal gas behavior will affect the value of ρ significantly.

1.13. The suit worn by the Apollo astronauts during exploration of the lunar surface weighed 200 lbf on earth. Average lunar gravitational acceleration is one-sixth that of the earth. What was the suit weight on the moon in lbf and in N?

Since the suit weighs 200 lbf where gravitational acceleration is standard, its mass is 200 lbm; i.e. by (1.19), with $F = $ wt.,

$$m = \frac{(\text{wt.}) g_c}{a} = \frac{(200 \text{ lbf})(32.17 \text{ ft-lbm/lbf-sec}^2)}{32.17 \text{ ft/sec}^2} = 200 \text{ lbm}$$

Hence, on the moon,

$$\text{wt.} = \frac{1}{g_c} ma = \frac{(200 \text{ lbm})[(32.17/6) \text{ ft/sec}^2]}{32.17 \text{ ft-lbm/lbf-sec}^2} = 33.33 \text{ lbf}$$

In SI units (by Table A-1),

$$\text{wt.} = 33.33 \text{ lbf} \times 4.4482 \frac{\text{N}}{\text{lbf}} = 148.273 \text{ N}$$

1.14. During a dive in an airplane a force-measuring instrument indicates a weight of 2.3 lbf for an object whose mass is 6.0 lbm. What is the downward acceleration of the airplane at this instant?

The object experiences a downward gravitational force $(1/g_c)mg$ and an upward force w, where w is numerically equal to the measured weight. Then, by Newton's second law,

$$\frac{1}{g_c} mg - w = \frac{1}{g_c} ma$$

or

$$a = g - \frac{wg_c}{m} = 32.17 \text{ ft/sec}^2 - \frac{2.3 \text{ lbf}}{6.0 \text{ lbm}} \left(32.17 \frac{\text{ft-lbm}}{\text{lbf-sec}^2} \right) = 19.838 \text{ ft/sec}^2, \text{ downward}$$

1.15. Repeat Problem 1.1 using SI units throughout.

Use Appendix A to convert k, T_1, T_2, and Δx.

$$k = \left(\frac{0.11 \text{ Btu}}{\text{hr-ft-}°\text{F}} \right) \left(\frac{1.7296 \text{ J/m-s-K}}{\text{Btu/hr-ft-}°\text{F}} \right) = 0.1903 \text{ J/m-s-K}$$

$$T_1 = \frac{5}{9}(100 + 459.67) = 310.93 \text{ K}$$

$$T_2 = \frac{5}{9}(70 + 459.67) = 294.26 \text{ K}$$

$$\Delta x = \left(\frac{1.5}{12} \text{ ft} \right) \left(\frac{0.3048 \text{ m}}{\text{ft}} \right) = 0.0381 \text{ m}$$

$$\frac{q}{A} = - \frac{0.1903 \text{ J}}{\text{m-s-K}} \left[\left(\frac{294.26 - 310.93}{0.0381} \right) \frac{\text{K}}{\text{m}} \right] = +83.227 \frac{\text{J}}{\text{s-m}^2} \quad \text{or} \quad +83.227 \frac{\text{W}}{\text{m}^2}$$

1.16. Verify the conversion factors

(a) 1 Btu/ft^2-hr = 3.1525 W/m^2
(b) 1 Btu/hr = 0.292875 W
(c) 1 Btu/hr-ft-°F = 1.729 W/m-K

(a) $$\left(1 \frac{\text{Btu}}{\text{ft}^2\text{-hr}} \right) \left(\frac{1054.35 \text{ J}}{\text{Btu}} \right) \left(\frac{\text{ft}^2}{(0.3048 \text{ m})^2} \right) \left(\frac{\text{hr}}{3600 \text{ sec}} \right) = 3.1525 \frac{\text{J}}{\text{s-m}^2} = 3.1525 \frac{\text{W}}{\text{m}^2}$$

(b) $$\left(1 \frac{\text{Btu}}{\text{hr}} \right) \left(\frac{1054.35 \text{ J}}{\text{Btu}} \right) \left(\frac{\text{hr}}{3600 \text{ sec}} \right) = 0.292875 \text{ W}$$

(c) $$\left(1 \frac{\text{Btu}}{\text{hr-ft-}°\text{F}} \right) \left(\frac{1054.35 \text{ J}}{\text{Btu}} \right) \left(\frac{\text{ft}}{0.3048 \text{ m}} \right) \left(\frac{\text{hr}}{3600 \text{ sec}} \right) \left(\frac{9 \text{ }°\text{F}}{5 \text{ K}} \right) = 1.72958 \frac{\text{W}}{\text{m-K}}$$

1.17.   Verify the conversion factors

$$(a)\ 1\ \text{ft/sec} = 0.3048\ \text{m/s}$$
$$(b)\ 1\ \text{ft/hr} = 8.4666 \times 10^{-5}\ \text{m/s}$$
$$(c)\ 1\ \text{km/hr} = 0.2777\ \text{m/s}$$

(a) $\left(1\dfrac{\text{ft}}{\text{sec}}\right)\left(\dfrac{0.3048\ \text{m}}{\text{ft}}\right) = 0.3048\ \text{m/s}$

(b) $\left(1\dfrac{\text{ft}}{\text{hr}}\right)\left(\dfrac{0.3048\ \text{m}}{\text{ft}}\right)\left(\dfrac{\text{hr}}{3600\ \text{sec}}\right) = 8.4666 \times 10^{-5}\ \text{m/s}$

(c) $\left(1\dfrac{\text{km}}{\text{hr}}\right)\left(\dfrac{1000\ \text{m}}{\text{km}}\right)\left(\dfrac{\text{hr}}{3600\ \text{sec}}\right) = 0.2777\ \text{m/s}$

Supplementary Problems

1.18.   A plane wall 0.5 ft thick, of a homogeneous material with $k = 0.25$ Btu/hr-ft-°F, has steady and uniform temperatures $T_1 = 70$ °F and $T_2 = 160$ °F (see Fig. 1-5).   Determine the heat transfer rate in the positive x-direction per square foot of surface area.
       Ans. -45 Btu/hr-ft^2

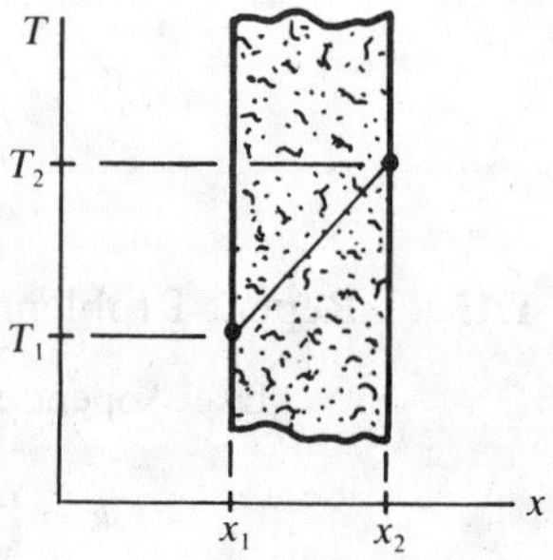

Fig. 1-5

1.19.   Forced air flows over a convective heat exchanger in a room heater, resulting in a convective heat-transfer coefficient $h = 200$ Btu/hr-ft^2-°F.   The surface temperature of the heat exchanger may be considered constant at 150 °F, and the air is at 65 °F.   Determine the heat exchanger surface area required for 30,000 Btu/hr of heating.
       Ans.   1.765 ft^2

1.20.   Asphalt pavements on hot summer days exhibit surface temperatures of approximately 120 °F.   Consider such a surface to emit as a blackbody and calculate the emitted radiant energy per unit surface area.   *Ans.*   193.56 Btu/hr-ft^2

1.21.   Plot thermal conductivity versus temperature for copper and cast iron over the range of values given in Appendix B.   Which of these is the better thermal conductor?

1.22.   Plot thermal conductivity versus temperature for saturated liquid ammonia and saturated liquid carbon dioxide over the ranges of temperature given in Table B-3.   Does the behavior of either of these depart from the general rule that k for liquids decreases with increasing temperature?

1.23.   Determine the thermal conductivity for gaseous oxygen at 530 °F and 100 psia.
       Ans.   0.0261 Btu/hr-ft-°F $(k/k_1 \approx 1.0)$

1.24.   Determine the thermal conductivity of helium gas at 33.15 K and 6.87×10^4 N/m^2.
       Ans.   0.0353 W/m-K $(k/k_1 \approx 1.0)$

1.25.   Determine the thermal conductivity of carbon monoxide gas at 25 psia and 200 °F.
       Ans.   0.0173 Btu/hr-ft-°F

1.26. Approximate the density of gaseous carbon monoxide at 530 °F and 120 psia.   *Ans.*   0.316 lbm/ft^3

1.27. Using the density of mercury at 50 °C, determine its specific gravity.
Ans.   13.51 (based on density of water equal to 62.4 lbm/ft^3)

1.28. Determine the thermal conductivity of helium gas at 15.78 K and 13.74×10^5 N/m^2. Use linear interpolation to obtain k_1 from Table B-4.
Ans.   0.0270 W/m-K (pressure correction required)

1.29. Determine the thermal diffusivity of air at 80 °F and 1000 psia. Note that k depends on pressure, and ρ deviates slightly from ideal gas prediction at this pressure.
Ans. $k \approx 0.01713$ Btu/hr-ft-°F; $\rho \approx 5.123$ lbm/ft^3; $c_p \approx 0.2402$ Btu/lbm-°F; $\alpha \approx 0.0138$ ft^2/hr

1.30. A 1.5-in.-diameter shaft rotates in a sleeve bearing which is 2.5 in. long. The radial clearance between the shaft and the sleeve is 0.0015 in., and this is filled with oil having $\mu_m = 1.53 \times 10^{-2}$ lbm/ft-sec. Assuming a rotational speed of 62 rpm and a linear velocity gradient in the lubricating oil, determine the resistive torque due to viscous shear at the shaft-lubricant interface.
Ans.   0.0079 ft-lbf

1.31. Plot the kinematic viscosity of steam at 300 psia for the range of temperature of steam properties in Table B-4.   *Ans.* $\nu = 2.4814 \times 10^{-5}$ ft^2/sec at 530 °F, $\rho = 0.5102$ lbm/ft^3 at 530 °F

1.32. During launch a NASA Space Shuttle Vehicle had a maximum acceleration of approximately 3.0 g, where g is standard gravitational acceleration. What total weight in lbf and in N should be used for a 180-lbm crew member?   *Ans.*   540 lbf; 2402.04 N

1.33. The Lunar Rover Vehicle used in the Apollo series manned exploration of the moon weighed 493.81 lbf on earth. This was loaded weight minus crew. Assume an average lunar gravity equal to one-sixth standard earth gravity. (*a*) What is the vehicle mass in kg? (*b*) What is the vehicle weight on earth in N? (*c*) What is the vehicle weight on the moon in lbf?   *Ans.*   223.988 kg; 2196.57 N; 82.302 lbf

1.34. Verify the viscosity conversion factors
(*a*) 1 stoke $= 1.0 \times 10^4$ m^2/s
(*b*) 1 ft^2/sec $= 9.2903 \times 10^{-2}$ m^2/s
(*c*) 1 ft^2/hr $= 2.5806 \times 10^{-5}$ m^2/s

1.35. Verify the following pressure conversion factors:
(*a*) 1 lbf/ft$^2 = 47.8803$ N/m^2
(*b*) 1 lbf/in$^2 = 6.8947 \times 10^3$ N/m^2
(*c*) 1 mmHg $= 1.3332 \times 10^2$ N/m^2

1.36. The lunar temperature range on the side facing the earth is approximately -400 °F to $+200$ °F. These correspond to the lunar night and day. What is this range in (*a*) °C, (*b*) °R, (*c*) K?
Ans. -240 °C to 93.33 °C; 59.67 °R to 659.67 °R; 33.15 K to 366.48 K

Chapter 2

One-Dimensional Steady-State Conduction

2.1 INTRODUCTORY REMARKS

The conductive heat-transfer rate at a point within a medium is related to the local temperature gradient by Fourier's law, (1.1). In many one-dimensional problems we can write the temperature gradient simply by inspection of the physical situation. However, more complex cases—and the multidimensional problems to be treated in later chapters—require the formation' of an energy equation which governs the temperature distribution in general. From the temperature distribution, the temperature gradient at any desired location within the medium can be formed, and consequently the heat transfer rate may be calculated.

2.2 GENERAL CONDUCTIVE ENERGY EQUATION

Consider a control volume consisting of a small parallelepiped, as shown in Fig. 2-1. This may be an element of material from a homogeneous solid or a homogeneous fluid so long as there is no relative motion between the macroscopic material particles. Heating of the material results in an energy flux per unit area within the control volume. This flux is, in general, a three-dimensional vector. For simplicity, only one component, q_x, is shown in Fig. 2-1.

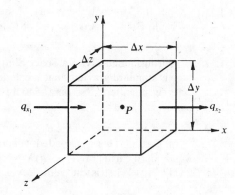

Fig. 2-1

Application of the first law of thermodynamics to the control volume, as carried out in Problem 2.1, yields the *general conduction equation*

$$\blacksquare \qquad \frac{\partial}{\partial x}\left(k\frac{\partial T}{\partial x}\right) + \frac{\partial}{\partial y}\left(k\frac{\partial T}{\partial y}\right) + \frac{\partial}{\partial z}\left(k\frac{\partial T}{\partial z}\right) + q''' = \rho c\frac{\partial T}{\partial t} \qquad (2.1)$$

for the temperature T as a function of x, y, z and t. (The $\blacksquare$ designates an important equation.) Here, k is the thermal conductivity, ρ is the density, c is the specific heat per unit mass, and q''' is the *rate of internal energy conversion* ("*heat generation*") per unit volume. A common instance of q''' is provided by resistance heating in an electrical conductor.

In most engineering problems k can be taken as constant, and (2.1) reduces to

$$\frac{\partial^2 T}{\partial x^2} + \frac{\partial^2 T}{\partial y^2} + \frac{\partial^2 T}{\partial z^2} + \frac{q'''}{k} = \frac{1}{\alpha}\frac{\partial T}{\partial t} \qquad (2.2)$$

where α is given by (1.16).

Special Cases of the Conduction Equation

1. *Fourier equation* (no internal energy conversion)

$$\frac{\partial^2 T}{\partial x^2} + \frac{\partial^2 T}{\partial y^2} + \frac{\partial^2 T}{\partial z^2} = \frac{1}{\alpha}\frac{\partial T}{\partial t} \qquad (2.3)$$

2. **Poisson equation**   (steady state with internal energy conversion)

$$\frac{\partial^2 T}{\partial x^2} + \frac{\partial^2 T}{\partial y^2} + \frac{\partial^2 T}{\partial z^2} + \frac{q'''}{k} = 0 \tag{2.4}$$

3. **Laplace equation**   (steady state and no internal energy conversion)

$$\frac{\partial^2 T}{\partial x^2} + \frac{\partial^2 T}{\partial y^2} + \frac{\partial^2 T}{\partial z^2} = 0 \tag{2.5}$$

Cylindrical and Spherical Coordinate Systems

The general conduction equation for constant thermal conductivity can be written as

$$\nabla^2 T + \frac{q'''}{k} = \frac{1}{\alpha} \frac{\partial T}{\partial t} \tag{2.6}$$

in which ∇^2 denotes the Laplacian operator. In cartesian coordinates,

$$\nabla^2 T = \nabla \cdot \nabla T = \left(\mathbf{i} \frac{\partial}{\partial x} + \mathbf{j} \frac{\partial}{\partial y} + \mathbf{k} \frac{\partial}{\partial z} \right) \cdot \left(\mathbf{i} \frac{\partial T}{\partial x} + \mathbf{j} \frac{\partial T}{\partial y} + \mathbf{k} \frac{\partial T}{\partial z} \right)$$

$$= \frac{\partial^2 T}{\partial x^2} + \frac{\partial^2 T}{\partial y^2} + \frac{\partial^2 T}{\partial z^2}$$

Forming $\nabla^2 T$ in cylindrical coordinates as given in Fig. 2-2 results in

$$\nabla^2 T = \frac{\partial^2 T}{\partial r^2} + \frac{1}{r} \frac{\partial T}{\partial r} + \frac{1}{r^2} \frac{\partial^2 T}{\partial \phi^2} + \frac{\partial^2 T}{\partial z^2} \tag{2.7}$$

and the result for the spherical coordinate system of Fig. 2-3 is

$$\nabla^2 T = \frac{1}{r} \frac{\partial^2}{\partial r^2} (rT) + \frac{1}{r^2 \sin \psi} \frac{\partial}{\partial \psi} \left(\sin \psi \frac{\partial T}{\partial \psi} \right) + \frac{1}{r^2 \sin^2 \psi} \frac{\partial^2 T}{\partial \phi^2} \tag{2.8}$$

In the following sections the general conduction equation will be used to obtain the temperature gradient only when such cannot be found by inspection or simple integration of Fourier's law.

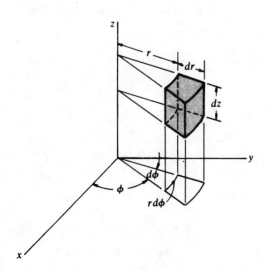

Fig. 2-2

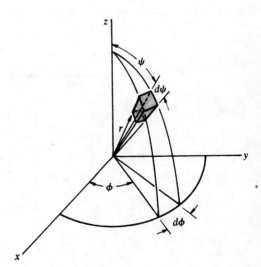

Fig. 2-3

2.3 PLANE WALL: FIXED SURFACE TEMPERATURES

The simplest heat transfer problem is that of one-dimensional, steady-state conduction in a plane wall of homogeneous material having constant thermal conductivity and with each face held at a constant uniform temperature, as shown in Fig. 2-4.

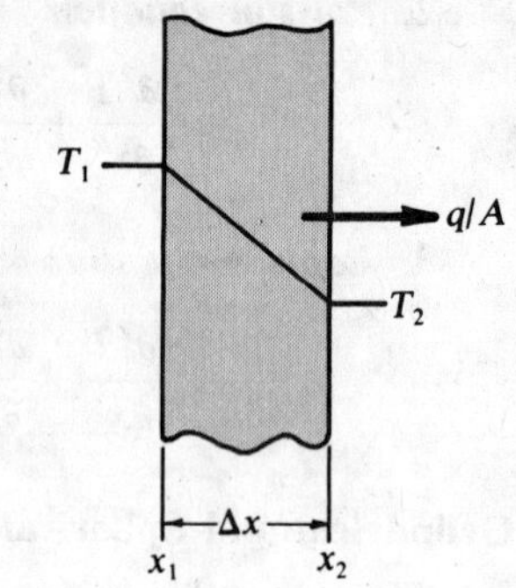

Separation of variables and integration of (1.1) where the gradient direction is x results in

$$q \int_{x_1}^{x_2} dx = -kA \int_{T_1}^{T_2} dT$$

or

$$q = -kA \frac{T_2 - T_1}{x_2 - x_1} = -kA \frac{T_2 - T_1}{\Delta x} \tag{2.9}$$

This equation can be rearranged as

$$q = \frac{T_1 - T_2}{\Delta x / kA} = \frac{\text{thermal potential difference}}{\text{thermal resistance}} \tag{2.10}$$

Fig. 2-4

Notice that the resistance to the heat flow is directly proportional to the material thickness, inversely proportional to the material thermal conductivity, and inversely proportional to the area normal to the direction of heat transfer.

These principles are readily extended to the case of a composite plane wall as shown in Fig. 2-5(a). In the steady state the heat transfer rate entering the left face is the same as that leaving the right face. Thus,

$$q = \frac{T_1 - T_2}{\Delta x_a / k_a A} \qquad \text{and} \qquad q = \frac{T_2 - T_3}{\Delta x_b / k_b A}$$

Together these give:

$$q = \frac{T_1 - T_3}{(\Delta x_a / k_a A) + (\Delta x_b / k_b A)} \tag{2.11}$$

Equations (2.10) and (2.11) illustrate the analogy between conductive heat transfer and electrical current flow, an analogy that is rooted in the similarity between Fourier's and Ohm's

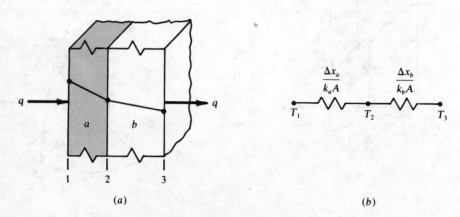

(a)   (b)

Fig. 2-5

laws.   It is convenient to express Fourier's law as

$$\text{conductive heat flow} = \frac{\text{overall temperature difference}}{\text{summation of thermal resistances}} \qquad (2.12)$$

In the case of the composite two-layered plane wall the total thermal resistance is simply the sum of the two resistances in series shown in Fig. 2-5(*b*).   The extension to three or more layers is obvious.

2.4   RADIAL SYSTEMS: FIXED SURFACE TEMPERATURES

Figure 2-6 depicts a single-layer cylindrical wall of a homogeneous material with constant thermal conductivity and uniform inner and outer surface temperatures.   At a given radius the area normal to radial heat flow by conduction is $2\pi rL$, where L is the cylinder length.   Substituting this into (*1.1*) and integrating with q constant gives

$$T_2 - T_1 = -\frac{q}{2\pi kL} \ln \frac{r_2}{r_1} \qquad (2.13)$$

or

$$q = \frac{2\pi kL(T_1 - T_2)}{\ln \dfrac{r_2}{r_1}} \qquad (2.14)$$

From (*2.14*) the thermal resistance of the single cylindrical layer is $[\ln (r_2/r_1)]/2\pi kL$.   For a two-layered cylinder (Fig. 2-7) the heat transfer rate is, by (*2.12*),

$$q = \frac{2\pi L(T_1 - T_3)}{\dfrac{1}{k_a} \ln \dfrac{r_2}{r_1} + \dfrac{1}{k_b} \ln \dfrac{r_3}{r_2}} \qquad (2.15)$$

and this too is readily extended to three or more layers.

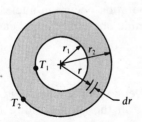

Fig. 2-6

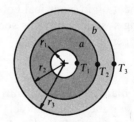

Fig. 2-7

For radial conductive heat transfer in a spherical wall the area at a given radius is $4\pi r^2$.   Substituting this into Fourier's law and integrating with q constant yields

$$q = \frac{4\pi k(T_1 - T_2)}{\dfrac{1}{r_1} - \dfrac{1}{r_2}} \qquad (2.16)$$

From this the thermal resistance afforded by a single spherical layer is $(1/r_1 - 1/r_2)/4\pi k$.   In a multilayered spherical problem the resistances of the individual layers are linearly additive, and (*2.12*) applies.

2.5   PLANE WALL: VARIABLE THERMAL CONDUCTIVITY

From Section 1.4 we recall that material thermal conductivity is temperature dependent, and rather strongly so for many engineering materials. It is common to express this dependence by the linear relationship (*1.9*). Then, as shown in Problem 2.11, (*2.10*) is replaced by

$$q = \frac{T_1 - T_2}{\Delta x / k_m A} \qquad (2.17)$$

where

$$k_m = k_0 \left(1 + b \frac{\theta_1 + \theta_2}{2} \right) = k_0 (1 + b\theta_m) \qquad (2.18)$$

is the thermal conductivity evaluated at the mean temperature of the wall,

$$\theta_m = \frac{\theta_1 + \theta_2}{2} = \frac{T_1 + T_2}{2} - T_{\text{ref}}$$

Often, however, the average material temperature is unknown at the start of the problem. This is generally true for multilayer walls, where only the overall temperature difference is initially specified. In such cases, if the data warrant an attempt at precision, the problem is attacked by assuming reasonable values for the interfacial temperatures, obtaining k_m's for each material, and then determining the heat flux per unit area by (*2.12*). Using this result, the assumed values of the interfacial temperatures may be improved by application of Fourier's law to each layer, beginning with a known surface temperature. This procedure can be repeated until satisfactory agreement between previous interfacial temperatures and the next set of computed values is obtained.

The temperature distribution for the plane wall having thermal conductivity which is linearly dependent upon temperature is obtained analytically in Problem 2.12, and the treatment for a cylindrical wall with linear dependence of k upon temperature is considered in Problem 2.13.

2.6   HEAT GENERATION SYSTEMS

Besides I^2R heating in electrical conductors, heat generation occurs in nuclear reactors and in chemically reacting systems. In this section we will examine one-dimensional cases with constant and uniform heat generation.

Plane Wall

Consider the plane wall with uniform internal conversion of energy, Fig. 2-8. Assuming constant thermal conductivity and very large dimensions in the y- and z-directions so that the temperature gradient is significant in the x-direction only, the Poisson equation, (*2.4*), reduces to

$$\frac{d^2 T}{dx^2} + \frac{q'''}{k} = 0 \qquad (2.19)$$

which is a second-order, ordinary linear differential equation. Two boundary conditions are sufficient for determination of the specific solution for $T(x)$. These are [Fig. 2-8(*a*)]:

$$T = T_1 \quad \text{at} \quad x = 0 \qquad \text{and} \qquad T = T_2 \quad \text{at} \quad x = 2L$$

Integrating (*2.19*) twice with respect to x results in

$$T = -\frac{q'''}{2k} x^2 + C_1 x + C_2$$

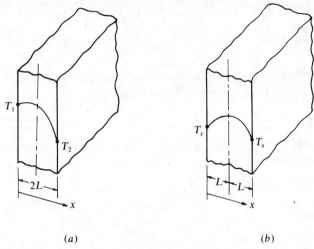

(a)   (b)

Fig. 2-8

The boundary conditions are then used to give

$$C_2 = T_1 \qquad C_1 = \frac{T_2 - T_1}{2L} + \frac{q'''L}{k}$$

Hence,

$$T = \left[\frac{T_2 - T_1}{2L} + \frac{q'''}{2k}(2L - x)\right]x + T_1 \tag{2.20}$$

The heat flux is dependent upon x-location; see Problem 2.16.

For the simpler case where $T_1 = T_2 = T_s$ [Fig. 2-8(b)], (2.20) reduces to

$$T = T_s + \frac{q'''}{2k}(2L - x)x \tag{2.21}$$

Differentiating (2.21) yields

$$\frac{dT}{dx} = \frac{q'''L}{k} - \frac{q'''x}{k} = \frac{q'''}{k}(L - x)$$

so that the heat flux out of the left face is

$$q = -kA\frac{dT}{dx}\Big|_{x=0} = -kA\frac{q'''L}{k} = -q'''AL \tag{2.22}$$

The minus sign indicates that the heat transfer is in the minus x-direction (for positive q'''); the product AL is one-half the plate volume. Thus (2.22) can be interpreted as indicating that the heat generated in the left half of the wall is conducted out of the left face, etc.

Cylinder

Consider a long circular cylinder of constant thermal conductivity having uniform internal energy conversion per unit volume, q'''. If the surface temperature is constant, the azimuthal gradient $\partial T/\partial \phi$ is zero during steady state, and the length precludes a significant temperature gradient along the axis, $\partial T/\partial z$. For this case, (2.7) simplifies to

$$\nabla^2 T = \frac{d^2T}{dr^2} + \frac{1}{r}\frac{dT}{dr}$$

and (2.6) becomes for steady state

$$\frac{d^2T}{dr^2} + \frac{1}{r}\frac{dT}{dr} + \frac{q'''}{k} = 0 \tag{2.23}$$

a second-order, ordinary differential equation requiring two boundary conditions on $T(r)$ to effect a solution. Usually the surface temperature is known and thus

$$T = T_s \quad \text{at} \quad r = r_s \tag{2.24}$$

A second boundary condition is provided by the physical requirement that the temperature be finite on the axis of the cylinder, i.e. $dT/dr = 0$ at $r = 0$.

Rewriting (2.23) as

$$\frac{d}{dr}\left(r\frac{dT}{dr}\right) = -\frac{rq'''}{k}$$

and then performing a first integration gives

$$\frac{dT}{dr} = -\frac{r}{2}\frac{q'''}{k} + \frac{C_1}{r}$$

Integrating again, we obtain

$$T = -\frac{r^2}{4}\frac{q'''}{k} + C_1 \ln r + C_2$$

The finiteness condition at $r = 0$ requires that $C_1 = 0$. Application of the remaining boundary condition, (2.24), yields

$$C_2 = T_s + \frac{r_s^2}{4}\frac{q'''}{k}$$

and consequently

$$T - T_s = \frac{r_s^2 q'''}{4k}\left[1 - \left(\frac{r}{r_s}\right)^2\right] \tag{2.25}$$

A convenient dimensionless form of the temperature distribution is obtained by denoting the centerline temperature of the rod as T_c and forming the ratio

$$\frac{T - T_s}{T_c - T_s} = 1 - \left(\frac{r}{r_s}\right)^2 \tag{2.26}$$

2.7 CONVECTIVE BOUNDARY CONDITIONS

Newton's law of cooling, (1.4), may be conveniently rewritten as

$$q = hA\,\Delta T \tag{2.27}$$

where h = convective heat-transfer coefficient, Btu/hr-ft^2-°F or W/m^2-K
         A = area normal to the direction of the heat flux, ft^2 or m^2
         ΔT = temperature difference between the solid surface and the fluid, °F or K

In this section we will consider problems wherein values of h are known or specified, and we will direct our attention to the solution of combined conductive-convective problems.

Overall Heat-Transfer Coefficient

It is often convenient to express the heat transfer rate for a combined conductive-convective problem in the form (2.27), with h replaced by an overall heat-transfer coefficient U. We now determine U for plane and cylindrical wall systems.

Plane wall. A plane wall of a uniform, homogeneous material a having constant thermal conductivity and exposed to fluid i at temperature T_i on one side and fluid o at temperature T_o on the other side is shown in Fig. 2-9(a). Frequently the fluid temperatures sufficiently far from the wall to be unaffected by the heat transfer are known, and the surface temperatures T_1 and T_2 are not specified.

Applying (2.27) at the two surfaces yields

$$\frac{q}{A} = \bar{h}_i(T_i - T_1) = \bar{h}_o(T_2 - T_o)$$

or

$$q = \frac{T_i - T_1}{1/\bar{h}_i A} = \frac{T_2 - T_o}{1/\bar{h}_o A} \tag{2.28}$$

where the overbar on h denotes an average value for the entire surface.

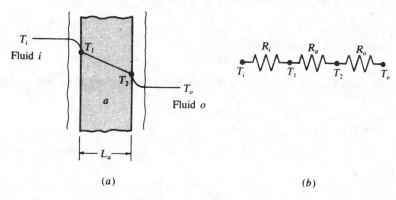

(a)   (b)

Fig. 2-9

In agreement with the electrical analogy of Section 2.3, $1/\bar{h}A$ can be thought of as a *thermal resistance due to the convective boundary*. Thus, the electrical analog to this problem is that of three resistances in series, Fig. 2-9(b). Here, $R_a = L_a/k_a A$ is the conductive resistance due to the homogeneous material a. Since the conductive heat flow within the solid must exactly equal the convective heat flow at the boundaries, (2.12) gives

$$\frac{q}{A} = \frac{T_i - T_o}{1/\bar{h}_i + L_a/k_a + 1/\bar{h}_o} = \frac{(\Delta T)_{\text{overall}}}{A \Sigma R_{\text{th}}} \tag{2.29}$$

Defining the overall heat-transfer coefficient U by

$$U \equiv \frac{1}{A \Sigma R_{\text{th}}} \tag{2.30}$$

for any geometry, we see that

$$\frac{q}{A} = U(\Delta T)_{\text{overall}} \tag{2.31}$$

and for the plane wall of Fig. 2-9(a),

$$U = \frac{1}{1/\bar{h}_i + L_a/k_a + 1/\bar{h}_o} \tag{2.32}$$

For a multilayered plane wall consisting of layers $a, b, \ldots,$

$$U = \frac{1}{1/\bar{h}_i + L_a/k_a + L_b/k_b + \cdots + 1/\bar{h}_o} \tag{2.33}$$

Radial systems.   Consider the cylindrical system consisting of a single material layer having an inner and an outer convective fluid flow as shown in Fig. 2-10(a).   If T_2 is the temperature at r_2, etc., then (2.12) gives

$$q = \frac{(\Delta T)_{\text{overall}}}{\Sigma R_{\text{th}}} = \frac{T_i - T_o}{\Sigma R_{\text{th}}} \tag{2.34}$$

where the thermal resistances are:

$$R_i = \text{inside convective } R_{\text{th}} = \frac{1}{2\pi r_1 L \bar{h}_i}$$

$$R_a = \text{conductive } R_{\text{th}} \text{ due to material } a = \frac{\ln(r_2/r_1)}{2\pi k_a L}$$

$$R_o = \text{outside convective } R_{\text{th}} = \frac{1}{2\pi r_2 L \bar{h}_o}$$

In these expressions L is the length of the cylindrical system.   Summing the thermal resistances,

$$\Sigma R_{\text{th}} = \frac{1}{2\pi r_1 L \bar{h}_i} + \frac{\ln(r_2/r_1)}{2\pi k_a L} + \frac{1}{2\pi r_2 L \bar{h}_o}$$

Now by definition $U = 1/(A \Sigma R_{\text{th}})$, and for A it is customary to use the outer surface area, $A_o = 2\pi r_2 L$, so that

$$U_o = \frac{1}{\dfrac{r_2}{r_1 \bar{h}_i} + \dfrac{r_2 \ln(r_2/r_1)}{k_a} + \dfrac{1}{\bar{h}_o}}$$

where the subscript o denotes that U_o is based on the outside surface area of the cylinder.   For a

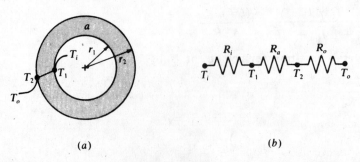

(a)   (b)

Fig. 2-10

multilayered cylindrical system having $n - 1$ material layers,

$$U_o = \cfrac{1}{\cfrac{r_n}{r_1 \bar{h}_i} + \cfrac{r_n \ln (r_2/r_1)}{k_{1,2}} + \cdots + \cfrac{r_n \ln (r_n/r_{n-1})}{k_{n-1,n}} + \cfrac{1}{\bar{h}_o}} \qquad (2.35)$$

where the subscripts on k denote the bounding radii of a layer (e.g., for a two-layered system with the outer layer of material b, $k_b = k_{2,3}$).

Critical Thickness of Cylindrical Insulation

In many cases, the thermal resistance offered by a metal pipe or duct wall is negligibly small in comparison with that of the insulation layer (see Problem 2.23). Also, the pipe wall temperature is often very nearly the same as that of the fluid inside the pipe. For a single layer of insulation material, the heat transfer rate per unit length is given by

$$\frac{q}{L} = U_o \frac{A}{L} \Delta T = \cfrac{2\pi(T_i - T_o)}{\cfrac{\ln (r/r_i)}{k} + \cfrac{1}{hr}} \qquad (2.36)$$

where the nomenclature is defined in Fig. 2-11.

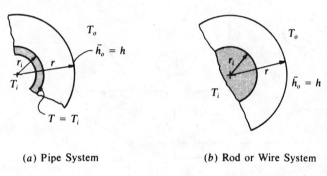

(a) Pipe System   (b) Rod or Wire System

Fig. 2-11

As a function of r, q/L has a *maximum* at

$$r = r_{\text{crit}} = \frac{k}{h} \qquad (2.37)$$

Thus, if $r_i < r_{\text{crit}}$, the heat loss rate *increases* with addition of insulation until $r = r_{\text{crit}}$, and then decreases with further addition of insulation. On the other hand, if $r_i > r_{\text{crit}}$, the heat loss rate decreases for any addition of insulation.

2.8 HEAT TRANSFER FROM FINS

Extended surfaces or fins are used to increase the effective surface area for convective heat transfer in heat exchangers, internal combustion engines, transistor designs, etc.

Uniform Cross Section

Two common designs, the rectangular fin and the rodlike fin, have uniform cross section and lend themselves to a common analysis.

Rectangular fin.   Figure 2-12 shows a rectangular fin having temperature T_b at the base and surrounded by a fluid at temperature T_∞.   Applying the first law of thermodynamics to an element of the fin of thickness Δx gives, in the steady state,

(energy conducted in at x) = (energy conducted out at $x + \Delta x$)

+ (energy out by convection)

Assuming no temperature variation in the y- or z-directions, the three energy terms are respectively

$$q\Big|_x = -kA\frac{dT}{dx}\Big|_x \qquad q\Big|_{x+\Delta x} = -kA\frac{dT}{dx}\Big|_{x+\Delta x} \qquad q_{\text{conv}} = \bar{h}(P\,\Delta x)(T - T_\infty)$$

where P is the perimeter, $2(w + t)$.   Substituting these three expressions, dividing by Δx, and taking the limit as $\Delta x \to 0$ results in

$$\frac{d^2T}{dx^2} - \frac{\bar{h}P}{kA}(T - T_\infty) = 0 \tag{2.38}$$

if the thermal conductivity k is constant.

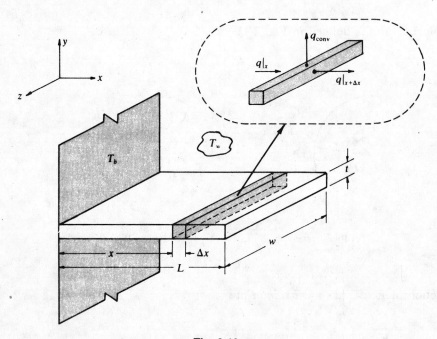

Fig. 2-12

Letting $\theta = T - T_\infty$ and $n = \sqrt{\bar{h}P/kA}$, (2.38) becomes

$$\frac{d^2\theta}{dx^2} - n^2\theta = 0 \tag{2.39}$$

which has the general solution

$$\theta(x) = C_1 e^{nx} + C_2 e^{-nx} \tag{2.40}$$

One boundary condition on (2.40) is $\theta(0) = T_b - T_\infty = \theta_b$, which requires that $C_1 + C_2 = \theta_b$.   Table 2-1 presents the solutions corresponding to three useful choices for the second boundary condition. For Case 3, $\bar{h}_L$ is the average heat-transfer coefficient for the end area; it may differ from $\bar{h}$ along the sides.

Table 2-1

Rectangular Fin	Second Boundary Condition	θ/θ_b
Case 1. Very long, with end at temperature of surrounding fluid	$\theta(+\infty) = 0$	e^{-nx}
Case 2. Finite length, insulated end	$\left.\dfrac{d\theta}{dx}\right|_{x=L} = 0$	$\dfrac{\cosh\,[n(L-x)]}{\cosh\,nL}$
Case 3. Finite length, heat loss by convection at end	$-k\left.\dfrac{d\theta}{dx}\right|_{x=L} = \bar{h}_L\,\theta(L)$	$\dfrac{\cosh\,[n(L-x)] + (\bar{h}_L/nk)\sinh\,[n(L-x)]}{\cosh\,nL + (\bar{h}_L/nk)\sinh\,nL}$

In all three cases the heat transfer from the fin is most easily found by evaluating the conductive flux into the fin at its base:

$$q = -kA\left.\frac{dT}{dx}\right|_{x=0} = -kA\left.\frac{d\theta}{dx}\right|_{x=0}$$

where $A = wt$ and where the gradient at $x = 0$ is derived from Table 2-1.　We have:

Case 1:　$q = kAn\theta_b$ 　　　　　　　　　　　　　　　　　　　　　　　(2.41)

Case 2:　$q = kAn\theta_b \tanh nL$ 　　　　　　　　　　　　　　　　　　(2.42)

Case 3:　$q = kAn\theta_b\left[\dfrac{\sinh nL + (\bar{h}_L/nk)\cosh nL}{\cosh nL + (\bar{h}_L/nk)\sinh nL}\right]$ 　　　　　(2.43)

Remark (a).　For a thin fin, $w \gg t$ and $P \approx 2w$, so

thin fin:　$n \approx \sqrt{\dfrac{2\bar{h}}{kt}}$ 　　　　　　　　　　　　　　　　　　　(2.44)

Remark (b).　The preceding solutions for temperature distributions and heat fluxes would be unchanged for a solid cylindrical rod or pin-type fin other than in the expressions for P and n.　If r is the radius of the rod,

pin or rod:　$n = \sqrt{\dfrac{2\bar{h}}{kr}}$ 　　　　　　　　　　　　　　　　　　(2.45)

Remark (c).　The insulated-end solution (2.42) is often used even when the end of the fin is exposed, the heat loss along the sides being typically much larger than that from the exposed end.　In that case, L in (2.42) is replaced by a corrected length only in evaluation of q:

rectangular fin:　$L_c = L + \dfrac{t}{2}$

cylindrical pin or rod:　$L_c = L + \dfrac{r}{2}$

Nonuniform Cross Section

The differential equation for the temperature distribution is now

$$\frac{d^2\theta}{d\xi^2} + \frac{1}{A}\frac{dA}{d\xi}\frac{d\theta}{d\xi} - \frac{\bar{h}}{k}\left(\frac{1}{A}\frac{dS}{d\xi}\right)\theta = 0 \qquad\qquad (2.46)$$

where $A = A(\xi)$ and $S = S(\xi)$ are respectively the variable cross-sectional area and variable surface area.

Annular fin of uniform thickness. Consider the annular fin shown in Fig. 2-13. For no circumferential temperature variation and for t small compared with $r_2 - r_1$, the temperature is a function of r only ($\xi = r$). The cross-sectional area and the surface area are $A(r) = 2\pi r t$ and $S(r) = 2\pi(r^2 - r_1^2)$ so that (2.46) becomes

$$\frac{d^2\theta}{dr^2} + \frac{1}{r}\frac{d\theta}{dr} - \frac{2\bar{h}}{kt}\theta = 0 \qquad (2.47)$$

This is a form of Bessel's differential equation of zero order, and it has the general solution

$$\theta = C_1 I_0(nr) + C_2 K_0(nr) \qquad (2.48)$$

where $\quad n = \sqrt{2\bar{h}/kt}$

$\qquad I_0 =$ modified Bessel function of the 1st kind

$\qquad K_0 =$ modified Bessel function of the 2nd kind

The constants C_1 and C_2 are determined by the boundary conditions, which are:

$$\theta(r_1) = T_b - T_\infty = \theta_b \qquad \left.\frac{d\theta}{dr}\right|_{r=r_2} = 0$$

The second of these conditions assumes no heat loss from the end of the fin. This is generally more realistic for the annular fin than for the rectangular case because of rapidly increasing surface area with increasing r.

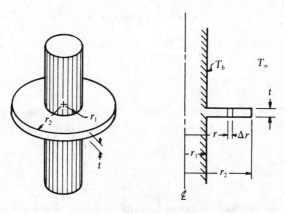

Fig. 2-13

With C_1 and C_2 evaluated, (2.48) becomes

$$\frac{\theta}{\theta_b} = \frac{I_0(nr)K_1(nr_2) + K_0(nr)I_1(nr_2)}{I_0(nr_1)K_1(nr_2) + K_0(nr_1)I_1(nr_2)} \qquad (2.49)$$

Determining the heat loss from the fin by evaluating the conductive heat transfer-rate into the base, we obtain

$$q = 2\pi kt\theta_b(nr_1)\frac{K_1(nr_1)I_1(nr_2) - I_1(nr_1)K_1(nr_2)}{I_0(nr_1)K_1(nr_2) + K_0(nr_1)I_1(nr_2)} \qquad (2.50)$$

A table of Bessel functions sufficiently accurate for most engineering applications is included in

Jahnke, E., F. Emde, and F. Lösch, *Tables of Higher Functions*, 6th ed., McGraw-Hill, New York, 1960.

Straight triangular fin. The solution of (2.46) for the fin shown in Fig. 2-14 [for $t \ll L$, the temperature will be a function of x alone] is

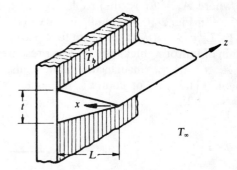

$$\frac{\theta}{\theta_b} = \frac{I_0(2px^{1/2})}{I_0(2pL^{1/2})} \qquad (2.51)$$

where

$$p = \sqrt{\frac{2f\bar{h}L}{kt}} \qquad f = \sqrt{1 + (t/2L)^2}$$

The heat loss from the fin per unit width (z-direction) may be found by application of Fourier's law at the base of the fin; this results in

Fig. 2-14

$$q = \frac{-kt\theta_b p}{L^{1/2}} \frac{I_1(2pL^{1/2})}{I_0(2pL^{1/2})} \qquad (2.52)$$

Fin Efficiency

The primary purpose of fins is to increase the effective heat-transfer surface area exposed to a fluid in a heat exchanger. The performance of fins is often expressed in terms of the *fin efficiency*, η_f, defined by

$$\eta_f = \frac{\text{actual heat transfer}}{\text{heat transfer if entire fin were at the base temperature}} \qquad (2.53)$$

In terms of η_f, the heat transfer rate is given by the simple expression

$$q = \bar{h}(A_b + \eta_f A_f)\,\theta_b \qquad (2.54)$$

where A_f is the total surface area of fins and A_b is the surface area of the wall, tube, etc., between fins (Fig. 2-15).

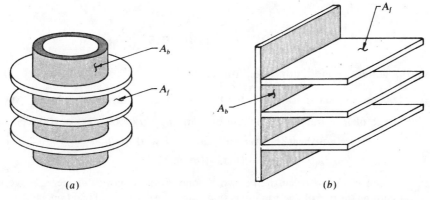

(a)   (b)

Fig. 2-15

Analytical expressions for η_f are readily obtained for several common configurations. Consider, for example, the simple case of a rectangular fin with no end heat loss. The efficiency is, from (2.42),

$$\eta_f = \frac{\sqrt{\bar{h}PkA}\,\theta_b \tanh nL}{\bar{h}PL\theta_b} = \frac{1}{nL}\tanh nL \qquad (2.55)$$

If the fin is thin, (2.44) gives

$$nL \approx \left(\frac{2\bar{h}}{kt}\right)^{1/2} L = L^{3/2}\left(\frac{2\bar{h}}{kLt}\right)^{1/2} = L^{3/2}\left(\frac{2\bar{h}}{kA_p}\right)^{1/2} \tag{2.56}$$

where $A_p = Lt$ is the profile area of the rectangular fin.

In Fig. 2-16 the fin efficiency of (2.55) is plotted against $nL/2^{1/2}$ of (2.56), in which L is replaced by $L_c = L + (t/2)$ to account for tip loss.   Similar graphs for the straight triangular fin and the annular fin of uniform thickness are also presented in this figure.   Note that the product $L_c t$ is the profile area A_p for the rectangular and annular fins, whereas A_p is one-half of the product Lt for the triangular fin.   There is, of course, no length correction for the triangular fin.   For the annular fin, $r_{2c} = r_2 + (t/2)$.

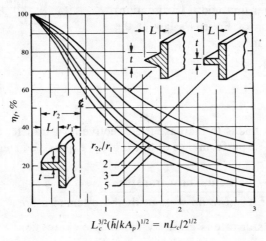

$$L_c^{3/2}(\bar{h}/kA_p)^{1/2} = nL_c/2^{1/2}$$

Fig. 2-16

Solved Problems

2.1.   Derive the general conduction equation, (2.1).

For the control volume in Fig. 2-1, the first law of thermodynamics may be expressed as

(rate of heat transfer in) + (rate of work in) + (rate of other energy conversion)

= (rate of heat transfer out) + (rate of work out)

+ (rate of internal energy storage) (1)

For an incompressible substance the net work done on the control volume is converted to internal energy.   Denoting the rate of energy conversion (from work, chemical reaction, etc.) as q''' (1) becomes

$$q_{x_1} + q_{y_1} + q_{z_1} + q''' \Delta x \, \Delta y \, \Delta z = q_{x_2} + q_{y_2} + q_{z_2} + \frac{\partial U}{\partial t} \tag{2}$$

Examine the heat transfer terms in (2).   In the x-direction the two terms may be grouped to form

$$q_{x_1} - q_{x_2} = -\Delta y \, \Delta z \left[\left(k\frac{\partial T}{\partial x}\right)_{x_1} - \left(k\frac{\partial T}{\partial x}\right)_{x_2}\right] \tag{3}$$

by application of Fourier's law.   Notice that k may be temperature dependent and hence spatially dependent.   By a Taylor's series expansion about the center point P,

$$\left(k\frac{\partial T}{\partial x}\right)\Bigg|_{x_1} = k\frac{\partial T}{\partial x} + \left(-\frac{\Delta x}{2}\right)\frac{\partial}{\partial x}\left(k\frac{\partial T}{\partial x}\right) + \cdots$$

$$\left(k\frac{\partial T}{\partial x}\right)\Bigg|_{x_2} = k\frac{\partial T}{\partial x} + \left(\frac{\Delta x}{2}\right)\frac{\partial}{\partial x}\left(k\frac{\partial T}{\partial x}\right) + \cdots$$

so that (3) becomes

$$q_{x_1} - q_{x_2} = \Delta y\,\Delta z\left[\Delta x\frac{\partial}{\partial x}\left(k\frac{\partial T}{\partial x}\right) + \cdots\right] \tag{4}$$

Similarly,

$$q_{y_1} - q_{y_2} = \Delta x\,\Delta z\left[\Delta y\frac{\partial}{\partial y}\left(k\frac{\partial T}{\partial y}\right) + \cdots\right] \tag{5}$$

$$q_{z_1} - q_{z_2} = \Delta x\,\Delta y\left[\Delta z\frac{\partial}{\partial z}\left(k\frac{\partial T}{\partial z}\right) + \cdots\right] \tag{6}$$

Finally, the internal energy storage per unit volume and per unit temperature is the product of density and specific heat, so

$$\frac{\partial U}{\partial t} = \rho c(\Delta x\,\Delta y\,\Delta z)\frac{\partial T}{\partial t} \tag{7}$$

Substituting expressions (4) through (7) into (2), dividing by the volume $\Delta x\,\Delta y\,\Delta z$, and taking the limit as $\Delta x, \Delta y,$ and Δz simultaneously approach zero yields (2.1).

2.2.   Beginning with the general conduction equation, show that the linear temperature distribution of Fig. 2-4 is correct.

The appropriate coordinate system is cartesian and hence the equation to be used is (2.2):

$$\frac{\partial^2 T}{\partial x^2} + \frac{\partial^2 T}{\partial y^2} + \frac{\partial^2 T}{\partial z^2} + \frac{q'''}{k} = \frac{1}{\alpha}\frac{\partial T}{\partial t}$$

Assumptions:

  1. The plane wall is very large in the y- and z-directions, hence

$$\frac{\Delta T}{\Delta y} \approx \frac{\Delta T}{\Delta z} = \frac{\text{small number}}{\text{very large number}}$$

Further, the rate of change of, say, $\Delta T/\Delta y$ with y will be even smaller, so

$$\frac{\partial^2 T}{\partial y^2} \approx 0 \qquad \frac{\partial^2 T}{\partial z^2} \approx 0$$

  2. No internal energy conversion, hence $q''' = 0$.
  3. The problem is steady-state, so $\partial T/\partial t = 0$.

Under assumptions 1, 2 and 3, the conduction equation reduces to

$$\frac{d^2 T}{dx^2} = 0$$

Integrating twice gives $T = C_1 x + C_2$, which is a linear temperature distribution.   The constants C_1 and C_2 are then chosen to satisfy the boundary conditions

$$T_1 = C_1 x_1 + C_2 \qquad T_2 = C_1 x_2 + C_2$$

2.3. Obtain dT/dr for the single-layered cylinder of Fig. 2-6 directly from the appropriate steady-state conduction equation. Substitute this into Fourier's law and obtain (2.13).

Since the problem is cylindrical and steady-state, and there is no internal energy conversion, the appropriate conduction equation is

$$\nabla^2 T = \frac{\partial^2 T}{\partial r^2} + \frac{1}{r}\frac{\partial T}{\partial r} + \frac{1}{r^2}\frac{\partial^2 T}{\partial \phi^2} + \frac{\partial^2 T}{\partial z^2} = 0$$

Assumptions: Very long cylinder (or negligible z-direction temperature change) and no angular temperature variation; hence,

$$\frac{\partial^2 T}{\partial z^2} = 0 \qquad \frac{\partial^2 T}{\partial \phi^2} = 0$$

Thus, the equation reduces to

$$\frac{d^2 T}{dr^2} + \frac{1}{r}\frac{dT}{dr} = 0 \qquad \text{or} \qquad \frac{1}{r}\frac{d}{dr}\left(r\frac{dT}{dr}\right) = 0$$

A first integration gives

$$r\frac{dT}{dr} = B \qquad \text{or} \qquad dT = B\frac{dr}{r}$$

and a second integration yields

$$T = B \ln r + c$$

The boundary conditions, $T(r_1) = T_1$ and $T(r_2) = T_2$, determine

$$B = \frac{T_1 - T_2}{\ln \dfrac{r_1}{r_2}}$$

Fourier's law becomes

$$q = -k(2\pi rL)\frac{dT}{dr} = -k(2\pi L)B = 2\pi kL\frac{T_1 - T_2}{\ln \dfrac{r_2}{r_1}}$$

which is the desired result.

2.4. An industrial furnace wall is constructed of 0.7-ft-thick fireclay brick having $k = 0.6$ Btu/hr-ft-°F. This is covered on the outer surface with a 0.1-ft-thick layer of insulating material having $k = 0.04$ Btu/hr-ft-°F. The innermost surface is at 1800 °F and the outermost is at 100 °F. Calculate the steady-state heat transfer per square foot.

Equation (2.11) applies, with the brick denoted a and the insulation b (see Fig. 2-5). Thus,

$$\frac{q}{A} = \frac{T_1 - T_3}{\dfrac{\Delta x_a}{k_a} + \dfrac{\Delta x_b}{k_b}} = \frac{(1800 - 100)\,°F}{\left(\dfrac{0.7}{0.6} + \dfrac{0.1}{0.04}\right)\dfrac{ft}{Btu/hr\text{-}ft\text{-}°F}} = 464\ \frac{Btu}{hr\text{-}ft^2}$$

2.5. A frequently encountered engineering problem is the determination of the thickness of insulation that will result in a specified heat flux. If the maximum allowable heat transfer rate for the furnace of Problem 2.4 is 300 Btu/hr-ft², the brick wall is unchanged and the same insulation material is to be used, how thick must the insulation be?

Solving

$$\frac{q}{A} = \frac{T_1 - T_3}{\dfrac{\Delta x_a}{k_a} + \dfrac{\Delta x_b}{k_b}}$$

for Δx_b, we have

$$\frac{300 \text{ Btu}}{\text{hr-ft}^2} = \frac{1700 \text{ °F}}{\left(\dfrac{0.7 \text{ ft}}{0.6} + \dfrac{\Delta x_b}{0.04}\right) \dfrac{\text{hr-ft-°F}}{\text{Btu}}}$$

$$\Delta x_b = (0.04)\left(\frac{1700}{300} - 1.166\right) \text{ ft} \approx 0.18 \text{ ft}$$

2.6. The ceilings of many American homes consist of a 5/8-in.-thick sheet of Celotex board supported by ceiling joists, with the space between the joists filled with 4-lbm/ft^3 loose rock wool insulation (Fig. 2-17). Neglecting the effect of the wooden joists, determine the heat transfer rate per unit area for a ceiling lower surface temperature of 85 °F and a rock wool upper surface temperature of 45 °F.

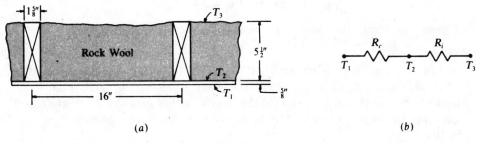

(a)   (b)

Fig. 2-17

From Table B-2, the thermal conductivity of rock wool ($\rho = 4$ lbm/ft^3) at 65 °F average temperature is $k_i \approx 0.0192$ Btu/hr-ft-°F and that of the Celotex board is $k_c = 0.028$ Btu/hr-ft-°F. (The latter value is at 90 °F, which should be acceptable for this problem.) By the electrical analogy [Fig. 2-17(b)]

$$\frac{q}{A} = \frac{T_1 - T_3}{R_c + R_i}$$

The thermal resistances per unit area of the Celotex and the insulation (rock wool) are

$$R_c = \frac{(0.625/12) \text{ ft}}{0.028 \dfrac{\text{Btu}}{\text{hr-ft-°F}}} = 1.86 \frac{\text{hr-ft}^2\text{-°F}}{\text{Btu}}$$

$$R_i = \frac{(5.5/12) \text{ ft}}{0.0192 \dfrac{\text{Btu}}{\text{hr-ft-°F}}} = 23.87 \frac{\text{hr-ft}^2\text{-°F}}{\text{Btu}}$$

Hence,

$$\frac{q}{A} = \frac{(85 - 45) \text{ °F}}{(1.86 + 23.87) \text{ hr-ft}^2\text{-°F/Btu}} = 1.55 \text{ Btu/hr-ft}^2$$

2.7. A composite three-layered wall is formed of a 0.5-cm-thick aluminum plate, a 0.25-cm-thick layer of sheet asbestos, and a 2.0-cm-thick layer of rock wool (density = 64 kg/m^3); the asbestos is the center layer. The outer aluminum surface is at 500 °C, and the outer rock wool surface is at 50 °C. Determine the heat flow per unit area.

From Table B-1 at 500 °C, $k_{al} = 268.08$ W/m-K.
From Table B-2 at 51 °C, $k_{asb} = 0.1660$ W/m-K.
From Table B-3 at 93 °C, $k_{rw} = 0.0548$ W/m-K.

Note that the asbestos sheet average temperature is certainly greater than 51 °C, but this is the only k value available in the Appendix. Also, the other two thermal conductivities were taken at reasonable temperatures for this problem. By (2.12),

$$\frac{q}{A} = \frac{(500 - 50)\,°C}{\left(\dfrac{0.5 \times 10^{-2}}{268.08} + \dfrac{0.25 \times 10^{-2}}{0.1660} + \dfrac{2.0 \times 10^{-2}}{0.0548}\right)\dfrac{m}{W/m\text{-}K}} = \frac{450\,K}{0.38004\,m^2\text{-}K/W} = 1184.08\,\frac{W}{m^2}$$

2.8. Repeat Problem 2.7 for a two-layer composite wall consisting of the asbestos sheet and the rock wool, with the same overall temperature difference.

Using the thermal conductivities given in the solution of Problem 2.7 and applying (2.12),

$$\frac{q}{A} = \frac{(500 - 50)\,K}{\left(\dfrac{0.25 \times 10^{-2}}{0.1660} + \dfrac{2.0 \times 10^{-2}}{0.0548}\right)\dfrac{m}{W/m\text{-}K}} = 1184.14\,\frac{W}{m^2}$$

Clearly, the thermal resistance of the aluminum sheet is negligibly small.

2.9. A 3-in.-o.d. steel pipe is covered with a 1/2-in. layer of asbestos ($\rho = 36$ lbm/ft^3) which is covered in turn with a 2-in. layer of glass wool ($\rho = 4$ lbm/ft^3). Determine (a) the steady-state heat transfer per lineal foot and (b) the interfacial temperature between the asbestos and the glass wool if the pipe outer surface temperature is 400 °F and the glass wool outer temperature is 100 °F.

From Table B-2 at 392 °F and 200 °F for asbestos and glass wool, respectively,

$$k_{asb} = 0.120\,\frac{Btu}{hr\text{-}ft\text{-}°F} \qquad k_{gw} = 0.0317\,\frac{Btu}{hr\text{-}ft\text{-}°F}$$

These temperatures should be reasonably close to the average values for these two materials.
(a) By (2.15),

$$\frac{q}{L} = \frac{2\pi(400 - 100)\,°F}{\left(\ln\dfrac{2.0}{1.5}\right)\Big/k_{asb} + \left(\ln\dfrac{4.0}{2.0}\right)\Big/k_{gw}}$$

$$= \frac{2\pi(300)\,°F}{\dfrac{0.288}{0.120\ Btu/hr\text{-}ft\text{-}°F} + \dfrac{0.693}{0.0317\ Btu/hr\text{-}ft\text{-}°F}}$$

$$= \frac{1884.96\,°F}{24.261\ hr\text{-}ft\text{-}°F/Btu} = 77.69\,\frac{Btu}{hr\text{-}ft}$$

(b) Since the heat transfer per ft is now known, the single-layer equation, (2.13), can be used to determine the interfacial temperature. Thus, considering the glass wool layer,

$$T_2 - (100\,°F) = \left[\frac{77.69}{2\pi(0.0317)}\ln\frac{4.0}{2.0}\right]°F$$

$$T_2 = 270.37 + 100 = 370.37\,°F$$

We could have instead used the asbestos layer to find T_2, since q/L is the same for either layer in steady state. Note that the average temperature of the glass wool is about 235 °F, which is reasonably close to the temperature at which the thermal conductivity was chosen.

2.10. Show that a positive value of b in (1.9) results in a temperature curve of type (1), i.e. concave downward, in Fig. 2-18 for a plane wall with steady, uniform surface temperatures.

From (1.9),

$$\frac{dk}{dT} = k_0 b > 0$$

Consequently, k increases with increasing temperature or decreases with decreasing temperature. Now examine Fourier's law,

$$\frac{q}{A} = k\left(-\frac{dT}{dx}\right)$$

As x increases, T decreases and so k decreases. Then, to keep q/A constant, $-dT/dx$ must increase; i.e. dT/dx must decrease, as it does for curve (1).

2.11. Derive (2.17).

For a linear dependence of thermal conductivity upon temperature, $k = k_0(1 + b\theta)$, Fourier's law becomes (in terms of the variable θ)

$$q = -k_0(1 + b\theta)A\frac{d\theta}{dx} \qquad \text{or} \qquad \frac{q}{A}dx = -k_0(1 + b\theta)d\theta$$

Integrating with q/A constant (steady state) gives

$$\frac{q}{A}\int_{x_1}^{x_2} dx = -k_0\int_{\theta_1}^{\theta_2}(1 + b\theta)d\theta$$

$$\frac{q}{A}(x_2 - x_1) = k_0\left[\theta_1 - \theta_2 + \frac{b}{2}(\theta_1^2 - \theta_2^2)\right]$$

$$= k_0(\theta_1 - \theta_2)\left(1 + b\frac{\theta_1 + \theta_2}{2}\right)$$

Writing $x_2 - x_1 = \Delta x$, $\theta_2 - \theta_1 = T_2 - T_1 = \Delta T$ and defining k_m as in (2.18), the last equation becomes

$$\frac{q}{A}\Delta x = -k_m\,\Delta T$$

which is equivalent to (2.17).

2.12. Obtain an analytical expression for the temperature distribution $T(x)$ in the plane wall of Fig. 2-18 having uniform surface temperatures T_1 and T_2 at x_1 and x_2, respectively, and a thermal conductivity which varies linearly with temperature.

Separating variables in Fourier's law (see Problem 2.11) and integrating from x_1 to arbitrary x yields

$$\frac{q}{A}\int_{x_1}^{x} dx = -k_0\int_{\theta_1}^{\theta}(1 + b\theta)d\theta$$

$$\frac{q}{A}(x - x_1) = -k_0\left[\left(\theta + \frac{b}{2}\theta^2\right) - \left(\theta_1 + \frac{b}{2}\theta_1^2\right)\right]$$

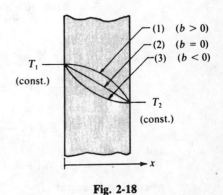

Fig. 2-18

But, as shown in Problem 2.11, the heat flux per unit area is

$$\frac{q}{A} = -k_m \frac{T_2 - T_1}{x_2 - x_1} = -k_m \frac{\theta_2 - \theta_1}{x_2 - x_1}$$

which coupled with the preceding equation results in

$$\frac{b}{2}\theta^2 + \theta - \left[\theta_1 + \frac{b}{2}\theta_1^2 + \frac{k_m(x - x_1)}{k_0(x_2 - x_1)}(\theta_2 - \theta_1) \right] = 0$$

This last is a quadratic equation, which may be solved to give θ, and hence T, explicitly in terms of x.

2.13. A hollow cylinder having inner and outer radii r_1 and r_2, respectively, is subjected to a steady heat transfer resulting in constant surface temperatures T_1 and T_2 at r_1 and r_2. If the thermal conductivity can be expressed as $k = k_0(1 + b\theta)$, obtain an expression for the heat transfer per unit length of the cylinder.

In terms of θ, Fourier's law is

$$q = -kA_r \frac{d\theta}{dr}$$

where A_r is the area normal to r. Substituting $k = k_0(1 + b\theta)$ and $A_r = 2\pi rL$, where L is cylinder length, results in

$$q = -k_0(1 + b\theta)(2\pi rL)\frac{d\theta}{dr} \qquad \text{or} \qquad \frac{q}{2\pi L}\frac{dr}{r} = -k_0(1 + b\theta)d\theta$$

In the steady state $q/2\pi L$ is constant. Hence, integrating through the cylinder wall yields, after rearrangement,

$$\frac{q}{L} = -2\pi k_0 \left[1 + \frac{b}{2}(\theta_2 + \theta_1) \right] \frac{\theta_2 - \theta_1}{\ln(r_2/r_1)}$$

Using (2.18) to define the mean thermal conductivity, we can simplify this result to

$$\frac{q}{L} = -2\pi k_m \frac{\theta_2 - \theta_1}{\ln(r_2/r_1)} = -2\pi k_m \frac{T_2 - T_1}{\ln(r_2/r_1)}$$

2.14. In a single experiment with a 2-cm-thick sheet of pure copper having one face maintained at 500 °C and the other at 300 °C, the measured heat flux per unit area is 3.633 MW/m² (1 MW = 10^6 W). A reported value of k for this material at 150 °C is 371.9 W/m-K. Determine an expression for $k(T)$ of form (1.9).

For the experiment, (2.17) yields

$$\frac{q}{A} = 3.633 \times 10^6 \frac{W}{m^2} = k_m \left| \frac{\Delta T}{\Delta x} \right| = k_m \left| \frac{200\ K}{2 \times 10^{-2}\ m} \right|$$

$$k_m = 363.3\ \text{W/m-K}$$

Using $k_0 = 371.9$ W/m-K and $\theta = T - 150$ °C in (2.18),

$$363.3 \frac{W}{m\text{-}K} = \left(371.9 \frac{W}{m\text{-}K} \right) \left[1 + b\frac{(500 - 150) + (300 - 150)}{2} \right]$$

$$b = \left(\frac{363.3}{371.9} - 1 \right) \times \frac{1}{250} = -9.25 \times 10^{-5}\ K^{-1}$$

and

$$k = (371.9)[1 - 9.25 \times 10^{-5}(T - 150)]\ \text{W/m-K}$$

where T is in degrees C.

The accuracy of this expression may be checked by comparison with tabulated values in Appendix B.   At 300 °C,

$$k = (371.9)[1 - 9.25 \times 10^{-5}(300 - 150)] = 366.74 \text{ W/m-K} = 212.04 \text{ Btu/hr-ft-°F}$$

At 500 °C,

$$k = (371.9)[1 - 9.25 \times 10^{-5}(500 - 150)] = 359.86 \text{ W/m-K} = 208.06 \text{ Btu/hr-ft-°F}$$

These values are in reasonable agreement with the tabulated data.   The lack of agreement is simply due to the actual nonlinearity of the $k(T)$ relationship, and the linear representation of $k(T)$ obtained should not be applied over an excessive temperature range.

2.15.   A thick-walled copper cylinder has an inside radius of 1 cm and an outside radius of 1.8 cm.   The inner and outer surface temperatures are held at 305 °C and 295 °C, respectively.   Assume k varies linearly with temperature, with k_0 and b the same as in Problem 2.14.   Determine the heat loss per unit length.

From Problem 2.13,

$$\frac{q}{L} = -2\pi k_m \frac{T_2 - T_1}{\ln(r_2/r_1)}$$

where $k_m = k_0[1 + b\theta_m]$.   For this problem,

$$\theta_m = \frac{\theta_2 + \theta_1}{2} = \frac{(305 - 150) + (295 - 150)}{2} = 150°$$

and from Problem (2.14),

$$k_m = (371.9)[1 - 9.25 \times 10^{-5}(150)] = 366.74 \text{ W/m-K}$$

Hence,

$$\frac{q}{L} = -2\pi\left(366.74 \frac{\text{W}}{\text{m-K}}\right)\frac{(295 - 305) \text{ K}}{\ln(1.8/1)} = 39.203 \text{ kW/m}$$

where $1 \text{ kW} = 10^3 \text{ W}$.

2.16.   Consider a plate with uniform heat generation q''' as shown in Fig. 2-8(a).   For $k = 200$ W/m-K, $q''' = 40$ MW/m^3, $T_1 = 160$ °C (at $x = 0$), $T_2 = 100$ °C (at $x = 2L$), and a plate thickness of 2 cm, determine (a) $T(x)$, (b) q/A at the left face, (c) q/A at the right face, and (d) q/A at the plate center.

(a)   By (2.20),

$$T = \left[\frac{100 - 160}{0.02} + \frac{(4 \times 10^7)(0.02 - x)}{2(200)}\right]x + 160 = 160 - 10^3 x - 10^5 x^2$$

where T is in °C and x is in m.

(b)   Obtain dT/dx at $x = 0$ and substitute into Fourier's law.

$$\frac{dT}{dx} = [-10^3 - (2)(10^5)x] \text{ K/m} \qquad \left.\frac{dT}{dx}\right|_{x=0} = -10^3 \text{ K/m}$$

$$\left.\frac{q}{A}\right|_0 = -k\left.\frac{dT}{dx}\right|_0 = -\left(\frac{200 \text{ W}}{\text{m-K}}\right)\left(-10^3 \frac{\text{K}}{\text{m}}\right) = +200 \text{ kW/m}^2$$

The + sign signifies a heat flux into the left surface.

(c)
$$\left.\frac{dT}{dx}\right|_{2L} = -10^3 - 2(10)^5(0.02) = -5(10)^3 \text{ K/m}$$

$$\left.\frac{q}{A}\right|_{2L} = -k\left.\frac{dT}{dx}\right|_{2L} = -\left(\frac{200 \text{ W}}{\text{m-K}}\right)\left(\frac{-5(10)^3 \text{ K}}{\text{m}}\right) = 1 \text{ MW/m}^2$$

An energy balance on the plate,

$$\left.\frac{q}{A}\right|_{2L} = \left.\frac{q}{A}\right|_0 + \frac{q''' \times \text{volume}}{A}$$

can be used to check the above results.

(d)
$$\left.\frac{dT}{dx}\right|_{L} = -10^3 - 2(10)^5(0.01) = -3 \times 10^3 \text{ K/m}$$

$$\left.\frac{q}{A}\right|_{L} = -\left(\frac{200 \text{ W}}{\text{m-K}}\right)\left(\frac{-3 \times 10^3 \text{ K}}{\text{m}}\right) = +600 \text{ kW/m}^2$$

2.17. For the heat generation problem depicted in Fig. 2-8(b), determine an analytical expression for the dimensionless temperature $(T - T_s)/(T_c - T_s)$, where T_c is the temperature at the center of the wall.

From (2.21),

$$T - T_s = \frac{q'''}{2k}(2L - x)x \quad \text{and} \quad T_c - T_s = \frac{q'''}{2k}(2L - L)L = \frac{q'''L^2}{2k}$$

so

$$\frac{T - T_s}{T_c - T_s} = \frac{(2L - x)x}{L^2} = 2\left(\frac{x}{L}\right) - \left(\frac{x}{L}\right)^2$$

This shows the parabolic form of the nondimensional temperature distribution.

2.18. For the heat generation problem of Fig. 2-8(b), show that the temperature at the center, T_c, is the maximum temperature when q''' is positive and the minimum temperature when q''' is negative.

By differentiating (2.21),

$$\frac{dT}{dx} = \frac{q'''}{k}(L - x)$$

and at $x = L$, which is the center plane, dT/dx vanishes. This is the condition for an extremal; T is either a maximum or a minimum at this location. To determine which, examine the second derivative,

$$\frac{d^2T}{dx^2} = -\frac{q'''}{k}$$

This is negative for positive q''', the condition for a maximum, and positive for negative q''', the condition for a minimum.

2.19. An electrical resistance heater wire has a 0.08-in. diameter. The electrical resistivity is $\rho = 80 \times 10^{-6}$ ohm-cm, and the thermal conductivity is 11 Btu/hr-ft-°F. For a steady-state current of 150 A passing through the wire, determine the centerline temperature rise above the surface temperature, in °F.

It will be convenient to solve this problem in SI units, since we will deal with electrical power dissipation and electrical resistivity, and then to convert the temperature difference from K to °F. Using Appendix A,

$$d = (0.08 \text{ in.})\left(0.0254\,\frac{\text{m}}{\text{in.}}\right) = 2.032 \times 10^{-3} \text{ m}$$

$$\rho = (80 \times 10^{-6} \text{ ohm-cm})\left(0.01\,\frac{\text{m}}{\text{cm}}\right) = 8 \times 10^{-7} \text{ ohm-m}$$

$$k = \left(11\,\frac{\text{Btu}}{\text{hr-ft-°F}}\right)\left(1.729577\,\frac{\text{W/m-K}}{\text{Btu/hr-ft-°F}}\right) = 19.0253\,\frac{\text{W}}{\text{m-K}}$$

Assuming uniform energy conversion within the wire, $q'''(\pi r_s^2)L = I^2R$, where r_s is the wire outside radius; I is the electric current; and R is the electrical resistance of the wire, which is

$$R = \rho\frac{L}{A} = \rho\frac{L}{\pi r_s^2}$$

Thus,

$$q''' = \frac{I^2R}{(\pi r_s^2)L} = \frac{I^2\rho}{(\pi r_s^2)^2} = \frac{(150 \text{ A})^2(8 \times 10^{-7} \text{ ohm-m})}{\pi^2\left(\dfrac{2.032 \times 10^{-3}}{2}\,\text{m}\right)^4} = 1.7116 \times 10^9\,\frac{\text{W}}{\text{m}^3}$$

By (2.25),

$$T - T_s = \frac{r_s^2 q'''}{4k}\left[1 - \left(\frac{r}{r_s}\right)^2\right] = \frac{(1.016 \times 10^{-3} \text{ m})^2(1.7116 \times 10^9 \text{ W/m}^3)}{4(19.0253) \text{ W/m-K}}\left[1 - \left(\frac{0}{r_s}\right)^2\right] = 23.22 \text{ K}$$

Since $(\Delta T)_{\text{°F}} = \frac{9}{5}(\Delta T)_{\text{K}}$, $T - T_s = 41.79$ °F.

2.20. A 6-in.-thick concrete wall, having thermal conductivity $k = 0.50$ Btu/hr-ft-°F, is exposed to air at 70 °F on one side and air at 20 °F on the opposite side. The heat transfer coefficients are $\bar{h}_i = 2.0$ Btu/hr-ft²-°F on the 70 °F side and $\bar{h}_o = 10$ Btu/hr-ft²-°F on the 20 °F side. Determine the heat transfer rate and the two surface temperatures of the wall.

By (2.29) and with reference to Fig. 2-9,

$$\frac{q}{A} = \frac{T_i - T_o}{\dfrac{1}{\bar{h}_i} + \dfrac{L_a}{k_a} + \dfrac{1}{\bar{h}_o}} = \frac{(70 - 20)\,°\text{F}}{\left(\dfrac{1}{2} + \dfrac{0.5}{0.50} + \dfrac{1}{10}\right)\dfrac{\text{hr-ft}^2\text{-°F}}{\text{Btu}}} = 31.25\,\frac{\text{Btu}}{\text{hr-ft}^2}$$

The surface temperatures can be determined from (2.28):

$$\frac{q}{A} = \frac{T_i - T_1}{1/\bar{h}_i}$$

$$T_1 = T_i - \frac{q}{A}\frac{1}{\bar{h}_i} = 70\,°\text{F} - \left(31.25\,\frac{\text{Btu}}{\text{hr-ft}^2}\right)\left(\frac{\text{hr-ft}^2\text{-°F}}{2.0 \text{ Btu}}\right) = 54.375\,°\text{F}$$

$$\frac{q}{A} = \frac{T_2 - T_o}{1/\bar{h}_o}$$

$$T_2 = T_o + \frac{q}{A}\frac{1}{\bar{h}_o} = 20\,°\text{F} + \left(31.25\,\frac{\text{Btu}}{\text{hr-ft}^2}\right)\left(\frac{\text{hr-ft}^2\text{-°F}}{10 \text{ Btu}}\right) = 23.125\,°\text{F}$$

2.21. A 12-in.-thick brick outer wall is used in an office building in a southern city with no insulation or added internal finish. On a winter day the following temperatures were measured: inside air temperature, $T_i = 70$ °F; outside air temperature, $T_o = 15$ °F; inside surface temperature, $T_1 = 56\,°F$; outside surface temperature, $T_2 = 19.5\,°F$. Using $k = 0.76\ Btu/hr\text{-}ft\text{-}°F$ from Table B-2, estimate the average values of the inner and outer heat-transfer coefficients, $\bar{h}_i$ and $\bar{h}_o$.

With reference to Fig. 2-9, the heat transfer per unit area may be determined by applying Fourier's law to the solid brick wall. Thus

$$\frac{q}{A} = -k\frac{\Delta T}{\Delta x} = \frac{-0.76\ Btu}{hr\text{-}ft\text{-}°F}\frac{(19.5 - 56)\,°F}{1\ ft} = 27.74\ \frac{Btu}{hr\text{-}ft^2}$$

Now, from (2.28),

$$\frac{q}{A} = 27.74\ \frac{Btu}{hr\text{-}ft^2} = \bar{h}_i[(70 - 56)\,°F] \quad \text{or} \quad \bar{h}_i = 1.981\ \frac{Btu}{hr\text{-}ft^2\text{-}°F}$$

$$\frac{q}{A} = 27.74\ \frac{Btu}{hr\text{-}ft^2} = \bar{h}_o[(19.5 - 15)\,°F] \quad \text{or} \quad \bar{h}_o = 6.164\ \frac{Btu}{hr\text{-}ft^2\text{-}°F}$$

2.22. Determine U for the situation of Problem 2.21, using SI units throughout. From U and the overall temperature difference, determine q/A in W/m^2. Compare this with the result of Problem 2.21.

Using the conversion factors of Appendix A,

$$\bar{h}_i = \left(1.981\ \frac{Btu}{hr\text{-}ft^2\text{-}°F}\right)\left(\frac{3.1524\ W/m^2}{Btu/hr\text{-}ft^2}\right)\left(\frac{9\,°F}{5\ K}\right) = 11.241\ W/m^2\text{-}K$$

$$\bar{h}_o = \left(6.164\ \frac{Btu}{hr\text{-}ft^2\text{-}°F}\right)\left(\frac{3.1524\ W/m^2}{Btu/hr\text{-}ft^2}\right)\left(\frac{9\,°F}{5\ K}\right) = 34.976\ W/m^2\text{-}K$$

$$L_a = (12\ in.)(0.0254\ m/in.) = 0.3048\ m$$

$$k_a = \left(0.76\ \frac{Btu}{hr\text{-}ft\text{-}°F}\right)\left(\frac{1.729577\ W/m\text{-}K}{Btu/hr\text{-}ft\text{-}°F}\right) = 1.314\ W/m\text{-}K$$

Applying (2.32),

$$U = \frac{1}{\left(\dfrac{1}{11.241} + \dfrac{0.3048}{1.314} + \dfrac{1}{34.976}\right)\dfrac{m^2\text{-}K}{W}} = 2.861\ W/m^2\text{-}K$$

Now

$$(\Delta T)_{overall} = [(70 - 15)\,°F] \times \frac{5\ K}{9\,°F} = 30.555\ K$$

so

$$\frac{q}{A} = \frac{2.861\ W}{m^2\text{-}K}(30.555\ K) = 87.421\ W/m^2$$

Converting to British Engineering units,

$$\frac{q}{A} = \left(87.421\ \frac{W}{m^2}\right)\left(\frac{Btu/hr\text{-}ft^2}{3.1524\ W/m^2}\right) = 27.73\ \frac{Btu}{hr\text{-}ft^2}$$

which agrees well with the result in Problem 2.21. Note that the conversion $(9\,°F)/(5\ K)$ may *not* be used to convert a specific temperature but may be used for ΔT conversions and for unit conversions.

2.23. Steam at $250\,°F$ flows in an insulated pipe. The pipe is mild steel and has an inside radius of 2.0 in. and an outside radius of 2.25 in. The pipe is covered with a one-inch layer of 85% magnesia. The inside heat-transfer coefficient, h_i, is 15 Btu/hr-ft²-°F, and the outside coefficient, h_o, is 2.2 Btu/hr-ft²-°F.   Determine the overall heat-transfer coefficient U_o and the heat transfer rate from the steam per foot of pipe length, if the surrounding air temperature is 65 °F.

In terms suitable for (2.35),

$$n = 3 \qquad\qquad k_{1,2} = k_{\text{steel}} = 26 \text{ Btu/hr-ft-°F}$$
$$r_1 = 2.0 \text{ in.} \qquad k_{2,3} = k_{\text{mag}} = 0.041 \text{ Btu/hr-ft-°F}$$
$$r_2 = 2.25 \text{ in.} \qquad h_i = 15 \text{ Btu/hr-ft}^2\text{-°F}$$
$$r_3 = r_n = 3.25 \text{ in.} \qquad h_o = 2.2 \text{ Btu/hr-ft}^2\text{-°F}$$

where thermal conductivity data are from Tables B-1 and B-2 at temperatures reasonably close to the expected average material temperatures.   By (2.35),

$$U_o = \cfrac{1}{\cfrac{3.25}{2.0 \times 15} + \cfrac{(3.25/12) \ln (2.25/2.0)}{26} + \cfrac{(3.25/12) \ln (3.25/2.25)}{0.041} + \cfrac{1}{2.2}}$$

where each term in the denominator has units hr-ft²-°F/Btu.   Thus,

$$U_o = \frac{1}{0.1083 + 0.0012 + 2.4291 + 0.4545} = \frac{1}{2.9932} = 0.3341 \text{ Btu/hr-ft}^2\text{-°F}$$

Clearly, the thermal resistance of the steel pipe wall is negligibly small in this problem.   The heat transfer per unit length of the pipe is

$$\frac{q}{L} = U_o \frac{A}{L} (\Delta T)_{\text{overall}} = \frac{0.3341 \text{ Btu}}{\text{hr-ft}^2\text{-°F}} (2\pi) \left(\frac{3.25}{12} \text{ ft} \right) [(250 - 65) \text{ °F}] = 105.18 \text{ Btu/hr-ft}$$

2.24.   In Problem 2.23 the thermal conductivity of the magnesia was taken at 200 °F.   Determine the two surface temperatures of the magnesia using $q/L = 105.18$ Btu/hr-ft, and evaluate k_{mag} at the average temperature.

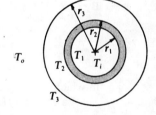

Fig. 2-19

Refer to Fig. 2-19.   Since $\bar{h}_o$ and $\bar{h}_i$ are specified together with T_o and T_i, we need to determine T_2 beginning with T_i and working from the inside out; and we should calculate T_3 beginning with T_o and working inward.   From the steam to the inner wall of the steel pipe:

$$\frac{q}{L} = 2\pi r_1 \bar{h}_i (T_i - T_1)$$

$$T_1 = T_i - \frac{q/L}{2\pi r_1 \bar{h}_i} = 250 - \frac{105.18}{2\pi(2.0/12)(15)} = 243.30 \text{ °F}$$

Through the steel pipe:

$$\frac{q}{L} = \frac{2\pi k_{\text{st}} (T_1 - T_2)}{\ln (r_2/r_1)}$$

$$T_2 = T_1 - \frac{(q/L) \ln (r_2/r_1)}{2\pi k_{\text{st}}} = 243.30 - \frac{105.18 \ln (2.25/2.0)}{2\pi(26)} = 243.23 \text{ °F}$$

From the ambient air to the magnesia outer surface:

$$\frac{q}{L} = 2\pi r_3 \bar{h}_o (T_3 - T_o)$$

$$T_3 = 65 + \frac{105.18}{2\pi(3.25/12)(2.2)} = 93.09 \text{ °F}$$

Hence, the average temperature of the magnesia is

$$T_{avg} = \frac{243.23 + 93.09}{2} = 168.16\ °F$$

and by linear interpolation of the data of Table B-2, $k_{mag} \approx 0.0404$ Btu/hr-ft-°F. However, it is questionable that Table B-2 is sufficiently accurate to justify recalculation of U_o and q/L for Problem 2.23.

2.25. Determine the critical radius in cm for an abestos-covered pipe ($k_{asb} = 0.208$ W/m-K) if the external heat-transfer coefficient is 1.5 Btu/hr-ft²-°F.

First, we need to convert h to SI units. Using Appendix A,

$$h = (1.5\ \text{Btu/hr-ft}^2\text{-°F})\left(\frac{3.15248\ \text{W/m}^2}{\text{Btu/hr-ft}^2}\right)\left(\frac{9\ °F}{5\ K}\right) = 8.51\ \text{W/m}^2\text{-K}$$

By (2.37)

$$r_{crit} = \frac{k}{h} = \frac{0.208\ \text{W/m-K}}{8.51\ \text{W/m}^2\text{-K}} = 0.0244\ \text{m} = 2.44\ \text{cm}$$

2.26. Plot q/L (Btu/hr-ft) versus r (in.) for the situation of Problem 2.25 if $r_i = 0.5$ in., $T_i = 250$ °F, and $T_o = 70$ °F. Consider the range $r = r_i$ to $r = 1.5$ in.

Converting units,

$$k = \left(0.208\ \frac{W}{m\text{-}K}\right)\left(\frac{\text{Btu/hr-ft-°F}}{1.729577\ \text{W/m-K}}\right) = 0.120\ \frac{\text{Btu}}{\text{hr-ft-°F}}$$

By (2.36),

$$\frac{q}{L} = \frac{2\pi(T_i - T_o)}{\dfrac{\ln(r/r_i)}{k} + \dfrac{1}{hr}} = \frac{2\pi(250 - 70)\ °F}{\left[\dfrac{\ln(r/0.5)}{0.120} + \dfrac{12}{1.5r}\right]\dfrac{\text{hr-ft-°F}}{\text{Btu}}} = \frac{1130.97}{\dfrac{\ln(r/0.5)}{0.12} + \dfrac{8}{r}}\ \text{Btu/hr-ft}$$

in which r is in inches.

Values of q/L are displayed in Table 2-2 and Fig. 2-20. Observe that 1.0 in. of insulation

Table 2-2

r, in.	q/L, Btu/hr-ft	r	q/L
0.5	70.69	1.0	82.10
0.6	76.15	1.1	81.70
0.7	79.46	1.2	81.00
0.8	81.27	1.3	80.12
0.9	82.03	1.4	79.12
0.961	82.14	1.5	78.06

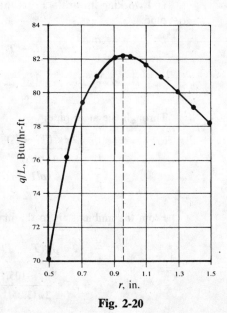

Fig. 2-20

($r = 1.5$ in.) still results in a higher q/L than the bare or uninsulated pipe! This illustrates the necessity of checking the critical radius prior to specification of an insulation thickness, especially for small pipes, ducts, or wires. As a final comment, the critical radius is strongly dependent upon h, which was arbitrarily specified to be constant in this problem. Accurate determination of h can be made using the methods of Chapters 6, 7 and 8.

2.27. An aluminum cylindrical rod ($k = 132$ Btu/hr-ft-°F), having a diameter of 0.375 in. and a length of 4 in., is attached to a surface having a temperature of 200 °F. The rod is exposed to ambient air at 70 °F, and the heat transfer coefficient along the length and at the end is 1.5 Btu/hr-ft²-°F. Determine the temperature distribution and the heat flux (*a*) neglecting the heat transfer at the end and (*b*) accounting for the heat transfer at the end.

For a cylindrical rod, (*2.45*) gives

$$n = \sqrt{\frac{2\bar{h}}{kr}} = \sqrt{\frac{2(1.5)}{132[0.375/2(12)]}} = 1.2060 \text{ ft}^{-1}$$

(*a*) Using the solution for θ/θ_b given in Table 2-1 (Case 2),

$$\frac{T - 70}{200 - 70} = \frac{\cosh\left[1.2060\left(\frac{4}{12} - x\right)\right]}{\cosh\left[(1.2060)\left(\frac{4}{12}\right)\right]}$$

$$T = 70\,°\text{F} + (130\,°\text{F})\frac{\cosh\left[1.2060\left(\frac{4}{12} - x\right)\right]}{1.0819}$$

Evaluating T at 1/2-in. intervals, we obtain Table 2-3.
The heat transfer rate is, by (*2.42*),

$$q = kAn\theta_b \tanh nL$$

$$= 132\left(\frac{\pi}{4}\right)\left(\frac{0.375}{12}\right)^2 (1.2060)(200 - 70)\tanh\left[(1.2060)\left(\frac{4}{12}\right)\right] = 6.058 \text{ Btu/hr}$$

(*b*) The appropriate solution (Case 3) from Table 2-1, with $\bar{h}_L/nk = \bar{h}/nk = 0.00942$, gives

$$T = 70\,°\text{F} + (130\,°\text{F})\left\{\frac{\cosh\left[1.2060\left(\frac{4}{12} - x\right)\right] + 0.00942\sinh\left[1.2060\left(\frac{4}{12} - x\right)\right]}{\cosh\left[(1.2060)\left(\frac{4}{12}\right)\right] + 0.00942\sinh\left[(1.2060)\left(\frac{4}{12}\right)\right]}\right\}$$

Evaluating T at 1/2-in. intervals yields Table 2-4.

Table 2-3

x, in.	T, °F	x	T
0.0	200.00	2.5	191.53
0.5	197.67	3.0	190.77
1.0	195.66	3.5	190.31
1.5	193.97	4.0	190.16
2.0	192.59		

Table 2-4

x, in.	T, °F	x	T
0.0	200.00	2.5	191.26
0.5	197.63	3.0	190.45
1.0	195.56	3.5	189.94
1.5	193.81	4.0	189.73
2.0	192.38		

The values in Tables 2-3 and 2-4 are only slightly different, the largest difference being near the rod end ($x = 4$).
The heat transfer rate from the rod is, by (*2.43*),

$$q = 132\left(\frac{\pi}{4}\right)\left(\frac{0.375}{12}\right)^2(1.2060)(200-70)$$

$$\times \frac{\sinh\left[(1.2060)(\frac{4}{12})\right]+0.00942\cosh\left[(1.2060)(\frac{4}{12})\right]}{\cosh\left[(1.2060(\frac{4}{12})\right]+0.00942\sinh\left[(1.2060)(\frac{4}{12})\right]}$$

$$= 6.185 \text{ Btu/hr}$$

If, instead of this exact solution for q, we use the corrected length $L_c = (4 + 0.375/4)$ in. in (2.42), we find

$$q = 132\left(\frac{\pi}{4}\right)\left(\frac{0.375}{12}\right)^2(1.2060)(130)\tanh\left[(1.2060)\left(\frac{4.0938}{12}\right)\right] = 6.185 \text{ Btu/hr}$$

The solution for q in part (a) is 2.04 percent low, whereas the L_c-approach yields a result essentially the same as the exact solution.

2.28. A thin fin of length L has its two ends attached to two parallel walls which have temperatures T_1 and T_2 (Fig. 2-21). The fin loses heat by convection to the ambient air at T_∞. Obtain an analytical expression for the one-dimensional temperature distribution along the length of the fin.

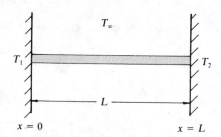

Fig. 2-21

The general solution for the rectangular fin,

$$\theta = C_1 e^{nx} + C_2 e^{-nx}$$

where $\theta = T - T_\infty$ applies to this problem. The two boundary conditions are

$$\theta(0) = T_1 - T_\infty = \theta_1 \qquad \theta(L) = T_2 - T_\infty = \theta_2$$

Applying these gives

$$C_1 = \frac{\theta_2 - \theta_1 e^{-nL}}{e^{nL} - e^{-nL}} \qquad C_2 = \theta_1 - C_1$$

Substituting,

$$\theta = \left(\frac{\theta_2 - \theta_1 e^{-nL}}{e^{nL} - e^{-nL}}\right)(e^{nx} - e^{-nx}) + \theta_1 e^{-nx}$$

$$= \theta_2\left(\frac{e^{nx} - e^{-nx}}{e^{nL} - e^{-nL}}\right) + \theta_1\left[\frac{e^{-nx}(e^{nL} - e^{-nL}) - e^{-nL}(e^{nx} - e^{-nx})}{e^{nL} - e^{-nL}}\right]$$

Since $\sinh y \equiv (e^y - e^{-y})/2$,

$$\theta = \theta_1\frac{\sinh n(L-x)}{\sinh nL} + \theta_2\frac{\sinh nx}{\sinh nL}$$

This result is, of course, applicable to cylindrical rods of small diameter as well as to rectangular fins. Also, the total heat loss from such a fin or rod with both ends fixed could be obtained by applying Fourier's law at each end and algebraically summing the two conductive heat-transfer rates into the fin. See Problem 2.47.

2.29. Consider a thin cylindrical rod with its ends fixed to two parallel surfaces (similar to Fig. 2-21). Let the dimensions, properties, and temperatures be such that $n = 3.0 \text{ ft}^{-1}$, $\theta_1 = 20\,°F$, $\theta_2 = 20\,°F$, and $L = 1.0 \text{ ft}$. (a) Determine the temperature at $x = 0.4 \text{ ft}$ using the result found in Problem 2.28, if the ambient temperature is $75\,°F$. (b) Obtain the same numerical result by considering the symmetry of the problem and using the temperature distribution given in Table 2-1 for the insulated-end case.

(a) From Problem 2.28

$$\theta(0.4) = 20\frac{\sinh\,[3(1.0 - 0.4)]}{\sinh\,[3(1.0)]} + 20\frac{\sinh\,[(3)(0.4)]}{\sinh\,[(3)(1.0)]}$$

$$= 5.8738 + 3.0135 = 8.8873\,°\text{F}$$

so that $T(0.4) = 75 + 8.8873 = 83.89\,°\text{F}$.

(b) As the problem is symmetrical about the midpoint of the rod, the insulated-end solution applies with $L = 0.5\,\text{ft}$. So,

$$\theta(0.4) = 20\frac{\cosh\,[3(0.5 - 0.4)]}{\cosh\,(3 \times 0.5)} = 20\frac{\cosh\,(0.3)}{\cosh\,(1.5)} = 8.8874$$

as before.

2.30. A very long, 1-cm-diameter copper rod ($k = 377\,\text{W/m-K}$) is exposed to an environment at $22\,°\text{C}$. The base temperature of the rod is maintained at $150\,°\text{C}$. The heat transfer coefficient between the rod and the surrounding air is $11\,\text{W/m}^2\text{-K}$. Determine the heat transfer rate from the rod to the surrounding air.

Since the rod is very long, (2.41) may be used. We have

$$n = \sqrt{\frac{2h}{kr}} = \left[\frac{2(11\,\text{W/m}^2\text{-K})}{(377\,\text{W/m-K})(0.005\,\text{m})}\right]^{1/2} = 3.416\,\text{m}^{-1}$$

Thus,

$$q = kAn\theta_b = (377\,\text{W/m-K})\left(\frac{\pi}{4}\right)(0.01\,\text{m})^2(3.416\,\text{m}^{-1})[(150 - 22)\,\text{K}] = 12.948\,\text{W}$$

2.31. Repeat Problem 2.30 for finite lengths $2, 4, 8, \ldots, 128\,\text{cm}$, assuming heat loss at the end, i.e. Case 3. Assume $h_L = 11\,\text{W/m}^2\text{-K}$ also.

The heat loss from the rod may be calculated from (2.43). We need the parameters

$$n = 3.416\,\text{m}^{-1}\quad\text{(from Problem 2.30)}$$

$$\frac{h_L}{nk} = \frac{11\,\text{W/m}^2\text{-K}}{(3.416\,\text{m}^{-1})(377\,\text{W/m-K})} = 0.00854$$

$$kAn\theta_b = 12.948\,\text{W}\quad\text{(from Problem 2.30)}$$

For $L = 2\,\text{cm}$,

$$nL = 3.416\,\text{m}^{-1} \times 0.02\,\text{m} = 0.06832$$

$$\sinh\,nL = 0.06837\qquad\cosh\,nL = 1.00233$$

and by (2.43)

$$q = (12.948\,\text{W})\left[\frac{0.06837 + 0.00854(1.00233)}{1.00233 + 0.00854(0.06837)}\right] = 0.993\,\text{W}$$

Repeating for lengths of 4, 8, 16, 32, 64, and $128\,\text{cm}$ we obtain the results plotted in Fig. 2-22. This problem illustrates that when k is large there are significant differences between the finite-length and infinite-length cases.

2.32. An annular aluminum alloy fin ($k = 90\,\text{Btu/hr-ft-}°\text{F}$) is mounted on a 1-in.-o.d. heated tube. The fin is of constant thickness equal to $1/64\,\text{in.}$ and has an outer radius of $1.5\,\text{in.}$ The tube wall temperature is $300\,°\text{F}$, the surrounding temperature is $70\,°\text{F}$, and the average convective heat-transfer coefficient is $5\,\text{Btu/hr-ft}^2\text{-}°\text{F}$. Calculate the heat loss from the fin.

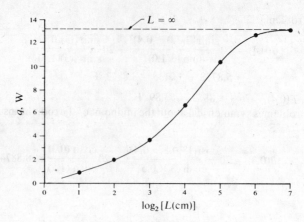

Fig. 2-22

Equation (2.50) may be used, with the parameters

$$nr_1 = \left[\frac{2\bar{h}}{kt}\right]^{1/2} r_1 = \left[\frac{2(5)(12)}{90(1/64)}\right]^{1/2}\left(\frac{0.5}{12}\right) = 0.3849$$

$$nr_2 = \left[\frac{2(5)(12)}{90(1/64)}\right]^{1/2}\left(\frac{1.5}{12}\right) = 1.1547$$

Thus,

$$q = 2\pi(90)\left[\frac{1}{64(12)}\right](300-70)(0.3849)\left[\frac{K_1(0.385)I_1(1.155) - I_1(0.385)K_1(1.155)}{I_0(0.385)K_1(1.155) + K_0(0.385)I_1(1.155)}\right]$$

The modified Bessel functions are obtained from standard tables of higher functions, and thus

$$q = (65.18)\left[\frac{(2.2860)(0.6793) - (0.1961)(0.4667)}{(1.0374)(0.4667) + (1.1483)(0.6793)}\right] = 75.35 \text{ Btu/hr}$$

The main drawback to the use of (2.50) is the effort required to obtain accurate values of the Bessel functions.

2.33. Repeat Problem 2.32 using the fin-efficiency approach with (a) no length correction for the tip loss and (b) a length correction.

(a)
$$L = \frac{1}{12} \text{ ft} \qquad A_p = \frac{1}{12}\left[\frac{1}{(64)(12)}\right] \text{ ft}^2$$

$$L^{3/2}\left(\frac{\bar{h}}{kA_p}\right)^{1/2} = \left(\frac{1}{12}\right)^{3/2}\left[\frac{5(64)(144)}{(90)(1)}\right]^{1/2} = 0.5443$$

$$\frac{r_2}{r_1} = \frac{1.5}{0.5} = 3.0$$

From Fig. 2-16, $\eta_f \approx 0.75$, so that (2.53) gives

$$q_{\text{act}} = \eta_f \bar{h} A_f \theta_b \approx (0.75)\left(\frac{5 \text{ Btu}}{\text{hr-ft}^2\text{-}°\text{F}}\right)\left[\left(\frac{2}{144}\right)(\pi)(1.5^2 - 0.5^2) \text{ ft}^2\right](230 °\text{F}) = 75.27 \text{ Btu/hr}$$

which is very close to the previous solution.

(b)
$$L_c = \frac{1}{12} + \frac{1}{2}\left[\frac{1}{(64)(12)}\right] = \frac{1 + 1/128}{12} \text{ ft} = 0.08398 \text{ ft}$$

$$A_p = \left(\frac{1 + 1/128}{12} \text{ ft}\right)\left[\frac{1}{(64)(12)} \text{ ft}\right] = 1.0935 \times 10^{-4} \text{ ft}^2$$

$$L_c^{3/2}\left(\frac{\bar{h}}{kA_p}\right)^{1/2} = (0.08398)^{3/2}\left[\frac{5 \times 10^4}{(90)(1.0935)}\right]^{1/2} = 0.5485$$

This difference is negligible when using the efficiency method in conjunction with Fig. 2-16.

2.34. A 1.0-in.-o.d. tube is fitted with 2.0-in.-o.d. annular fins spaced on 3/16-in. centers. The fins are aluminum alloy (k = 93 Btu/hr-ft-°F) and are of constant thickness 0.0009 in. The external free convective heat-transfer coefficient to the ambient air is 1.5 Btu/hr-ft²-°F. For a tube wall temperature of 330 °F and an ambient temperature of 80 °F, determine the heat loss per foot of length of finned tube.

We will first determine the fin efficiency and then apply (2.54).

$$L_c = \frac{2.0 - 1.0}{2 \times 12} + \frac{0.009}{2 \times 12} = \frac{0.5045}{12} \text{ ft}$$

$$(L_c)^{3/2}\left[\frac{\bar{h}}{kA_p}\right]^{1/2} = \left(\frac{0.5045}{12}\right)^{3/2}\left(\frac{1.5 \times 144}{93 \times 0.5045 \times 0.009}\right)^{1/2} = 0.1950$$

$$\frac{r_{2c}}{r_1} = \frac{1.0045}{0.5} = 2.009$$

From Fig. 2-16, $\eta_f \approx 0.94$. The number of fins per lineal foot of tube is

$$\text{no.} = \frac{12 \text{ in.}}{3/16 \text{ in./fin}} = 64 \text{ fins}$$

Area of fins:

$$A_f = (\text{no.})(2)(\pi)(r_{2c}^2 - r_1^2) = \frac{(64)(2\pi)}{144}(1.0045^2 - 0.5^2) = 2.1196 \text{ ft}^2$$

Area of exposed tube:

$$A_b = [\pi D - t(\text{no.})](1 \text{ ft}) = \left[\pi\left(\frac{1}{12}\right) - \frac{0.009}{12}(64)\right](1 \text{ ft}) = 0.2138 \text{ ft}^2$$

The heat transfer per lineal foot is

$$q = \bar{h}(A_b + \eta_f A_f)\theta_b = (1.5)[0.2138 + (0.94)(2.1196)](330 - 80) = 827.3 \text{ Btu/hr}$$

Supplementary Problems

2.35.   Begin with an appropriate form of the general conduction equation for the steady-state, radial flow in a single-layered, spherical shell. Obtain the temperature gradient dT/dr by solution of the equation and substitute this into Fourier's law to obtain (*2.16*).

2.36.   In Problem 2.4 what is the steady-state interfacial temperature T_2 between the brick and the insulation?   *Ans.*   1258.7 °F

2.37.   For the ceiling of Problem 2.6, what is the temperature at the center of the rock wool insulation ($2\frac{3}{4}$ in. below upper surface)?   *Ans.*   63.6 °F

2.38.   Steam at 120 °C flows in an insulated steel pipe. The pipe inner radius is 10 cm and the outer radius is 11 cm. This is covered with a 3-cm-thick layer of asbestos having a density of 577 kg/m³. The outer asbestos surface is at 45 °C. Using mild steel thermal-conductivity data at 100 °C and asbestos thermal-conductivity data at 70 °C (interpolate if necessary), determine the heat transfer from the steam per meter of pipe length. Use SI units throughout.   *Ans.*   351 W/m

2.39.   Problem 2.14 presents an expression of the form $k = k_0[1 + b(T - T_{ref})]$ for the thermal conductivity of copper in SI units, where $k_0 = 371.9$ W/m-K at $T_{ref} = 150$ °C and $b = -9.25 \times 10^{-5}$ K⁻¹. Convert this expression to British Engineering units. Compare k from the resulting expression with the four values in Table B-1. Comment on any lack of agreement.

     Ans. $k = (215.02)[1 - 5.14 \times 10^{-5}(T - 302)]$ Btu/hr-ft-°F; $k_{32} = 218.0$, $k_{212} = 216.01$, $k_{572} = 212.04$,
         $k_{932} = 208.06$ Btu/hr-ft-°F; not highly accurate outside range of approximately 300 °F to 750 °F

2.40.   Estimate the conductive heat transfer in the insulated copper wire between the two liquid surfaces for two thermocouples located in boiling water and an ice bath (Fig. 2-23). The wire length between the surfaces is 14 in., the wire is A.W.G. no. 28 (0.0126 in. diameter) and is pure copper. Use linear interpolation of Table B-1 to obtain k_m.

     Ans. 2.95×10^{-2} Btu/hr

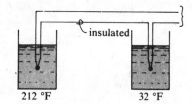

Fig. 2-23

2.41.   Determine a linear expression for $k(T)$ for 1% mild carbon steel in the temperature range between 0 °C and 300 °C. Comment on the applicability of the resulting expression outside of the stated temperature range.

     Ans. $k = (45.8338)[1 - 1.8868 \times 10^{-4} T]$ W/m-°C, where T is in °C
           $k = (26.5)[1 - 1.0482 \times 10^{-4}(T - 32)]$ Btu/hr-ft-°F, where T is in °F

2.42.   A wall has a freshly plastered layer which is 0.5 in. thick. If q''' due to the chemical reaction during "curing" of the plaster is approximately constant at 5000 Btu/hr-ft³, the outer surface is insulated (no heat transfer), and the inner surface is held at 90 °F, determine the steady-state temperature of the outer surface. Assume that k of the fresh plaster is 0.5 Btu/hr-ft-°F (which is higher than the values listed in Appendix B due to increased moisture content). (*Hint:* This problem is mathematically the same as that of a plaster wall 1 in. thick with both surfaces held at 90 °F.)   *Ans.*   98.68 °F

2.43.   An electric resistance wire 0.10 in. in diameter and 1.5 ft long has a measured voltage drop of 25 V for a current flow of 40 A. The material thermal conductivity is 14 Btu/hr-ft-°F. Determine (*a*) q''' in Btu/hr-ft³ and (*b*) the maximum temperature in the wire if the surface temperature is 1200 °F.

     Ans.   (*a*) 4.1705×10^7 Btu/hr-ft³; (*b*) 1212.93 °F

2.44.   A rectangular steel tank is filled with a liquid at 150 °F and exposed along the outside surface to air at 70 °F, the inner and outer convective heat-transfer coefficients being $\bar{h}_i = 4.0$ Btu/hr-ft^2-°F and $\bar{h}_o = 1.5$ Btu/hr-ft^2-°F. The tank wall is 1/4-in. mild steel ($k = 26$ Btu/hr-ft-°F), and this is covered with a 1-in. layer of glass wool ($k = 0.024$ Btu/hr-ft-°F). Determine (*a*) the overall heat-transfer coefficient U and (*b*) the heat transfer rate per sq ft.

Ans.   (*a*) 0.2278 Btu/hr-ft^2-°F; (*b*) 18.224 Btu/hr-ft^2
(Note that the thermal resistance due to the steel wall is negligible in this problem.)

2.45.   Determine the critical radius of insulation for asbestos felt, 40 laminations per inch, at 100 °F, if the external heat-transfer coefficient is 1.2 Btu/hr-ft^2-°F.   *Ans.*   0.33 in.

2.46.   Repeat Problem 2.31 for an aluminum alloy rod ($k = 161$ W/m-K). All other factors remain unchanged.

Ans.   Selected values of L and q are: $L = \infty$, $q = 8.465$ W; $L = 16$ cm, $q = 5.848$ W; $L = 32$ cm, $q = 7.903$ W

2.47.   Using the expression for the temperature distribution obtained in Problem 2.28, determine an analytical expression for the heat transfer from a fin or rod attached to two parallel walls as shown in Fig. 2-21.

Ans. $q = kA\left(\dfrac{d\theta}{dx}\bigg|_{x=L} - \dfrac{d\theta}{dx}\bigg|_{x=0}\right) = \dfrac{kAn}{\sinh nL}[\theta_2(\cosh nL - 1) + \theta_1(\cosh nL - 1)]$

(q is the sum of the two conductive heat-transfer rates *into* the fin at its ends.)

2.48.   Show that for the straight rectangular fin, (*2.46*) reduces to (*2.39*).

Chapter 3

Multidimensional Steady-State Conduction

3.1 INTRODUCTION

The steady-state temperature in a three-dimensional cartesian coordinate system obeys, when thermal conductivity is constant, the Laplace equation,

$$\frac{\partial^2 T}{\partial x^2} + \frac{\partial^2 T}{\partial y^2} + \frac{\partial^2 T}{\partial z^2} = 0 \tag{3.1}$$

The solution of this equation, $T(x, y, z)$, can be differentiated and combined with Fourier's equation to yield the components of the *vector heat-transfer rate*. These components are

$$q_x = -kA_x\frac{\partial T}{\partial x} \qquad q_y = -kA_y\frac{\partial T}{\partial y} \qquad q_z = -kA_z\frac{\partial T}{\partial z} \tag{3.2}$$

where A_x is the area normal to q_x, etc.

A number of methods of solving the Laplace equation are available, including analytical, numerical, graphical and analog techniques.

3.2 ANALYTICAL SOLUTIONS

Method of Separation of Variables

We illustrate this classical method of solution by means of an example.

EXAMPLE 3.1. A very long (z-direction) rectangular bar has three of its lateral sides held at a fixed temperature; the temperature distribution across the fourth side is sinusoidal (see Fig. 3-1). Find the temperature distribution within the bar.

By using the shifted temperature $\theta = T - T_0$ we may suppose the fixed temperature to be zero. Since there is no z-direction temperature gradient, the Laplace equation is

$$\frac{\partial^2 \theta}{\partial x^2} + \frac{\partial^2 \theta}{\partial y^2} = 0 \tag{3.3}$$

subject to the boundary conditions

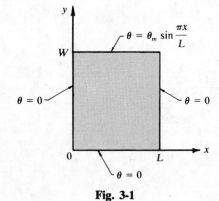

Fig. 3-1

    (1) $\theta(0, y) = 0$ $(0 < y < W)$
    (2) $\theta(L, y) = 0$ $(0 < y < W)$
    (3) $\theta(x, 0) = 0$ $(0 < x < L)$
    (4) $\theta(x, W) = \theta_m \sin\dfrac{\pi x}{L}$ $(0 < x < L)$

Assume a solution of the form $\theta(x, y) = X(x)\,Y(y)$. When substituted into the Laplace equation this yields

$$-\frac{1}{X}\frac{d^2 X}{dx^2} = \frac{1}{Y}\frac{d^2 Y}{dy^2}$$

48

The left side, a function of x alone, can equal the right side, a function of y alone, only if both sides have a constant value, say $\lambda^2 > 0$ (see Problem 3.1).   Then,

$$\frac{d^2X}{dx^2} + \lambda^2 X = 0 \qquad \frac{d^2Y}{dy^2} - \lambda^2 Y = 0$$

The general solutions to these separated equations are:

$$X = C_1 \cos \lambda x + C_2 \sin \lambda x \qquad Y = C_3 e^{-\lambda y} + C_4 e^{\lambda y}$$

so that

$$\theta = (C_1 \cos \lambda x + C_2 \sin \lambda x)(C_3 e^{-\lambda y} + C_4 e^{\lambda y})$$

Now, applying the boundary conditions, (1) gives $C_1 = 0$ and (3) gives $C_3 = -C_4$.   Using these together with (2) yields

$$0 = C_2 C_4 (\sin \lambda L)(e^{\lambda y} - e^{-\lambda y})$$

which requires that

$$\sin \lambda L = 0 \qquad \text{or} \qquad \lambda = \frac{n\pi}{L} \text{ (n an integer)}$$

Because the original differential equation (3.3) is linear, the sum of any number of solutions constitutes a solution.   Thus, θ can be written as the sum of an infinite series:

$$\theta = \sum_{n=1}^{\infty} C_n \sin \frac{n\pi x}{L} \sinh \frac{n\pi y}{L} \tag{3.4}$$

where the constants have been combined and where we have replaced $e^{\lambda y} - e^{-\lambda y}$ by $2 \sinh \lambda y$.
   Finally, boundary condition (4) gives

$$\theta_m \sin \frac{\pi x}{L} = \sum_{n=1}^{\infty} C_n \sin \frac{n\pi x}{L} \sinh \frac{n\pi W}{L} \tag{3.5}$$

which holds only if $C_2 = C_3 = C_4 = \cdots = 0$ and

$$C_1 = \frac{\theta_m}{\sinh \dfrac{\pi W}{L}}$$

Therefore,

■ $$\theta = \theta_m \frac{\sinh (\pi y/L)}{\sinh (\pi W/L)} \sin \frac{\pi x}{L} \tag{3.6}$$

which is the final expression for the temperature distribution in the plate.

EXAMPLE 3.2.   Example 3.1 is changed so that the shifted temperature along $y = W$ is given by the arbitrary function $f(x)$.   Find the temperature distribution within the bar.
   Everything in Example 3.1 through (3.4) remains valid for the present problem.   The new fourth boundary condition gives, instead of (3.5),

$$f(x) = \sum_{n=1}^{\infty} C_n \sin \frac{n\pi x}{L} \sinh \frac{n\pi W}{L} \qquad (0 < x < L)$$

Thus, the quantities $C_n \sinh (n\pi W/L)$ must be the coefficients of the Fourier sine series for $f(x)$ in the interval $0 < x < L$.   From the theory of Fourier series,

$$C_n \sinh \frac{n\pi W}{L} = \frac{2}{L} \int_0^L f(x) \sin \frac{n\pi x}{L} \, dx$$

and

$$\theta = \frac{2}{L} \sum_{n=1}^{\infty} \frac{\sinh (n\pi y/L)}{\sinh (n\pi W/L)} \sin \frac{n\pi x}{L} \int_0^L f(x) \sin \frac{n\pi x}{L} \, dx \tag{3.7}$$

For the special case $f(x) = \theta_c = $ constant, (3.7) reduces to

$$\theta = \theta_c \frac{2}{\pi} \sum_{n=1}^{\infty} \frac{(-1)^{n+1} + 1}{n} \frac{\sinh (n\pi y/L)}{\sinh (n\pi W/L)} \sin \frac{n\pi x}{L} \tag{3.8}$$

Note that only odd-order terms appear in (3.8).

Principle of Superposition

The linearity of the Laplace equation requires that a linear combination of solutions is itself a solution; this is the *principle of superposition*. Examples 3.1 and 3.2 dealt with problems having only one nonhomogeneous boundary condition (for present purposes, a nonhomogeneous boundary condition is one for which the dependent variable θ is not zero). A problem having more than one nonhomogeneous boundary condition can be resolved into a set of simpler problems each with the physical geometry of the original problem and each having only one nonhomogeneous boundary condition. The solutions to the simpler problems can be superposed (at the geometric point being considered) to yield the solution to the original problem.

EXAMPLE 3.3. Determine the temperature at the center of Fig. 3-2.

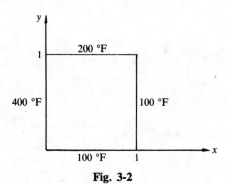

Fig. 3-2

Two of the nonhomogeneous boundary conditions can be removed by defining $\theta = T - (100\ °F)$. Then the resulting problem may be separated into the two subproblems of Fig. 3-3.

Either by use of (3.8) or by intuition (see Problem 3.19), the solutions are $\theta_1 = 75\ °F$ and $\theta_2 = 25\ °F$. Consequently,

$$\theta = 75 + 25 = 100\ °F \qquad \text{and} \qquad T = \theta + 100 = 200\ °F$$

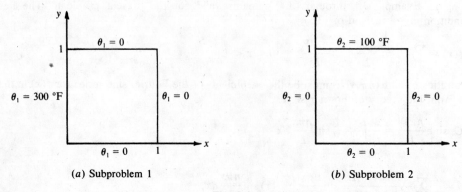

(a) Subproblem 1   (b) Subproblem 2

Fig. 3-3

3.3 CONDUCTIVE SHAPE FACTOR

Figure 3-4 represents a heated pipe with a thick layer of insulation.   The inner surface of the insulation is at uniform temperature T_i, the outer surface is at uniform temperature T_o, and there is a resulting outward heat flux for $T_i > T_o$.   Constructing *uniformly spaced lines* perpendicular to the isotherms results in a group of *heat flow lanes*.   In Fig. 3-4 there are four such lanes in the quadrant selected for study, and the other three quadrants would be similar due to the problem symmetry.   If we can determine the rate of heat transfer for a single lane, then we can easily find the total.   Note that there is no heat transfer across one of the radial lines, such as line a-b, because there is no angular temperature gradient.

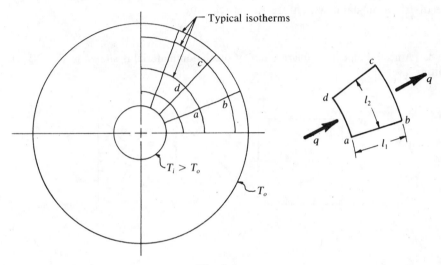

Fig. 3-4

By Fourier's law applied to the element a-b-c-d of a typical lane, the heat transfer per unit depth is

$$\frac{q}{L} = \frac{kl_2(T_{ad} - T_{bc})}{l_1} \tag{3.9}$$

For the case where $l_1 = l_2$ (then the element is called a *curvilinear square*) this simplifies to

$$\frac{q}{L} = k(T_{ad} - T_{bc}) \tag{3.10}$$

In this case, $l_1 = (ab + cd)/2$ and $l_2 = (ad + bc)/2$.

Now if the isotherms are uniformly spaced (with regard to temperature difference) and if there are M such curvilinear squares in the flow lane, then the temperature difference across one square is

$$\Delta T = \frac{T_i - T_o}{M} \tag{3.11}$$

Using this together with (3.10) and noting that the entire system consists of N flow lanes yields

$$\frac{q}{L} = Nk\left(\frac{T_i - T_o}{M}\right) = \frac{S}{L} k(T_i - T_o) \tag{3.12}$$

where the *conductive shape factor per unit depth* is

$$\frac{S}{L} = \frac{N}{M}$$

Note that although the configuration of Fig. 3-4 results in a one-dimensional conduction problem, the development just presented is equally applicable to two-dimensional problems (see Example 3.4).

Freehand Plotting

One method of obtaining the conductive shape factor is by freehand plotting. As illustrated in the preceding development, a graphical plot of equally spaced isotherms and adiabatics is sufficient to determine the shape factor. The graphical net formed by isotherms and adiabatics can frequently be obtained by freehand drawing to quickly yield heat transfer results, as well as temperature distribution data, which are quite accurate.

EXAMPLE 3.4. A heated pipe in a square block of insulation material is shown in Fig. 3-5(a). Find S/L by freehand plotting.

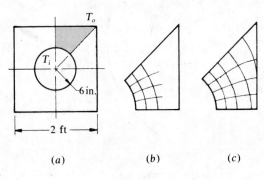

       (a)   (b)   (c)

Fig. 3-5

Since the inner and outer surfaces have constant temperatures T_i and T_o, respectively, the vertical and horizontal centerlines, as well as the corner-to-corner diagonals, are lines of geometrical and thermal symmetry. These are also adiabatics, so we need to construct the flux plot for only one of the typical one-eighth sections; the resulting shape factor for this section will be one-eighth of the overall shape factor.

Figure 3-5(b) illustrates the beginning of the freehand sketch. Notice that the plot was begun by fixing the number of heat flow lanes to be four; the choice is quite arbitrary, but the use of an integral number is advisable. The freehand work is continued in Fig. 3-5(c) by progressing outward by the formation of curvilinear squares. Upon completion the average number of squares per lane for this sketch is $M \approx 3.7$, whence $S/L \approx 8(4)/3.7 = 8.65$.

Suggested techniques for freehand plotting include:

1. Identify all known isotherms.

2. Apply symmetry (geometrical and thermal) to reduce the art work.

3. Flow lines (adiabatics) should bisect the corners of isothermal boundaries.

4. Begin, if possible, in a region where the adiabatics can be uniformly spaced.

5. Begin with a crude network sketch to find the approximate locations of isotherms and adiabatics.

6. Continuously modify the network by maintaining adiabatic lines normal to isothermal lines while forming curvilinear squares.

Electrical Analog

A second technique for determination of the conductive shape factor for a two-dimensional problem is afforded by the fact that the electric potential E also obeys the Laplace equation:

$$\frac{\partial^2 E}{\partial x^2} + \frac{\partial^2 E}{\partial y^2} = 0$$

for steady-state conditions. Consequently, if the boundary conditions for E are similar to those for temperature and if the physical geometry of the problem is the same as for the thermal problem, then lines of constant electric potential are also lines of constant temperature. This analogy leads to a more accurate grid of curvilinear squares than obtainable by freehand plotting and consequently to a somewhat better value of the conductive shape factor. (See Problem 3.26.)

Tabulated Values of S

A summary of useful conductive shape factors is given in Table 3-1, page 54.

3.4 NUMERICAL ANALYSIS

Consider a general two-dimensional body as shown in Fig. 3-6. The body has uniform thickness L in the z-direction and no temperature gradient in that direction. Choosing an appropriate Δx and Δy, the body is divided into a network of rectangles, each containing a single nodal point at its center. It is convenient to consider the heat transfer as occurring between nodal points only, these being connected by fictitious rods acting as conductors/resistors for the heat flow. Thermal energy is considered to be "stored" at the nodal points only. The horizontal and vertical conductances are given by

$$kA_h/\Delta x = kL(\Delta y)/\Delta x \qquad \text{and} \qquad kA_v/\Delta y = kL(\Delta x)/\Delta y$$

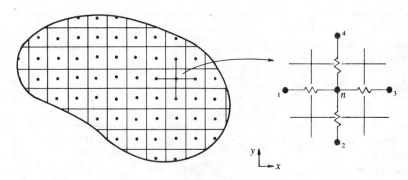

Fig. 3-6

A steady-state energy balance on an interior nodal point n (see the enlarged section of Fig. 3-6) is

$$q_{1\to n} + q_{2\to n} + q_{3\to n} + q_{4\to n} = 0 \qquad\qquad (3.13)$$

where q is taken positive for heat flow into n. Using the product of a conductance and a finite temperature difference for each conductive flux, (3.13) becomes

$$kL(\Delta y)\left(\frac{T_1 - T_n}{\Delta x}\right) + kL(\Delta x)\left(\frac{T_2 - T_n}{\Delta y}\right) + kL(\Delta y)\left(\frac{T_3 - T_n}{\Delta x}\right) + kL(\Delta x)\left(\frac{T_4 - T_n}{\Delta y}\right) = 0$$

Table 3-1. (summarized from J. E. Sunderland and K. R. Johnson, *Trans. ASHRAE*, **10**: 238–239, 1964)

Physical Description	Sketch	Conductive Shape Factor
Conduction through a material of uniform k from a horizontal isothermal cylinder to an isothermal surface		(a) Finite length $$S = \frac{2\pi L}{\cosh^{-1}(z/r)} \quad \frac{z}{L} \ll 1$$ (b) Infinite length (per unit length) $$\frac{S}{L} = \frac{2\pi}{\cosh^{-1}(z/r)}$$
Conduction in a medium of uniform k from a cylinder of length L to two parallel planes of infinite width and length L		$$S = \frac{2\pi L}{\ln(4z/r)}$$
Conduction from an isothermal sphere through a material of uniform k to an isothermal surface		$$S = \frac{4\pi r}{1 - (r/2z)}$$
Conduction between two long isothermal parallel cylinders in an infinite medium of constant k		$$\frac{S}{L} = \frac{2\pi}{\cosh^{-1}[(x^2 - r_1^2 - r_2^2)/2r_1 r_2]}$$ $L \gg r$ $L \gg x$
Conduction between a vertical isothermal cylinder in a medium of uniform k and a horizontal isothermal surface		$$S = \frac{2\pi L}{\ln(4L/d)}$$ $L \gg d$
Conduction through an edge formed by intersection of two plane walls, with inner wall temperature T_1 and outer wall temperature T_2 as shown*		$S = 0.54\,L$ $a > t/5$ $b > t/5$
Conduction through a corner at intersection of three plane walls, each of thickness t, with uniform inner temperature T_1 and outer temperature T_2		$S = 0.15\,t$ inside dimensions $> t/5$

*S for the plane wall is simply A/t, where A for the top wall shown is $A = aL$; for side wall, $A = bL$.

or, when we choose $\Delta x = \Delta y$,

$$T_1 + T_2 + T_3 + T_4 - 4T_n = 0 \qquad (3.14)$$

An equation of the form (3.14) can be written for each interior nodal point of a body, and the simultaneous solution of the resulting set of equations yields the temperatures at the nodal points. These temperatures can then be used to approximate temperature gradients for calculation of heat transfer rates over finite areas.

EXAMPLE 3.5. Write the set of nodal temperature equations for a six-inch square grid for the square chimney shown in Fig. 3-7. Assume the material to have uniform thermal conductivity, uniform inside temperature $T_i = 300$ °F, and uniform outside temperature $T_o = 100$ °F:

In the indicated quadrant of the chimney, the only unknown nodal temperatures are T_a, T_b, and T_c, since clearly $T_{b'} = T_b$, $T_{c'} = T_c$, etc. The nodal equations are, from (3.14),

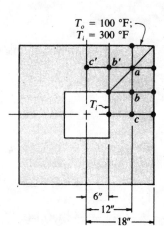

Fig. 3-7

node a:　$T_{b'} + T_b + 100 + 100 - 4T_a = 0$

　　or　　$T_b + 100 - 2T_a = 0$

node b:　$300 + T_c + 100 + T_a - 4T_b = 0$

　　or　　$400 + T_c + T_a - 4T_b = 0$

node c:　$300 + T_{b'} + 100 + T_b - 4T_c = 0$

　　or　　$200 + T_b - 2T_c = 0$

In conventional form, the set of equations is

$$2T_a - 9T_b + \quad 0 = 100$$
$$-T_a + 4T_b - \quad T_c = 400$$
$$- \quad T_b + 2T_c = 200$$

and the problem is reduced to solving this set of linear algebraic equations.

Computer Solution

The two most widely used methods for solving a set of linear algebraic equations with a digital computer are (1) matrix inversion and (2) Gaussian elimination. The latter method is generally more efficient with regard to computer time, and it will be emphasized here.

EXAMPLE 3.6. Solve the system

$$x_1 + 2x_2 + 3x_3 = 20 \qquad (a)$$
$$x_1 - 3x_2 + \quad x_3 = -3 \qquad (b)$$
$$2x_1 + \quad x_2 + \quad x_3 = 11 \qquad (c)$$

by Gaussian elimination.

First, triangularize the given set of equations. This can always be accomplished by repeated application of three basic row operations: (i) multiplication of a row by a constant, (ii) addition to a row of another row, (iii) interchange of two rows.

Thus, eliminate x_1 from (b) and (c) by respectively adding to these equations -1 times (a) and -2 times (a). The result is:

$$x_1 + 2x_2 + 3x_3 = \quad 20 \qquad (a')$$
$$- 5x_2 - 2x_3 = -23 \qquad (b')$$
$$- 3x_2 - 5x_3 = -29 \qquad (c')$$

Now eliminate x_2 from (c') by adding to it $-3/5$ times (b'):

$$x_1 + 2x_2 + 3x_3 = 20$$
$$5x_2 + 2x_3 = 23$$
$$x_3 = 4$$

This is the triangularized set of equations.

Finally, "back substitute," beginning with the bottom equation and working upward, to obtain successively $x_2 = 3$ and $x_1 = 2$. Thus, the final solution is

$$x_1 = 2 \qquad x_2 = 3 \qquad x_3 = 4$$

To solve large systems of equations by Gaussian elimination, it is convenient to express the

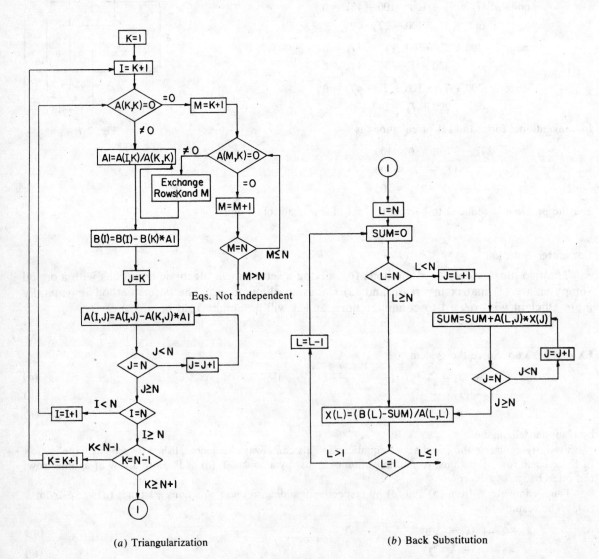

(a) Triangularization   (b) Back Substitution

Fig. 3-8

equations in matrix form:

$$
\begin{bmatrix}
A(1,1) & A(1,2)\ldots A(1,N) \\
A(2,1) & A(2,2)\ldots A(2,N) \\
\cdot & \cdot \qquad \cdot \\
\cdot & \cdot \qquad \cdot \\
\cdot & \cdot \qquad \cdot \\
A(N,1) & A(N,2)\ldots A(N,N)
\end{bmatrix}
\begin{bmatrix}
T(1) \\
T(2) \\
\cdot \\
\cdot \\
\cdot \\
T(N)
\end{bmatrix}
=
\begin{bmatrix}
B(1) \\
B(2) \\
\cdot \\
\cdot \\
\cdot \\
B(N)
\end{bmatrix}
\qquad (3.15)
$$

Figure 3-8 presents one possible flow diagram which is useful in formulating a computer program for solution of a set of linear algebraic equations such as (3.15).

Relaxation Technique

A method of solving a set of linear algebraic equations without the use of a digital computer will be considered next. This method is frequently useful for nonrepetitive-type problems involving a relatively small number of nodal points, say not more than 10, in a conduction problem.

EXAMPLE 3.7. Determine the steady-state temperatures at the 4 interior nodal points of Fig. 3-9.

The nodal equations, obtained with the aid of (3.14), are

node 1: $400 + 500 + T_2 + T_4 - 4T_1 = 0$ 　　(1)

node 2: $500 + 200 + T_1 + T_3 - 4T_2 = 0$ 　　(2)

node 3: $200 + 300 + T_2 + T_4 - 4T_3 = 0$ 　　(3)

node 4: $300 + 400 + T_1 + T_3 - 4T_4 = 0$ 　　(4)

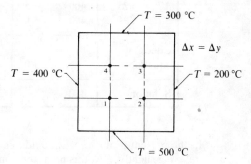

Fig. 3-9

which comprise a set of four linear algebraic equations containing the four unknown nodal temperatures. The *relaxation method* of solution proceeds as follows:

1. Assume (guess) values for the four unknown temperatures. Good initial guesses help to minimize the ensuing work.

2. Since the initial guesses will usually be in error, the right side of each nodal equation will differ from zero; a *residual* will exist due to inaccuracies in the assumed values. Consequently, we replace the zeros in equations (1) through (4) with R_1, R_2, R_3, and R_4, respectively:

$$900 + T_2 + T_4 - 4T_1 = R_1 \qquad (5)$$
$$700 + T_1 + T_3 - 4T_2 = R_2 \qquad (6)$$
$$500 + T_2 + T_4 - 4T_3 = R_3 \qquad (7)$$
$$700 + T_1 + T_3 - 4T_4 = R_4 \qquad (8)$$

3. Set up a "unit change" table such as Table 3-2, which shows the effect of a one-degree change of

Table 3-2

	ΔR_1	ΔR_2	ΔR_3	ΔR_4
$\Delta T_1 = +1$	-4	$+1$	0	$+1$
$\Delta T_2 = +1$	$+1$	-4	$+1$	0
$\Delta T_3 = +1$	0	$+1$	-4	$+1$
$\Delta T_4 = +1$	$+1$	0	$+1$	-4
Block Change $= +1$	-2	-2	-2	-2

temperature at one node upon the residuals. The fact that a "block" (overall) unit change has the same effect upon all residuals is unusual, this being due to the overall problem symmetry.

4. Calculate the initial residuals for the initially assumed temperatures using the "residual equations" (5) through (8).

5. Set up a *relaxation table* such as Table 3-3. Begin with the initially assumed temperatures and the resulting initial residuals. The left-hand column records the changes from the initially assumed temperature values. Notice that the procedure begins by "relaxing" the largest initial residual (or perhaps by making a block change, a technique useful when all residuals are of the same sign).

Table 3-3.   Relaxation Table

	T_1	R_1	T_2	R_2	T_3	R_3	T_4	R_4
Initial Values	400	−25	325	+75	275	+75	350	−25
$\Delta T_2 = +20$	400	−5	345	−5	275	+95	350	−25
$\Delta T_3 = +25$	400	−5	345	+20	300	−5	350	0
$\Delta T_2 = +5$	400	0	350	0	300	0	350	0
Check by equations		√ 0		√ 0		√ 0		√ 0
Solution	400		350		300		350	

In the present problem, we should begin by reducing R_2 or R_3. Arbitrarily choose R_2 and proceed by over-relaxing slightly. At this point the convenience afforded by Table 3-2 becomes evident; this facilitates rapid calculation of the changes in the residuals without recourse to the equations. Notice that the +20° change in T_2 reduced the residuals at nodes 1 and 2 but unfortunately increased R_3.

The first row in Table 3-3 shows the new residuals and temperatures; the only temperature changed is underlined. Proceeding, we next relax the largest resulting residual, this being R_3. Following a temperature change of +25° at node 3, we see that $R_4 = 0$. This does not necessarily mean that we have obtained the correct temperature at node 4, but rather that the set of as yet incorrect temperature values happens to satisfy equation (4) exactly. Proceeding, the largest residual is now R_2, which is reduced to 0 by a +5° change in T_2. This also reduces all remaining residuals to zero. A check is made by substituting the temperatures thus obtained into equations (1) through (4); this verifies the solution.

Exterior Nodal Points

Frequently the temperatures of exterior nodal points are not specified or known at the outset but must be determined as a part of the solution. For example, consider the boundary nodal point subjected to a convective heat transfer, as shown in Fig. 3-10. A steady-state energy balance on nodal point n is

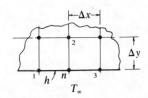

Fig. 3-10

$$kL\frac{\Delta y}{2}\left(\frac{T_1 - T_n}{\Delta x}\right) + kL(\Delta x)\left(\frac{T_2 - T_n}{\Delta y}\right)$$

$$+ kL\frac{\Delta y}{2}\left(\frac{T_3 - T_n}{\Delta x}\right) + hL(\Delta x)(T_\infty - T_n) = 0 \qquad (3.16)$$

where L is again the thickness of the body in the z-direction. Note that the effective horizontal conductance between nodes 1 and n, or that between 3 and n, involves only one-half the area that is associated with a horizontal conductance between a pair of adjacent interior nodal points.

For a square grid, (3.16) simplifies to

$$\frac{1}{2}(T_1 + 2T_2 + T_3) + \frac{h\,\Delta x}{k}(T_\infty) - \left(\frac{h\,\Delta x}{k} + 2\right)T_n = 0 \qquad (3.17)$$

For numerical solution using a digital computer, this last equation is suitable for each exterior nodal point along a plane boundary experiencing convective heat transfer. For solution using the relaxation method, we simply replace the zero on the right-hand side of (3.17) with a residual, R_n.

A summary of useful nodal equations (in residual form) is given in Table 3-4. Simply replace the R_n with zero to use any of these in a computer solution. Although not an exterior nodal point per se, the case of an interior nodal point near a curved boundary has been included in this table for convenience.

Table 3-4. Nodal Equations for Numerical Calculations, Square Grid

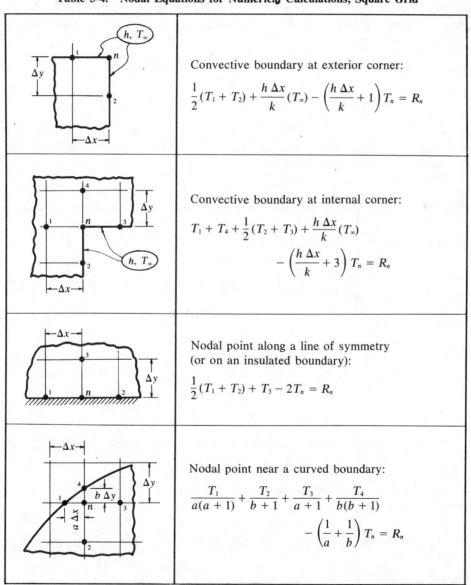

	Convective boundary at exterior corner: $\frac{1}{2}(T_1 + T_2) + \frac{h\,\Delta x}{k}(T_\infty) - \left(\frac{h\,\Delta x}{k} + 1\right)T_n = R_n$
	Convective boundary at internal corner: $T_1 + T_4 + \frac{1}{2}(T_2 + T_3) + \frac{h\,\Delta x}{k}(T_\infty)$ $\qquad - \left(\frac{h\,\Delta x}{k} + 3\right)T_n = R_n$
	Nodal point along a line of symmetry (or on an insulated boundary): $\frac{1}{2}(T_1 + T_2) + T_3 - 2T_n = R_n$
	Nodal point near a curved boundary: $\frac{T_1}{a(a+1)} + \frac{T_2}{b+1} + \frac{T_3}{a+1} + \frac{T_4}{b(b+1)}$ $\qquad - \left(\frac{1}{a} + \frac{1}{b}\right)T_n = R_n$

Solved Problems

3.1. Show that, in the solution of (3.3), the sine-function boundary condition (4) requires a positive separation constant λ^2.

The equation for $X(x)$,

$$\frac{d^2X}{dx^2} + \lambda^2 X = 0$$

has general solutions (for other than $\lambda^2 > 0$)

$$X = C_5 + C_6 x \quad (\lambda^2 = 0) \qquad X = C_7 e^{-|\lambda|x} + C_8 e^{|\lambda|x} \quad (\lambda^2 < 0)$$

In neither case could X fit a sine function along the edge $y = W$.

3.2. Consider a two-dimensional problem of the type shown in Fig. 3-1. The linear dimensions are $W = L = 2$ m, and the temperature on three sides is $T_0 = 280$ K. The temperature along the upper surface is $T_c = 320$ K. Determine the temperature at the center of the plate.

The answer is obtained from the infinite series (3.8), with

$$\frac{\pi y}{L} = \frac{\pi x}{L} = \frac{\pi(1)}{2} = \frac{\pi}{2} \qquad \frac{\pi W}{L} = \frac{\pi(2)}{2} = \pi$$

$$\theta_c = T_c - T_0 = 320 - 280 = 40 \text{ K}$$

Using six significant figures, we compute in Table 3-5 the quantities needed in the first three nonvanishing terms of the infinite series.

Table 3-5

n	$\dfrac{(-1)^{n+1}+1}{n}$	$\sinh\dfrac{n\pi}{2}$	$\sinh n\pi$	$\sin\dfrac{n\pi}{2}$
1	2	2.30130	11.5487	1.0
3	2/3	55.6544	6195.82	−1.0
5	2/5	1287.98	3.31781×10^6	1.0

Thus

$$\theta(1, 1) = 40\left(\frac{2}{\pi}\right)\left[2\frac{\sinh(\pi/2)}{\sinh\pi}\sin\left(\frac{\pi}{2}\right) + \frac{2}{3}\frac{\sinh(3\pi/2)}{\sinh 3\pi}\sin(3\pi/2)\right.$$

$$\left. + \frac{2}{5}\frac{\sinh(5\pi/2)}{\sinh 5\pi}\sin\left(\frac{5\pi}{2}\right) + \cdots\right]$$

$$= \frac{80}{\pi}[0.398538 - 0.005988 + 0.000155 - \cdots] = 10.0002 \text{ K}$$

and $T(1, 1) = 10.0002 + 280 \approx 290$ K.

It can be shown (Problem 3.19) that the exact answer is $\theta(1, 1) = 10$ K. The very slight error in the series solution is due to round-off and truncation errors.

3.3. For the situation shown in Fig. 3-11, determine the steady-state temperature at the point $(\frac{1}{2}, \frac{1}{2})$.

The problem in θ has only one nonhomogeneous boundary condition, and consequently (3.8)

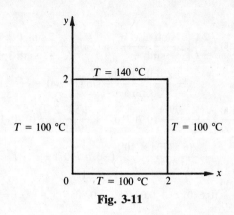

Fig. 3-11

gives the solution as

$$\theta(\tfrac{1}{2}, \tfrac{1}{2}) = (40\ °C)\left(\frac{2}{\pi}\right)\left[2\frac{\sinh(\pi/4)}{\sinh\pi}\sin(\pi/4) + \frac{2}{3}\frac{\sinh(3\pi/4)}{\sinh 3\pi}\sin(3\pi/4)\right.$$

$$\left. + \frac{2}{5}\frac{\sinh(5\pi/4)}{\sinh 5\pi}\sin(5\pi/4) + \frac{2}{7}\frac{\sinh(7\pi/4)}{\sinh 7\pi}\sin(7\pi/4) + \cdots\right]$$

$$= \frac{80}{\pi}\left[2\left(\frac{0.868671}{11.5487}\right)(0.707107) + \frac{2}{3}\left(\frac{5.22797}{6195.82}\right)(0.707107)\right.$$

$$\left. + \frac{2}{5}\left(\frac{25.3671}{3.31781\times 10^6}\right)(-0.707107) + \frac{2}{7}\left(\frac{122.073}{1.77666\times 10^9}\right)(-0.707107) + \cdots\right]$$

$$= \frac{80}{\pi}[0.106374 + 0.000398 - 0.000002 - \cdots] = 2.71888 \approx 2.72\ °C$$

and $T(\tfrac{1}{2}, \tfrac{1}{2}) \approx 102.72\ °C$.

3.4.   For the two-dimensional configuration of Fig. 3-2, determine the temperature at the point $(\tfrac{1}{4}, \tfrac{3}{4})$.

The problem can be separated into two simpler subproblems in θ, as shown in Fig. 3-3, and the final temperature is given by $\theta = \theta_1 + \theta_2$, where θ is $T - (100\ °F)$.   To solve the two subproblems, we first must reorient subproblem 1 to match the orientation used in the derivation of (3.8).   This is shown in Fig. 3-12(a); notice that in the new (x', y') coordinate system the point in question is $(\tfrac{3}{4}, \tfrac{3}{4})$, obtained by a 90° clockwise rotation of the original problem.

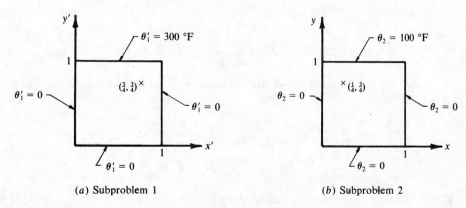

(a) Subproblem 1   (b) Subproblem 2

Fig. 3-12

Solving,

$$\theta_1'(\tfrac{3}{4}, \tfrac{3}{4}) = 300\left(\frac{2}{\pi}\right)\left[2\frac{\sinh(3\pi/4)}{\sinh\pi}\sin(3\pi/4) + \frac{2}{3}\frac{\sinh(9\pi/4)}{\sinh 3\pi}\sin(9\pi/4)\right.$$

$$\left. + \frac{2}{5}\frac{\sinh(15\pi/4)}{\sinh 5\pi}\sin(15\pi/4) + \cdots\right]$$

$$= \frac{600}{\pi}\left[\frac{2(5.22797)}{11.5487}(0.707107) + \frac{2(587.241)}{3(6195.82)}(0.707107)\right.$$

$$\left. + \frac{2(6.53704\times 10^4)}{5(3.3178\times 10^6)}(-0.707107) + \cdots\right]$$

$$= \frac{600}{\pi}(0.640199 + 0.044680 - 0.005573 + \cdots) = 129.7379\ °F = \theta_1(\tfrac{1}{4}, \tfrac{3}{4})$$

$$\theta_2(\tfrac{1}{4}, \tfrac{3}{4}) = 100\left(\frac{2}{\pi}\right)\left[2\frac{\sinh(3\pi/4)}{\sinh\pi}\sin(\pi/4) + \frac{2}{3}\frac{\sinh(9\pi/4)}{\sinh 3\pi}\sin(3\pi/4)\right.$$

$$\left. + \frac{2}{5}\frac{\sinh(15\pi/4)}{\sinh 5\pi}\sin(5\pi/4) + \cdots\right]$$

$$= \frac{200}{\pi}\left[\frac{2(5.22797)}{11.5487}(0.707107) + \frac{2(587.241)}{3(6195.82)}(0.707107)\right.$$

$$\left. + \frac{2(6.53704\times 10^4)}{5(3.31781\times 10^6)}(-0.707107) + \cdots\right]$$

$$= \frac{200}{\pi}(0.640199 + 0.044680 - 0.005573 + \cdots) \approx 43.2460\ °F$$

Therefore,

$$\theta(\tfrac{1}{4}, \tfrac{3}{4}) = \theta_1(\tfrac{1}{4}, \tfrac{3}{4}) + \theta_2(\tfrac{1}{4}, \tfrac{3}{4}) \approx 172.98\ °F \qquad \text{and} \qquad T(\tfrac{1}{4}, \tfrac{3}{4}) \approx 172.98 + 100 \approx 272.98\ °F$$

It should be noted that in the solution of the two subproblems, the infinite series for $\theta_1'(\tfrac{3}{4}, \tfrac{3}{4})$ is numerically the same as that for $\theta_2(\tfrac{1}{4}, \tfrac{3}{4})$; this is intuitively correct from a physical viewpoint.

3.5. Consider an 8.0-inch-o.d. pipe with a 12.3-inch-thick insulation blanket. By flux plotting determine the heat transfer per unit length if the inner surface of the insulation is at 300 °F, the outer surface is at 120 °F, and the thermal conductivity of the insulation is 0.35 Btu/hr-ft-°F.

Starting with an accurately scaled set of two concentric circles, as shown in Fig. 3-4, construct a network of curvilinear squares as explained in Section 3.3. Since this has been done in Fig. 3-4, it will not be repeated here. From Fig. 3-4, there are approximately $3\tfrac{1}{3}$ squares in each heat flow lane, and there are 4 flow lanes per quarter-section. So, $M \approx 3.33$, $N = 16$, and

$$\frac{S}{L} \approx \frac{16}{3.33} = 4.80$$

Thus,

$$\frac{q}{L} \approx 4.80\, k(T_i - T_o) = 4.80\left(\frac{0.35\ \text{Btu}}{\text{hr-ft-°F}}\right)[(300 - 120)\ °F] = 302.4\ \text{Btu/hr-ft}$$

Since this is actually a one-dimensional problem, we can readily check the result by use of (*2.14*). Hence

$$\frac{q}{L} = \frac{2\pi k(T_i - T_o)}{\ln(r_o/r_i)} = \frac{2\pi}{\ln(16.3/4)}(k)(T_i - T_o)$$

$$= 4.47\left(\frac{0.35\ \text{Btu}}{\text{hr-ft-°F}}\right)[(300 - 120)\ °F] = 281.61\ \frac{\text{Btu}}{\text{hr-ft}}$$

This exact result is 7% lower than the result by freehand plotting; this is an indication of the accuracy to be expected from freehand plotting. Note that in this case, the analytical representation for S/L is $2\pi/\ln(r_o/r_i)$.

3.6. Consider a 4 × 4-inch square block of fire clay having a 1-inch square hole at the center, as shown in Fig. 3-13. If the inner and outer surface temperatures are 150 °C and 30 °C, respectively, and the thermal conductivity is 1.00 W/m-K, determine the heat transfer from the inner surface to the outer surface per meter of length by freehand plotting.

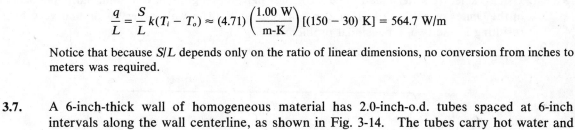

Since this problem has isothermal boundaries, freehand plotting is applicable. Due to symmetry, it is necessary to construct the freehand grid of curvilinear squares for only one-eighth of the body, as shown in Fig. 3-13. From the plot,

$$M_{av} \approx \frac{8.15 + 8.25 + 8.35 + 8.70 + 9.00}{5} \approx 8.5$$

and, for the entire block, $N = 8 \times 5 = 40$. Thus,

$$\frac{S}{L} \approx \frac{40}{8.5} = 4.71$$

Fig. 3-13

and

$$\frac{q}{L} = \frac{S}{L} k(T_i - T_o) \approx (4.71)\left(\frac{1.00 \text{ W}}{\text{m-K}}\right)[(150 - 30)\text{ K}] = 564.7 \text{ W/m}$$

Notice that because S/L depends only on the ratio of linear dimensions, no conversion from inches to meters was required.

3.7. A 6-inch-thick wall of homogeneous material has 2.0-inch-o.d. tubes spaced at 6-inch intervals along the wall centerline, as shown in Fig. 3-14. The tubes carry hot water and are at constant wall temperature T_i. Both sides of the wall may be assumed to be at the same constant wall temperature, T_o. Determine by freehand plotting the conductive shape factor for one tube.

With reference to Fig. 3-14, the wall longitudinal centerline, the centerlines for spacing of the tubes at 6-inch intervals, and the centerlines located midway between the tubes are lines of

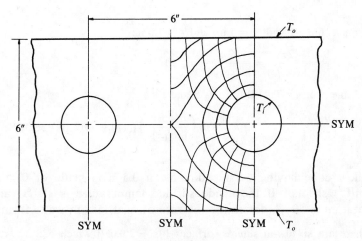

Fig. 3-14

geometrical and thermal symmetry marked by "SYM." Thus a freehand plot of isotherms and adiabatics for any quarter-section will suffice. Two such plots have been completed in Fig. 3-14 for the rightmost tube. For the upper plot,

$$M_{av} \approx \frac{3.9 + 4.1 + 4.7 + 5.2 + 5.9}{5} = 4.76$$

and S/L for the entire tube is

$$\frac{S}{L} \approx 4\left(\frac{N_{1/4}}{M_{av}}\right) = 4\left(\frac{5}{4.76}\right) = 4.20$$

Likewise, from the lower plot,

$$M_{av} \approx \frac{4.1 + 4.2 + 4.4 + 5.0 + 5.7}{5} = 4.68$$

and

$$\frac{S}{L} \approx 4\left(\frac{5}{4.68}\right) = 4.27$$

The agreement between the two plots is to within 2%. This, however, is more an indication of consistency in sketching than of accuracy; the accuracy is probably to within ±10%.

3.8. Using Table 3-1, determine the heat transfer per unit length from a 2.0-in.-o.d. pipe located in the center of a 10-in.-thick cinder concrete wall. The wall is very wide and very high, resulting in the two-dimensional problem of Fig. 3-15.

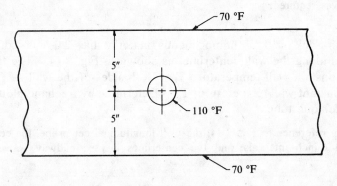

Fig. 3-15

From Table 3-1 with $z = 5$ in. and $r = 1.0$ in.,

$$\frac{S}{L} = \frac{2\pi}{\ln(4z/r)} = \frac{2\pi}{\ln[4(5)/1]} = 2.10$$

From Table B-1, $k = 0.44$ Btu/hr-ft-°F and thus

$$\frac{q}{L} = \frac{S}{L} k(T_2 - T_1) = (2.10)\left(\frac{0.44 \text{ Btu}}{\text{hr-ft-°F}}\right)[(110 - 70) \text{ °F}] \doteq 36.91 \text{ Btu/hr-ft}$$

3.9. A hollow cube having outside dimension 0.5 m is made of 0.05-m-thick asbestos ($\rho = 36$ lbm/ft^3) sheets. If the inside surface temperature is 150 °C and the outside surface temperature is 50 °C, determine the rate of heat loss in watts from the cube.

There are six 0.4-m square surfaces $[0.5 - 2(0.05)]$ which can be treated as one-dimensional conduction problems. There are twelve edges each 0.4 m long, as shown in the next-to-last sketch in

Table 3-1; these are two-dimensional problems. Finally, there are eight three-dimensional corners, as shown in the bottom sketch of Table 3-1.

From Appendix B at 100 °C,

$$k = \frac{0.111 \text{ Btu}}{\text{hr-ft-°F}} \left(1.72958 \frac{\text{W/m-K}}{\text{Btu/hr-ft-°F}} \right) = 0.192 \text{ W/m-K}$$

Square Surfaces:

$$q_s = 6kA_s \frac{\Delta T}{\Delta n}$$

$$= 6(0.192 \text{ W/m-K})(0.4 \text{ m})^2 \frac{(150 - 50) \text{ K}}{0.05 \text{ m}} = 368.64 \text{ W}$$

Edges: From Table 3-1, $S_e = (0.54)(0.4 \text{ m}) = 0.22 \text{ m}$ and

$$q_e = 12(S_e k \, \Delta T) = 12(0.22 \text{ m}) \left(\frac{0.192 \text{ W}}{\text{m-K}} \right) (100 \text{ K}) = 50.69 \text{ W}$$

Corners: From Table 3-1, $S_c = (0.15) \, t = (0.15)(0.05 \text{ m}) = 0.0075 \text{ m}$ and

$$q_c = 8(S_c k \, \Delta T) = 8(0.0075 \text{ m}) \left(\frac{0.192 \text{ W}}{\text{m-K}} \right) (100 \text{ K}) = 1.15 \text{ W}$$

The total heat-transfer rate is then

$$q_{\text{total}} = q_s + q_e + q_c = 368.64 + 50.69 + 1.15 = 420.48 \text{ W, say } 420 \text{ W}$$

since clearly S_c and S_e are approximations.

3.10.   Solve by Gaussian elimination the set of algebraic equations obtained in Example 3.5.

The nodal equations to be solved are

$$2T_a - \quad T_b + \quad \ \ 0 = 100 \tag{1a}$$
$$-T_a + 4T_b - \quad T_c = 400 \tag{2a}$$
$$-T_b + \quad 2T_c = 200 \tag{3a}$$

Multiplying (1a) by 1/2 and adding to (2a) yields

$$2T_a - \quad T_b + \quad \ \ 0 = 100 \tag{1b}$$
$$0 + \frac{7}{2} T_b - \quad T_c = 450 \tag{2b}$$
$$-T_b + \quad 2T_c = 200 \tag{3b}$$

Multiplying (2b) by 2/7 and adding to (3b) yields

$$2T_a - \quad T_b + \quad \ \ 0 = 100 \tag{1c}$$
$$\frac{7}{2} T_b - \quad T_c = 450 \tag{2c}$$
$$+ \frac{12}{7} T_c = \frac{2300}{7} \tag{3c}$$

Thus, by back substitution,

$$T_c = \frac{2300}{12} = 191.67 \text{ °F}$$

$$T_b = \frac{2}{7} (450 + 191.67) = 183.33 \text{ °F}$$

$$T_a = \frac{1}{2} (100 + 183.33) = 141.67 \text{ °F}$$

3.11.   Using the flow diagram of Fig. 3-8, write a computer program for Gaussian elimination in Fortran IV.   Allow for 10 equations in the DIMENSION and FORMAT statements.

The program given below is *one* suitable solution.   It can easily be modified to handle more equations by appropriate renumbering in the DIMENSION statement and suitable FORMAT statement(s) changes.   It is best not to heavily overdimension in the DIMENSION statement as this fixes computer storage requirements.

Part 1.   Triangularization

```
0060    DIMENSION A(10,10) , B(10) ,X(10)
0070    READ (5,1000) N
0080    N1 = N
0090    DO 100 I = 1 , N
0100    READ (5,1010) (A(I,J), J=1,N1)
0110    READ (5,1010) (B(I),I=1,N)
0120    WRITE (6,1020) N
0130    WRITE (6,1030)
0140    DO 200 I= 1,N
0200    WRITE (6,1040) (A(I,J),J=1,N1) , B(I)
0220    K = 1
0260    I = K+1
0270    IF (A(K,K).EQ.0) GO TO 410
0280    A1 = A(I,K)/A(K,K)
0290    B(I) = B(I)-B(K)*A1
0300    J = K
0310    A(I,J) = A(I,J)-A(K,J)*A1
0320    IF (J.GE.N) GO TO 350
0330    J = J+1
0340    GO TO 310
0350    IF (I.GE.N) GO TO 380
0360    I = I+1
0370    GO TO 270
0380    IF (K.GE.(N-1)) GO TO 550
0390    K = K+1
0400    GO TO 260
0410    M = K+1
0420    IF ((A(M,K)).NE.0) GO TO 460
0430    M = M+1
0440    IF (M.LE.N) GO TO 420
0450    WRITE (6,1050)
0455    GO TO 2000
0460    C1 = B(K)
0470    B(K) = B(M)
0480    B(M) = C1
0490    DO 520 J=1,N
0500    Z1=A(K,J)
0510    A(K,J) = A(M,J)
0520    A(M,J)=Z1
0530    GO TO 280
0550    WRITE (6,1060)
0560    DO 570 I=1,N
0570    WRITE (6,1040) (A(I,J),J=1,N1)
0575    WRITE (6,1070)
0580    WRITE (6,1040) (B(I),I=1,N)
```

Part 2.   Back Substitution

```
0585    WRITE(6,1090)
0590    L=N
0600    SUM=0
0610    IF (L.LT.N) GO TO 700
0620    X(L)=(B(L)-SUM)/A(L,L)
0640    IF (L.LE.1.0) GO TO 1200
0650    L=L-1
0660    GO TO 600
0700    J=L+1
0710    SUM=SUM+A(L,J)*X(J)
0720    IF(J.GE.N) GO TO 620
0730    J=J+1
0740    GO TO 710
```

```
1000   FORMAT(I2)
1010   FORMAT(10 F8.3)
1020   FORMAT('1','THERE ARE!',I3,2X,'EQUATIONS'//)
1030   FORMAT(' ','THE EQUATIONS ARE:'/)
1040   FORMAT('0',11F11.3)
1050   FORMAT('1','THE EQUATIONS ARE NOT INDEPENDENT')
1060   FORMAT('1',' MATRIX A TRIANGULARIZED'//)
1070   FORMAT('0',' MATRIX B TRIANGULARIZED'//)
1090   FORMAT('1',' THE EQUATION ROOTS ARE:'//)
1100   FORMAT('0',' ROOT #',I2,' = ',F10.3)
1200   DO 1210 L=1,N
1210   WRITE(6,1100) L, X(L)
2000   CONTINUE
       CALL EXIT
       END
```

3.12. Repeat Example 3.5 using a 3-inch square grid of nodal points (Fig. 3-16).

The nodal equations are, by (3.14),

$$\text{node } a: \quad 2T_b + 2T_f - 4T_a = 0$$
$$\text{node } b: \quad 300 + T_c + T_g + T_a - 4T_b = 0$$
$$\text{node } c: \quad 300 + T_d + T_h + T_b - 4T_c = 0$$
$$\text{node } d: \quad 300 + 2T_c + T_i - 4T_d = 0$$
$$\text{node } e: \quad 2T_f + 2T_k - 4T_e = 0$$
$$\text{node } f: \quad T_a + T_g + T_l + T_e - 4T_f = 0$$
$$\text{node } g: \quad T_b + T_h + T_m + T_f - 4T_g = 0$$
$$\text{node } h: \quad T_c + T_i + T_n + T_g - 4T_h = 0$$
$$\text{node } i: \quad T_d + 2T_h + T_o - 4T_i = 0$$
$$\text{node } j: \quad 2T_k + 2(100) - 4T_j = 0$$
$$\text{node } k: \quad T_e + T_l + 100 + T_j - 4T_k = 0$$
$$\text{node } l: \quad T_f + T_m + 100 + T_k - 4T_l = 0$$
$$\text{node } m: \quad T_g + T_n + 100 + T_l - 4T_m = 0$$
$$\text{node } n: \quad T_h + T_o + 100 + T_m - 4T_n = 0$$
$$\text{node } o: \quad T_i + 2T_n + 100 - 4T_o = 0$$

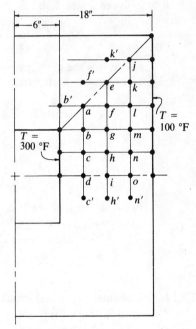

Fig. 3-16

where, due to thermal and physical symmetry, $T_b = T_{b'}$, etc. Putting these equations in the matrix form (3.15) for subsequent computer solution, we have

	a	b	c	d	e	f	g	h	i	j	k	l	m	n	o				
	−4	2				2											T_a		0
	1	−4	1				1										T_b		−300
		1	−4	1				1									T_c		−300
			2	−4					1								T_d		−300
					−4	2					2						T_e		0
	1				1	−4	1					1					T_f		0
		1				1	−4	1					1				T_g		0
			1				1	−4	1					1			T_h	=	0
				1				2	−4						1		T_i		0
										−4	2						T_j		−200
					1					1	−4	1					T_k		−100
						1					1	−4	1				T_l		−100
							1					1	−4	1			T_m		−100
								1					1	−4	1		T_n		−100
									1					2	−4		T_o		−100

The computer solution is (where units are °F):

$$T_a = 194.41 \qquad T_e = 140.34 \qquad T_i = 190.34 \qquad T_m = 137.58$$

$$T_b = 228.11 \qquad T_f = 160.71 \qquad T_j = 109.99 \qquad T_n = 142.23$$

$$T_c = 239.53 \qquad T_g = 178.51 \qquad T_k = 119.97 \qquad T_o = 143.70$$

$$T_d = 242.35 \qquad T_h = 187.65 \qquad T_l = 129.56$$

3.13. Derive the nodal temperature equation for the case of an exterior corner node with one adjacent side insulated and one adjacent side subject to a convective heat transfer, as shown in Fig. 3-17.

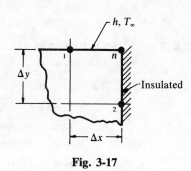

The rates of energy conducted between nodes 1 and n and between nodes 2 and n are, respectively,

$$kL\frac{\Delta y}{2}\left(\frac{T_1 - T_n}{\Delta x}\right) \qquad \text{and} \qquad kL\frac{\Delta x}{2}\left(\frac{T_2 - T_n}{\Delta y}\right)$$

where L is the depth perpendicular to the xy-plane. The rate of energy convected from T_∞ to T_n is

Fig. 3-17

$$hL\frac{\Delta x}{2}(T_\infty - T_n)$$

For steady state, the summation of energy transfer rates into node n must be zero, and thus

$$kL\frac{\Delta y}{2}\left(\frac{T_1 - T_n}{\Delta x}\right) + kL\frac{\Delta x}{2}\left(\frac{T_2 - T_n}{\Delta y}\right) + hL\frac{\Delta x}{2}(T_\infty - T_n) = 0$$

For a square grid, $\Delta x = \Delta y$ and this simplifies to

$$T_1 + T_2 + \frac{h\,\Delta x}{k}(T_\infty) - \left(\frac{h\,\Delta x}{k} + 2\right)T_n = 0$$

3.14. Consider a 1/4-inch-thick, rectangular, stainless-steel fin ($k = 8$ Btu/hr-ft-°F) which is 1.0 in. long in the x-direction and very wide in the direction perpendicular to the xy-plane of Fig. 3-18. The external convective heat-transfer coefficient is $h = 96$ Btu/hr-ft²-°F; the surrounding fluid temperature is $T_\infty = 80$ °F; the fin base temperature is $T_b = 200$ °F; and the end of the fin is insulated. Using the 1/8-inch square grid shown, determine the nodal temperatures T_1 through T_{16}.

Due to symmetry about the horizontal centerline, there are only 16 different nodal conditions. This set consists of interior nodal points, exterior nodal points with a convective boundary, an exterior nodal point with an insulated boundary, and an exterior corner nodal point having both an

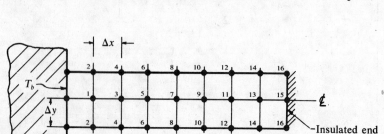

Fig. 3-18

insulated and a convective boundary.   Treating these by type:

Interior Nodes [by use of (3.14)]

node 1: $200 + 2T_2 + T_3 - 4T_1 = 0$

node 3: $T_1 + 2T_4 + T_5 - 4T_3 = 0$

node 5: $T_3 + 2T_6 + T_7 - 4T_5 = 0$

node 7: $T_5 + 2T_8 + T_9 - 4T_7 = 0$

node 9: $T_7 + 2T_{10} + T_{11} - 4T_9 = 0$

node 11: $T_9 + 2T_{12} + T_{13} - 4T_{11} = 0$

node 13: $T_{11} + 2T_{14} + T_{15} - 4T_{13} = 0$

Exterior Nodes with Convective Boundary [by use of (3.17)]

node 2: $\dfrac{1}{2}(200 + 2T_1 + T_4) + \dfrac{h\,\Delta x}{k}(80) - \left(\dfrac{h\,\Delta x}{k} + 2\right)T_2 = 0$

$$\frac{h\,\Delta x}{k} = \frac{(96\ \text{Btu/hr-ft}^2\text{-}^\circ\text{F})[(0.125/12)\ \text{ft}]}{8\ \text{Btu/hr-ft-}^\circ\text{F}} = 0.125$$

thus, $200 + 2T_1 + T_4 + 0.25(80) - 4.25\,T_2 = 0$

or, $220 + 2T_1 + T_4 - 4.25\,T_2 = 0$

node 4: $T_2 + 2T_3 + T_6 + 20 - 4.25\,T_4 = 0$

node 6: $T_4 + 2T_5 + T_8 + 20 - 4.25\,T_6 = 0$

node 8: $T_6 + 2T_7 + T_{10} + 20 - 4.25\,T_8 = 0$

node 10: $T_8 + 2T_9 + T_{12} + 20 - 4.25\,T_{10} = 0$

node 12: $T_{10} + 2T_{11} + T_{14} + 20 - 4.25\,T_{12} = 0$

node 14: $T_{12} + 2T_{13} + T_{16} + 20 - 4.25\,T_{14} = 0$

Exterior Node with Insulated Boundary (by use of Table 3-4)

node 15: $\dfrac{1}{2}(T_{16} + T_{16}) + T_{13} - 2T_{15} = 0$

or, $T_{16} + T_{13} - 2T_{15} = 0$

Exterior Corner Node (by the result of Problem 3.13)

node 16: $T_{14} + T_{15} + \dfrac{h\,\Delta x}{k}(T_\infty) - \left(\dfrac{h\,\Delta x}{k} + 2\right)T_{16} = 0$

or, $T_{14} + T_{15} + 10 - 2.125\,T_{16} = 0$

Solving the system on the computer gives (units are °F):

$T_1 = 167.178$	$T_5 = 124.928$	$T_9 = 104.533$	$T_{13} = 96.366$
$T_2 = 163.118$	$T_6 = 122.242$	$T_{10} = 103.049$	$T_{14} = 95.375$
$T_3 = 142.474$	$T_7 = 112.753$	$T_{11} = 99.232$	$T_{15} = 95.432$
$T_4 = 138.896$	$T_8 = 110.776$	$T_{12} = 98.115$	$T_{16} = 94.497$

3.15.   For the two-dimensional conduction problem of Fig. 3-19, determine the steady-state nodal temperatures T_1 through T_6 using the relaxation technique.   Consider the answers to be satisfactorily accurate for this problem when the absolute value of the largest residual is equal to or less than 1.0.

The nodal equations obtained with the aid of (3.14) are, in residual form:

node 1: $300 + T_2 + T_4 - 4T_1 = R_1$

node 2: $200 + T_1 + T_3 + T_5 - 4T_2 = R_2$

node 3: $400 + T_2 + T_6 - 4T_3 = R_3$

node 4: $200 + T_1 + T_5 - 4T_4 = R_4$

node 5: $100 + T_2 + T_4 + T_6 - 4T_5 = R_5$

node 6: $300 + T_3 + T_5 - 4T_6 = R_6$

Table 3-6 shows the effects of unit temperature changes.

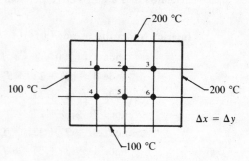

Fig. 3-19

Table 3-6

	ΔR_1	ΔR_2	ΔR_3	ΔR_4	ΔR_5	ΔR_6
$\Delta T_1 = +1$	-4	1	0	1	0	0
$\Delta T_2 = +1$	1	-4	1	0	1	0
$\Delta T_3 = +1$	0	1	-4	0	0	1
$\Delta T_4 = +1$	1	0	0	-4	1	0
$\Delta T_5 = +1$	0	1	0	1	-4	1
$\Delta T_6 = +1$	0	0	1	0	1	-4
Block $= +1$	-2	-1	-2	-2	-1	-2

Table 3-7

	T_1	R_1	T_2	R_2	T_3	R_3	T_4	R_4	T_5	R_5	T_6	R_6
Initial Guesses	160	-20	180	-20	190	-15	140	-50	150	-15	165	-20
Block $= -10$	_150_	0	_170_	-10	_180_	$+5$	_130_	-30	_140_	-5	_155_	0
$\Delta T_4 = -8$	150	-8	170	-10	180	$+5$	_122_	$+2$	140	-13	155	0
$\Delta T_5 = -3$	150	-8	170	-13	180	$+5$	122	-1	_137_	-1	155	-3
$\Delta T_2 = -3$	150	-11	_167_	-1	180	$+2$	122	-1	137	-4	155	-3
$\Delta T_1 = -3$	_147_	$+1$	167	-4	180	$+2$	122	-4	137	-4	155	-3
Check by equations	147	√ $+1$	167	√ -4	180	√ $+2$	122	√ -4	137	√ -4	155	√ -3
$\Delta T_2 = -1$	147	0	_166_	0	180	$+1$	122	-4	137	-5	155	-3
$\Delta T_5 = -1$	147	0	166	-1	180	$+1$	122	-5	_136_	-1	155	-4
$\Delta T_4 = -1$	147	-1	166	-1	180	$+1$	_121_	-1	136	-2	155	-4
$\Delta T_6 = -1$	147	-1	166	-1	180	0	121	-1	136	-3	_154_	0
$\Delta T_5 = -1$	147	-1	166	-2	180	0	121	-2	_135_	$+1$	154	-1
$\Delta T_2 = -\frac{1}{2}$	147	$-1\frac{1}{2}$	_165$\frac{1}{2}$_	0	180	$-\frac{1}{2}$	121	-2	135	$+\frac{1}{2}$	154	-1
$\Delta T_4 = -\frac{1}{2}$	147	-2	165$\frac{1}{2}$	0	180	$-\frac{1}{2}$	_120$\frac{1}{2}$_	0	135	0	154	-1
$\Delta T_1 = -\frac{1}{2}$	_146$\frac{1}{2}$_	0	165$\frac{1}{2}$	$-\frac{1}{2}$	180	$-\frac{1}{2}$	120$\frac{1}{2}$	$-\frac{1}{2}$	135	0	154	-1
Check by equations	146$\frac{1}{2}$	√ 0	165$\frac{1}{2}$	√ $-\frac{1}{2}$	180	√ $-\frac{1}{2}$	120$\frac{1}{2}$	√ $-\frac{1}{2}$	135	√ 0	154	√ -1

We next begin the relaxation effort. Setting up the relaxation table, Table 3-7, we make an initial guess for each nodal temperature. Using the six initial temperatures (guesses) and the set of six nodal equations, we calculate the initial residuals and record these in the relaxation table. The work proceeds by relaxing, in turn, the residual of largest absolute value. Since all initial residuals were of the same sign, a "block" 10-degree reduction in all temperatures was the first step. The block change is usually advantageous whenever all residuals are of like sign. Following this, the residual of largest absolute value was R_4; this was relaxed by reducing T_4, Table 3-6 showing the effect per degree of temperature change in T_4 upon all residuals.

Continuing, all residuals are within the specified tolerance for the bottom set of temperatures in the relaxation table. The work could have been reduced somewhat by over-relaxing. Notice that T_5, for example, was reduced three times. By reducing it more than apparently needed the first time, the work could have been shortened. Excessive over-relaxation, however, could increase the required effort.

3.16. Estimate the heat transfer rate from the horizontal 100 °C surface in Fig. 3-19. Use the nodal temperatures determined in Problem 3.15. The material is magnesite and the grid size is $\Delta x = \Delta y = 15$ cm.

From Appendix B, the thermal conductivity of magnesite at 205 °C (nearest listed temperature to conditions of this problem) is $k \approx 2.2 \times 1.7296$ W/m-K $= 3.81$ W/m-K. Dividing the body into heat flow lanes (Fig. 3-20) and assuming the lower right-hand corner to be at the average of the two adjacent surface temperatures, we proceed to approximate the heat transfer rate through each lane by one-dimensional methods. Thus for depth L perpendicular to the plane of the figure,

$$\text{Lane } a: \quad \left(\frac{q}{L}\right)_a \approx k\frac{A_a}{L}\frac{\Delta T}{\Delta S} = \frac{3.81 \text{ W}}{\text{m-K}}(7.5 \text{ cm})\left(\frac{100-100}{15}\frac{\text{K}}{\text{cm}}\right) \approx 0 \frac{\text{W}}{\text{m}}$$

$$\text{Lane } b: \quad \left(\frac{q}{L}\right)_b \approx \frac{3.81 \text{ W}}{\text{m-K}}(15 \text{ cm})\left(\frac{T_4-100}{15}\frac{\text{K}}{\text{cm}}\right)$$

$$\approx 3.81(120.5-100) = 78.11 \frac{\text{W}}{\text{m}}$$

$$\text{Lane } c: \quad \left(\frac{q}{L}\right)_c \approx \frac{3.81 \text{ W}}{\text{m-K}}(15 \text{ cm})\left(\frac{135-100}{15}\frac{\text{K}}{\text{cm}}\right) = 133.35 \frac{\text{W}}{\text{m}}$$

$$\text{Lane } d: \quad \left(\frac{q}{L}\right)_d \approx \frac{3.81 \text{ W}}{\text{m-K}}(15 \text{ cm})\left(\frac{154-100}{15}\frac{\text{K}}{\text{cm}}\right) = 205.74 \frac{\text{W}}{\text{m}}$$

$$\text{Lane } e: \quad \left(\frac{q}{L}\right)_e \approx \frac{3.81 \text{ W}}{\text{m-K}}(7.5 \text{ cm})\left(\frac{200-150}{15}\frac{\text{K}}{\text{cm}}\right) = 95.25 \frac{\text{W}}{\text{m}}$$

$$\text{Total:} \quad \frac{q}{L} = 78.11 + 133.35 + 205.74 + 95.25 \approx 512 \frac{\text{W}}{\text{m}}$$

The approximation could be improved by using a smaller grid size; in fact, almost any desired level of accuracy can be attained at the expense of computational time.

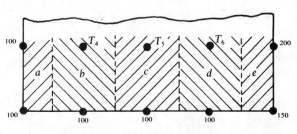

Fig. 3-20

Supplementary Problems

3.17. From (3.8) infer the solution to the steady-state problem illustrated in Fig. 3-21.

Ans. $\theta(x, y) = \theta_c \dfrac{2}{\pi} \displaystyle\sum_{n=1}^{\infty} \dfrac{(-1)^{n+1} + 1}{n} \dfrac{\sinh(n\pi x/W)}{\sinh(n\pi L/W)} \sin \dfrac{n\pi y}{W}$

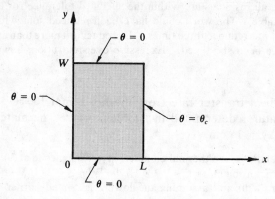

Fig. 3-21

3.18. Use the result of Problem 3.17 to obtain the temperature θ_1 at the point $(\frac{1}{4}, \frac{3}{4})$ for subproblem 1 of Fig. 3-3. (*Hint:* Rotate the figure 180°.)

Ans. 129.7379 °F, using 3 terms of the series

3.19. A square plate has three sides held at temperature $T_0 = 0$, while the fourth side is held at temperature T_c. Use the principle of superposition to show that the steady-state temperature at the center of the plate is $T_c/4$. (*Hint:* Break up the problem shown in Fig. 3-22 into four subproblems.)

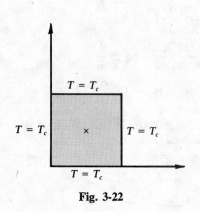

Fig. 3-22

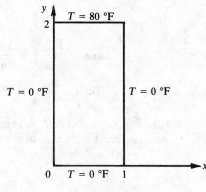

Fig. 3-23

3.20. Find the steady-state temperature at the center of the rectangular plate shown in Fig. 3-23.

Ans. $T(\frac{1}{2}, 1) \approx 4.39$ °F

3.21. Determine by freehand plotting the heat transfer per unit length for the square chimney configuration of Fig. 3-16 if $k = 0.5$ Btu/hr-ft-°F.

Ans. $S/L_{1/8} \approx 0.78$ (using 7 heat flow lanes); $q/L \approx 624$ Btu/hr-ft

3.22. Repeat Problem 3.21 using the method presented in Problem 3.16 and the numerical temperature values for nodes j, k, l, m, n and o from Problem 3.12.

Ans. $q/L \approx 644.72$ Btu/hr-ft, about 3% higher than obtained by flux plotting

3.23. Repeat Problem 3.8 for a 4-cm-o.d. pipe at 40 °C in the center of a 30-cm-thick cinder concrete wall with both surfaces at 25 °C.   *Ans.* $q/L = 21.1$ W/m

3.24. A long, $1\frac{1}{2}$-in.-o.d. tube at 120 °F and a long, 3/4-in.-o.d. tube at 40 °F are located parallel and 3.0 in. apart (center-to-center distance) in a service tunnel filled with rock wool ($k \approx 0.03$ Btu/hr-ft-°F).   Estimate the net heat transfer per foot of length from the hotter to the cooler tube, assuming the medium to be infinite.   *Ans.* $q/L \approx 4.46$ Btu/hr-ft

3.25. Derive the nodal temperature equation for the case of an exterior corner node as shown in Fig. 3-17, except both adjacent sides are insulated (i.e. the upper surface is insulated rather than subjected to a convective environment).   Assume a square grid.   *Ans.* $\frac{1}{2}(T_1 + T_2) - T_n = 0$

3.26. Using the electrical analogy discussed on page 53, devise an analog device for obtaining a grid of curvilinear squares to get the conductive shape factor for a two-dimensional, steady-state problem.

Ans.   The usual technique involves the use of a paper uniformly coated with a thin layer of conductive material.   A commercially available paper is known as "Teledeltos," but other papers such as those used in some types of strip-chart recorders work equally well.   The paper is cut into the exact shape of the two-dimensional model of the heat transfer problem, and a suitable d.c. electrical potential is impressed by the use of large electrodes at the boundaries.   The resistance of the paper is large compared with that of the electrodes, and lines of equal potential can easily be found by using a null detector system as shown in Fig. 3-24.   These lines are analogous to lines of constant temperature; constant heat-flow lines can be found by reversing the analog to apply the potential along the edges which are lines of symmetry in Fig. 3-24.   The null detector can be replaced with a sensitive voltmeter if desired.

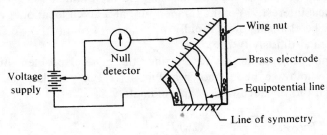

Fig. 3-24

Chapter 4

Time-Varying Conduction

4.1 INTRODUCTION

To this point we have considered conductive heat-transfer problems in which the temperatures are independent of time. In many applications, however, the temperatures are varying with time. Analysis of such transient problems can be undertaken with the general conduction equation, (2.2). In the present chapter we will deal primarily with one spatial dimension, in which case (2.2) reduces to

$$\frac{\partial^2 T}{\partial x^2} = \frac{1}{\alpha} \frac{\partial T}{\partial t}$$ (4.1)

For the solution of (4.1) we need two boundary conditions in the x-direction and one time condition. Boundary conditions are, as the name implies, frequently specified along the physical extremities of the body; they can, however, also be internal—e.g. a known temperature gradient at an internal line of symmetry. The time condition is usually the known initial temperature.

4.2 BIOT AND FOURIER MODULI

In some transient problems, the internal temperature gradients in the body may be quite small and of little practical interest. Yet the temperature at a given location, or the average temperature of the object, may be changing quite rapidly with time. From (4.1), we see that such could be the case for large thermal diffusivity α.

A more meaningful approach is to consider the general problem of transient cooling of an object, such as the hollow cylinder shown in Fig. 4-1. For very large r_i, the heat transfer rate by conduction through the cylinder wall is approximately

$$q \approx -k(2\pi r_s l)\left(\frac{T_s - T_i}{r_s - r_i}\right) = k(2\pi r_s l)\left(\frac{T_i - T_s}{L}\right)$$ (4.2)

where l is the length of the cylinder and L is the material thickness. The rate of heat transfer away from the outer surface by convection is

$$q = \bar{h}(2\pi r_s l)(T_s - T_\infty)$$ (4.3)

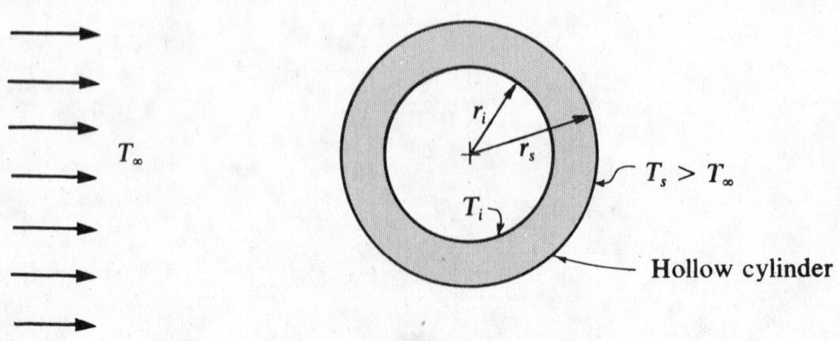

Fig. 4-1

where $\bar{h}$ is the average heat-transfer coefficient for convection from the entire surface. Equating (4.2) and (4.3) gives

$$\frac{T_i - T_s}{T_s - T_\infty} = \frac{\bar{h}L}{k} \equiv \text{Biot number} \qquad (4.4)$$

The Biot number is dimensionless, and it can be thought of as the ratio

$$\text{Bi} = \frac{\text{resistance to internal heat flow}}{\text{resistance to external heat flow}}$$

Whenever the Biot number is small, the internal temperature gradients are also small and a transient problem can be treated by the "lumped thermal capacity" approach wherein the object for analysis is considered to have a single mass-averaged temperature.

In the preceding derivation, the significant body dimension was the conductive path length, $L = r_s - r_i$. In general, a characteristic linear dimension may be obtained by dividing the volume of the solid by its surface area:

$$L = \frac{V}{A_s} \qquad (4.5)$$

Using this method to determine L, objects resembling a plate, a cylinder, or a sphere may be considered to have uniform temperature and the resulting error will be less than 5 percent if the Biot number is less than 0.1.

The *Fourier modulus* is the dimensionless time obtained by multiplying the dimensional time by the thermal diffusivity and dividing by the square of the characteristic length:

$$\text{dimensionless time} = \frac{\alpha t}{L^2} \equiv \text{Fo} \qquad (4.6)$$

4.3 LUMPED ANALYSIS

A typical transient problem which may be treated by a lumped analysis if the Biot modulus is suitably small is the cooling of a metal object after hot forming. In Fig. 4-2, $\bar{h}$ denotes the average heat-transfer coefficient for the entire surface area, A_s. Thermal energy is leaving the object from all elements of the surface; this is shown for simplicity by the single arrow.

The first law of thermodynamics applied to this problem is

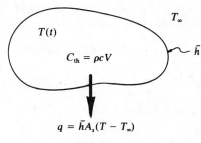

$$q = \bar{h}A_s(T - T_\infty)$$

Fig. 4-2

$$\left(\begin{array}{c}\text{heat out of object}\\\text{during time } dt\end{array}\right) = \left(\begin{array}{c}\text{decrease of internal thermal}\\\text{energy of object during time } dt\end{array}\right)$$

Now if the temperature of the object can be considered to be uniform, i.e. independent of location within the object, this equation may be written as

$$\bar{h}A_s[T(t) - T_\infty]dt = -\rho c V\, dT \qquad \text{or} \qquad \frac{dT}{T - T_\infty} = \frac{-\bar{h}A_s}{\rho c V}dt$$

Integrating and applying the initial condition $T(0) = T_i$ yields

$$\frac{T - T_\infty}{T_i - T_\infty} = \exp\left[-\left(\frac{\bar{h}A_s}{\rho c V}\right)t\right] \qquad (4.7)$$

The exponential temperature decay given in (4.7) is analogous to the voltage decay during discharge of an electrical capacitor, which is given by

$$\frac{E}{E_i} = e^{-t/(RC)_e}$$

To make the analogy complete, we define the *thermal time constant* by

$$(RC)_{th} \equiv \left(\frac{1}{\bar{h}A_s}\right)(\rho c V) \equiv \text{(thermal resistance)(thermal capacitance)} \tag{4.8}$$

so that

$$\frac{T - T_\infty}{T_i - T_\infty} = e^{-t/(RC)_{th}} = e^{-(Bi)(Fo)} \tag{4.9}$$

where the last equality follows from (4.4), (4.5) and (4.6).

Multiple Lumped Systems

A lumped analysis may be appropriate for each material lump of a system consisting of two or more parts such as two solids in good thermal contact or a liquid in a container. Assuming the Biot number is suitably small for each subsystem, we may treat each as having a single transient temperature; thus a two-body system would result in two time dependent temperatures. See Problems 4.9 and 4.10.

4.4 ONE-DIMENSIONAL SYSTEMS: FIXED SURFACE TEMPERATURE

Some transient conduction problems may be treated approximately by considering the body to be initially at a uniform temperature and suddenly having the temperature of part of the surface changed to and held at a known constant value different from the initial temperature.

Semi-Infinite Body

Consider a three-dimensional body that occupies the half-space $x \geq 0$. The body is initially at the uniform temperature T_i, including the surface at $x = 0$. The surface temperature at $x = 0$ is instantaneously changed to and held at T_s for all time greater than $t = 0$.

The temperature obeys (4.1), subject to

$$\text{boundary conditions:} \quad T(0, t) = T_s \quad \text{for } t > 0 \tag{4.10}$$
$$T(\infty, t) = T_i \quad \text{for } t > 0 \tag{4.11}$$
$$\text{time condition:} \quad T(x, 0) = T_i \tag{4.12}$$

The solution is

$$\frac{T(x, t) - T_s}{T_i - T_s} = \text{erf}\left(\frac{x}{\sqrt{4\alpha t}}\right) \tag{4.13}$$

where the *Gaussian error function* is defined by

$$\text{erf } u \equiv \frac{2}{\sqrt{\pi}} \int_0^u e^{-\eta^2} d\eta \tag{4.14}$$

Suitable values of the error function for arguments from 0 to 2 may be obtained from Fig. 4-3. More accurate values, which can be obtained from mathematical tables, are usually not necessary in conduction problems of this type, due to larger errors in thermal properties and other conditions.

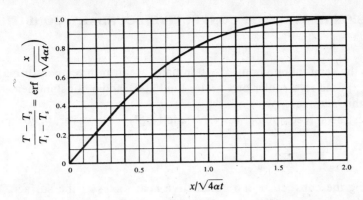

Fig. 4-3

Of interest is the heat transfer rate at a given x-location within the semi-infinite body at a specified time.   By (4.13),

$$T(x, t) = T_s + (T_i - T_s)\frac{2}{\sqrt{\pi}}\int_0^{x/\sqrt{4\alpha t}} e^{-\eta^2}d\eta$$

which may be differentiated by Leibniz' rule to yield

$$\frac{\partial T}{\partial x} = \frac{T_i - T_s}{\sqrt{\pi\alpha t}}e^{-x^2/4\alpha t}$$

Substituting this gradient in Fourier's law, the heat flux is

$$\frac{q}{A} = \frac{k(T_s - T_i)}{\sqrt{\pi\alpha t}}e^{-x^2/4\alpha t} \tag{4.15}$$

We are frequently interested in the heat flux at $x = 0$; it is

$$\left.\frac{q}{A}\right|_{x=0} = \frac{k(T_s - T_i)}{\sqrt{\pi\alpha t}} \tag{4.16}$$

and this heat flux clearly diminishes with time.

Finite Body

The preceding analytical solution for the temperature as a function of location and time was obtained under the condition that the temperature approaches T_i as x approaches infinity, for all values of time.   If the temperature at some point within a relatively thick, finite body is still unaffected by heat transfer, the temperature distribution in the portion of the body that *is* affected is identical with that in a semi-infinite body under otherwise identical circumstances.   The general criterion for the semi-infinite solution to apply to a body of finite thickness (slab) subjected to one-dimensional heat transfer is

$$\frac{2L}{\sqrt{4\alpha t}} \geq 0.5 \tag{4.17}$$

where $2L$ is the thickness of the body.

4.5 ONE-DIMENSIONAL SYSTEMS: CONVECTIVE BOUNDARY CONDITIONS

Semi-Infinite Body

Consider again the semi-infinite body $x \geq 0$ initially at uniform temperature T_i. The surface at $x = 0$ is suddenly exposed to a cooler or warmer fluid at T_∞ and a constant heat-transfer coefficient $\bar{h}$ is applicable to this surface.

Again, (4.1) must be solved subject to (4.11) and (4.12). However, (4.10) must be replaced by

$$\bar{h}(T_\infty - T)\Big|_{x=0} = -k\frac{\partial T}{\partial x}\Big|_{x=0} \tag{4.18}$$

obtained by equating the convective and conductive heat fluxes at the surface.

The solution can be shown to be

$$\frac{T - T_i}{T_\infty - T_i} = 1 - \mathrm{erf}\,\xi - \left[\exp\left(\frac{\bar{h}x}{k} + \frac{\bar{h}^2\alpha t}{k^2}\right)\right]\left[1 - \mathrm{erf}\left(\xi + \frac{\bar{h}\sqrt{\alpha t}}{k}\right)\right] \tag{4.19}$$

where $\xi = x/\sqrt{4\alpha t}$.

Slab of Finite Thickness

We will examine next the case of a large slab of x-thickness $2L$, as shown in Fig. 4-4(a). The slab is of infinite, or at least of very large, dimensions in the y- and z-directions. Initially at a uniform temperature T_i, it is suddenly exposed to the convective environment at T_∞. The Biot number is such that a lumped analysis is not suitable.

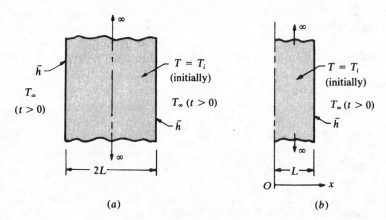

Fig. 4-4

The geometry, boundary conditions, and temperature distribution are symmetrical about the vertical centerline; consequently we may consider one-half of the problem as shown in Fig. 4-4(b). Also, the symmetry assures us that the centerline is adiabatic; thus one boundary condition is that $\partial T/\partial x$ is zero at the centerline.

The one-dimensional conduction equation appropriate for this problem is

$$\frac{\partial^2 \theta}{\partial x^2} = \frac{1}{\alpha}\frac{\partial \theta}{\partial t} \tag{4.20}$$

where $\theta \equiv T - T_\infty$. The boundary and time conditions are

$$\text{boundary conditions:} \quad \frac{\partial \theta}{\partial x} = 0 \qquad \text{at } x = 0 \tag{4.21}$$

$$\frac{\partial \theta}{\partial x} = -\frac{\bar{h}}{k}\theta \quad \text{at } x = L \tag{4.22}$$

$$\text{time condition:} \quad \theta = \theta_i \quad \text{at } t = 0 \tag{4.23}$$

where (4.22) results from equating the conductive and convective heat-transfer rates at the surface.

The classical separation-of-variables technique, which was used in Section 3.2, applied to equation (4.20) yields

$$\theta = e^{-\lambda^2 \alpha t}(C_1 \sin \lambda x + C_2 \cos \lambda x) \tag{4.24}$$

where the separation parameter, $-\lambda^2$, is chosen negative to ensure that $\theta \to 0$ as $t \to \infty$. Boundary condition (4.21) requires that $C_1 = 0$, and (4.22) then gives the transcendental equation

$$\cot \lambda L = \frac{\lambda}{\bar{h}/k} \quad \text{or} \quad \cot \lambda L = \frac{\lambda L}{\mathbf{Bi}} \tag{4.25}$$

for λ. Equation (4.25) has an infinite number of positive roots, $\lambda_1, \lambda_2, \ldots$, and in terms of these *eigenvalues* the solution for θ is

$$\theta = \sum_{n=1}^{\infty} C_n e^{-\lambda_n^2 \alpha t} \cos \lambda_n x \tag{4.26}$$

One method of determining the λ_n is by means of an accurate plot of the type shown in Fig. 4-5. Usually the first three or four roots are sufficient to yield an accurate answer.

Finally, (4.23) requires that

$$\theta_i = \sum_{n=1}^{\infty} C_n \cos \lambda_n x$$

which indicates that the C_n must be chosen so that θ_i is represented by an infinite series of cosine terms over the range $0 < x < L$. From the theory of orthogonal functions we find

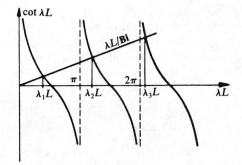

Fig. 4-5

$$C_n = \frac{2\theta_i \sin \lambda_n L}{\lambda_n L + (\sin \lambda_n L)(\cos \lambda_n L)} \tag{4.27}$$

so that (4.26) becomes

$$\frac{\theta}{\theta_i} = 2\sum_{n=1}^{\infty} e^{-\lambda_n^2 \alpha t} \frac{\sin \lambda_n L}{\lambda_n L + (\sin \lambda_n L)(\cos \lambda_n L)} \cos \lambda_n x \tag{4.28}$$

Using (4.28) for the temperature, one can show (see Problem 4.15) that the thermal energy transferred to or from the slab is given by

$$\frac{Q}{Q_i} = 2\sum_{n=1}^{\infty} \frac{1}{\lambda_n L}\left[\frac{\sin^2 \lambda_n L}{\lambda_n L + (\sin \lambda_n L)(\cos \lambda_n L)}\right](1 - e^{-\lambda_n^2 \alpha t}) \tag{4.29}$$

where $Q_i = V\rho c\theta_i = AL\rho c\theta_i$ is the thermal energy above the reference state, T_∞, initially stored in the half-slab per unit depth.

4.6 CHART SOLUTIONS: CONVECTIVE BOUNDARY CONDITIONS

The results obtained in Section 4.5 can be put in dimensionless form by use of the Biot and Fourier moduli. The resulting dimensionless equations have been solved for a wide range of values of **Bi** and **Fo**; these solutions are available in the form of graphs, obtained as follows

Slab of Finite Thickness

Letting $\delta_n = \lambda_n L$, (4.28) can be written as

$$\frac{\theta}{\theta_i} = 2 \sum_{n=1}^{\infty} \exp\left(-\delta_n^2 \mathbf{Fo}\right) \frac{\sin \delta_n}{\delta_n + (\sin \delta_n)(\cos \delta_n)} \cos\left(\delta_n x/L\right) \qquad (4.30)$$

and, at $x = 0$, the dimensionless centerline temperature is

$$\frac{T_c - T_\infty}{T_i - T_\infty} = \frac{\theta_c}{\theta_i} = 2 \sum_{n=1}^{\infty} \exp\left(-\delta_n^2 \mathbf{Fo}\right) \frac{\sin \delta_n}{\delta_n + (\sin \delta_n)(\cos \delta_n)} \qquad (4.31)$$

Furthermore, (4.29) becomes

$$\frac{Q}{Q_i} = 2 \sum_{n=1}^{\infty} \frac{1}{\delta_n} \frac{\sin^2 \delta_n}{\delta_n + (\sin \delta_n)(\cos \delta_n)} \left[1 - \exp\left(1 - \delta_n^2 \mathbf{Fo}\right)\right] \qquad (4.32)$$

Note that the δ_n are functions of **Bi** via (4.25):

$$\cot \delta_n = \delta_n/\mathbf{Bi}$$

Consequently, (4.30), (4.31) and (4.32) indicate that θ and Q are functions of **Bi** and **Fo**. Heisler's plot of the temperature at the centerline of a slab versus **Fo** and **Bi** is given in Fig. 4-6. The temperature at any other x-location in the body at a specified time (**Fo**) can be determined with the aid of Fig. 4-7 together with the centerline temperature at that **Fo**. Notice that Fig. 4-7 implies that θ/θ_c is independent of **Fo**, which is very nearly, but not exactly, the case. Finally, Gröber's chart for the dimensionless heat removal or addition is presented in Fig. 4-8. The use of these charts is illustrated in Problems 4.16 and 4.17.

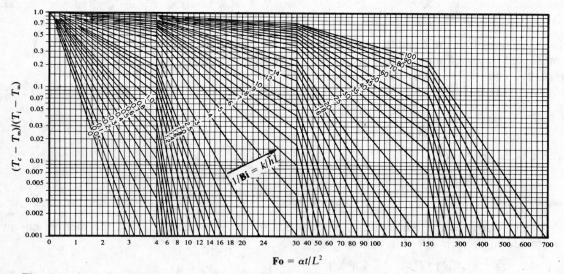

Fig. 4-6. Centerline temperature for a slab. [From M. P. Heisler, *Trans. ASME*, **69**: 227 (1947).]

Long Solid Cylinder

The temperature differential equation for purely radial heat transfer in a cylinder of radius R is

$$\frac{\partial^2 \theta}{\partial r^2} + \frac{1}{r} \frac{\partial \theta}{\partial r} = \frac{1}{\alpha} \frac{\partial \theta}{\partial t} \qquad (4.33)$$

where $\theta = T(r, t) - T_\infty$. The solution satisfying

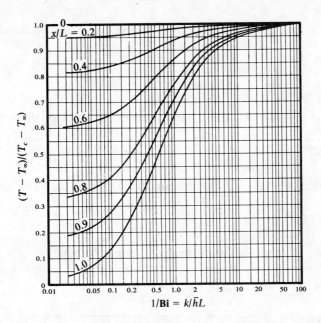

Fig. 4-7. Correction chart for a slab. [From M. P. Heisler, *Trans. ASME*, **69**: 227 (1947).]

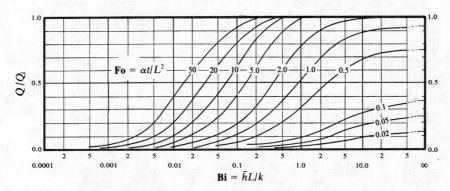

Fig. 4-8. Total heat flow, slab. (From H. Gröber, S. Erk, and U. Grigall, *Grundgesetze der Wärmeübertragung*, 3d ed., Springer-Verlag, Berlin, 1955, by permission.)

boundary conditions:
$$\frac{\partial \theta}{\partial r} = 0 \qquad \text{at } r = 0$$

$$\frac{\partial \theta}{\partial r} = -\frac{\bar{h}\theta}{k} \qquad \text{at } r = R$$

time condition: $\theta = \theta_i \qquad$ at $t = 0$

is given by

$$\frac{\theta}{\theta_i} = 2 \sum_{n=1}^{\infty} \frac{1}{\delta_n} e^{-\delta_n^2 \mathbf{Fo}} \frac{J_0(\delta_n r/R) J_1(\delta_n)}{J_0^2(\delta_n) + J_1^2(\delta_n)} \tag{4.34}$$

where J_0 and J_1 are the Bessel functions of the first kind of zero and first orders, respectively, and $\delta_n = \lambda_n R$.   Since $J_0(0) = 1$, the dimensionless temperature at the centerline is

$$\frac{\theta_c}{\theta_i} = 2 \sum_{n=1}^{\infty} \frac{1}{\delta_n} e^{-\delta_n^2 \mathbf{Fo}} \frac{J_1(\delta_n)}{J_0^2(\delta_n) + J_1^2(\delta_n)} \tag{4.35}$$

The dimensionless total thermal energy to or from the cylinder is obtained as it was for the slab (Problem 4.15); this is

$$\frac{Q}{Q_i} = 4 \sum_{n=1}^{\infty} \frac{1}{\delta_n^2} \frac{J_1^2(\delta_n)}{J_0^2(\delta_n) + J_1^2(\delta_n)} (1 - e^{-\delta_n^2 \, \mathbf{Fo}}) \qquad (4.36)$$

In all the equations for the cylinder, the δ_n are obtained as roots of the transcendental equation

$$\delta_n \frac{J_1(\delta_n)}{J_0(\delta_n)} = \mathbf{Bi^*}$$

and the Biot and Fourier numbers are based upon the radius of the cylinder, i.e.

$$\mathbf{Bi^*} = \frac{\bar{h}R}{k} \qquad \mathbf{Fo^*} = \frac{\alpha t}{R^2}$$

We use an asterisk to indicate a deviation from the general definition of a characteristic length, (4.5). The results of numerical solutions of (4.35) and (4.36) are presented in Figs. 4-9 and 4-10. A position correction chart is given in Fig. 4-11.

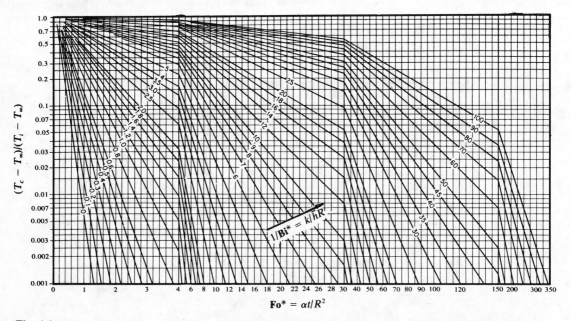

Fig. 4-9.   Centerline temperature for a solid cylinder.   [From M. P. Heisler, *Trans. ASME*, **69**: 227 (1947).]

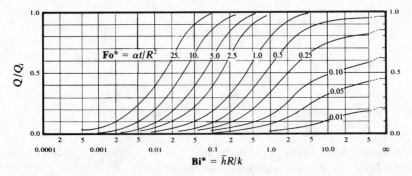

Fig. 4-10.   Total heat flow, solid cylinder.   (From H. Gröber, S. Erk, and U. Grigall, *Grundgesetze der Wärmeübertragung*, 3d ed., Springer-Verlag, Berlin, 1955, by permission.)

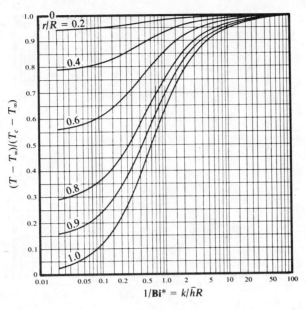

Fig. 4-11

Solid Sphere

The problem of determining the transient temperature distribution within a solid sphere of radius R which is initially at a uniform temperature and is suddenly exposed to a convective heat transfer at its surface has been studied extensively also. Dimensionless temperatures are presented in the Heisler charts of Figs. 4-12 and 4-13. Here the moduli are defined as

$$\mathbf{Bi^*} = \frac{\bar{h}R}{k} \qquad \mathbf{Fo^*} = \frac{\alpha t}{R^2}$$

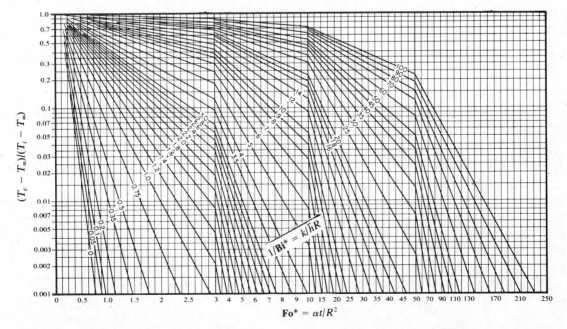

Fig. 4-12.   Centerline temperature of a solid sphere.   [From M. P. Heisler, *Trans. ASME*, **69**: 227 (1947).]

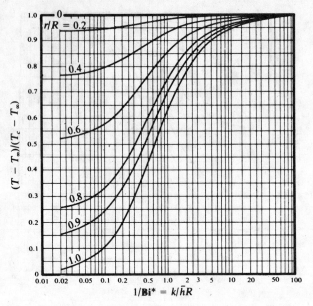

Fig. 4-13.   Correction chart for a solid sphere.   [From M. P. Heisler, *Trans. ASME*, **69**: 227 (1947).]

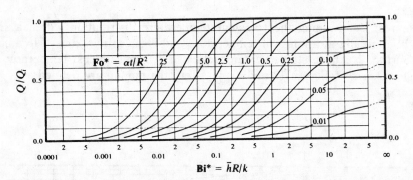

Fig. 4-14.   Total heat flow, solid sphere.   (From H. Gröber, S. Erk, and U. Grigall, *Grundgesetze der Wärmeübertragung*, 3d ed., Springer-Verlag, Berlin, 1955, by permission.)

and T_c denotes the temperature at the center.   The dimensionless total heat transfer is given by the Gröber chart of Fig. 4-14.

4.7   MULTIDIMENSIONAL SYSTEMS

Consider the temperature distribution during heat treatment of a long rectangular bar (Fig. 4-15).   If such a body is initially at a high temperature and is suddenly exposed to a cooler surrounding fluid, there will be two-dimensional heat transfer within the body, assuming the z-direction length is sufficient to result in negligible temperature gradients in that direction.

Applying the separation-of-variables method to the conduction equation for $\theta = T - T_\infty$,

$$\frac{\partial^2 \theta}{\partial x^2} + \frac{\partial^2 \theta}{\partial y^2} = \frac{1}{\alpha}\frac{\partial \theta}{\partial t}$$

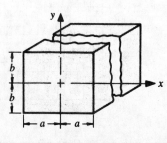

Fig. 4-15

we assume that $\theta(x, y, t) = X(x, t) \, Y(y, t)$ and realizing that the x- and y-solutions must have the same form, we easily reduce the problem to two one-dimensional problems of the type resulting in (4.28).   Thus,

$$\frac{\theta}{\theta_i} \equiv \left(\frac{T - T_\infty}{T_i - T_\infty}\right)_{\substack{\text{long} \\ \text{bar}}} = \left(\frac{T - T_\infty}{T_i - T_\infty}\right)_{\substack{2a- \\ \text{slab}}} \left(\frac{T - T_\infty}{T_i - T_\infty}\right)_{\substack{2b- \\ \text{slab}}} \tag{4.37}$$

where the $2a$-slab and the $2b$-slab solutions may be taken from Figs. 4-6 and 4-7.   It can also be shown that the transient temperature in a cylinder with both radial and axial heat transfer is given by

$$\left(\frac{T - T_\infty}{T_i - T_\infty}\right)_{\substack{\text{short} \\ \text{cyl}}} = \left(\frac{T - T_\infty}{T_i - T_\infty}\right)_{\substack{\text{inf} \\ \text{cyl}}} \left(\frac{T - T_\infty}{T_i - T_\infty}\right)_{\substack{2L- \\ \text{slab}}} \tag{4.38}$$

and that in a rectangular parallelepiped (box-shaped object) it is given by

$$\left(\frac{T - T_\infty}{T_i - T_\infty}\right)_{\text{box}} = \left(\frac{T - T_\infty}{T_i - T_\infty}\right)_{\substack{2a- \\ \text{slab}}} \left(\frac{T - T_\infty}{T_i - T_\infty}\right)_{\substack{2b- \\ \text{slab}}} \left(\frac{T - T_\infty}{T_i - T_\infty}\right)_{\substack{2c- \\ \text{slab}}} \tag{4.39}$$

where $2L$ is the total length of the cylinder and $2c$ is the total z-dimension of the parallelepiped.

4.8   NUMERICAL ANALYSIS

The fundamental approach to numerical analysis of a transient conduction problem lies in the replacement of the differential equation, which represents a continuous temperature distribution in both space and time, with a finite-difference equation, which can yield results only at discrete locations and at specified times within the body.

For the partial derivatives of $T(\xi, \tau)$ at the point (ξ, τ) we employ either the *forward finite-difference approximations*

$$\left(\frac{\partial T}{\partial \xi}\right)_{\xi, \tau, \text{fwd}} \approx \frac{1}{\delta \xi}[T(\xi + \delta \xi, \tau) - T(\xi, \tau)] \tag{4.40}$$

$$\left(\frac{\partial T}{\partial \tau}\right)_{\xi, \tau, \text{fwd}} \approx \frac{1}{\delta \tau}[T(\xi, \tau + \delta \tau) - T(\xi, \tau)] \tag{4.41}$$

or the *backward finite-difference approximations*

$$\left(\frac{\partial T}{\partial \xi}\right)_{\xi, \tau, \text{bkwd}} \approx \frac{1}{\delta \xi}[T(\xi, \tau) - T(\xi - \delta \xi, \tau)] \tag{4.42}$$

$$\left(\frac{\partial T}{\partial \tau}\right)_{\xi, \tau, \text{bkwd}} \approx \frac{1}{\delta \tau}[T(\xi, \tau) - T(\xi, \tau - \delta \tau)] \tag{4.43}$$

By definition of the derivative, these approximations become exact as $\delta \xi$ and $\delta \tau$ approach zero.

Second partial derivatives in the conduction equation are usually replaced by their *central finite-difference approximation*:

$$\left(\frac{\partial^2 T}{\partial \xi^2}\right)_{\xi, \tau, \text{cent}} \approx \frac{1}{(\delta \xi)^2}[T(\xi + \delta \xi, \tau) - 2T(\xi, \tau) + T(\xi - \delta \xi, \tau)] \tag{4.44}$$

Again, the approximation becomes exact as $\delta \xi \to 0$.

Explicit Finite-Difference Conduction Equation

Let us apply the above to the two-dimensional conduction equation

$$\frac{\partial^2 T}{\partial x^2} + \frac{\partial^2 T}{\partial y^2} = \frac{1}{\alpha} \frac{\partial T}{\partial t} \tag{4.45}$$

From (4.44), with $\xi = x$, $\delta\xi = \Delta x$, and $\tau = t$,

$$\left(\frac{\partial^2 T}{\partial x^2}\right)_{x,t,\text{cent}} \approx \frac{1}{(\Delta x)^2}[T(x + \Delta x, t) - 2T(x, t) + T(x - \Delta x, t)]$$

which for the general nodal point n of Fig. 4-16 results in

$$\left(\frac{\partial^2 T}{\partial x^2}\right)_{n,t,\text{cent}} \approx \frac{1}{(\Delta x)^2}[T_3^t - 2T_n^t + T_1^t] \tag{4.46}$$

where, for example, T_3^t is the temperature at node 3 at time t.   Similarly,

$$\left(\frac{\partial^2 T}{\partial y^2}\right)_{n,t,\text{cent}} \approx \frac{1}{(\Delta y)^2}[T_4^t - 2T_n^t + T_2^t] \tag{4.47}$$

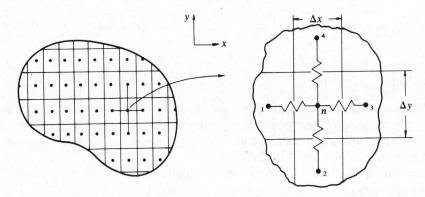

Fig. 4-16

The forward finite-difference formulation of the time derivative is obtained from (4.41) with $\tau = t$ and $\delta\tau = \Delta t$ at point n:

$$\left(\frac{\partial T}{\partial t}\right)_{n,t,\text{fwd}} \approx \frac{1}{\Delta t}[T_n^{t+1} - T_n^t] \tag{4.48}$$

where T_n^{t+1} is the value of T_n at time $t + \Delta t$.   Substituting (4.46), (4.47), and (4.48) into (4.45) yields

$$\frac{1}{(\Delta x)^2}[T_3^t - 2T_n^t + T_1^t] + \frac{1}{(\Delta y)^2}[T_4^t - 2T_n^t + T_2^t] = \frac{1}{\alpha \, \Delta t}[T_n^{t+1} - T_n^t] \tag{4.49}$$

Note that this equation expresses the *future* temperature at node n in terms of the *present* temperatures of node n and its adjacent nodes.   This gives rise to the notation "explicit formulation," since the future temperature is explicitly specified in terms of temperatures existing one time increment earlier.

To this point, the increments Δx and Δy may be different; and indeed, they may vary with location within the body.   For simplification, we now choose

$$\Delta x = \Delta y = \Delta s$$

and (4.49) reduces to

$$T_n^{t+1} = \frac{\alpha \, \Delta t}{(\Delta s)^2} (T_1^t + T_2^t + T_3^t + T_4^t) + \left[1 - 4 \frac{\alpha \, \Delta t}{(\Delta s)^2} \right] T_n^t \tag{4.50}$$

Defining the parameter M by

$$M \equiv \frac{(\Delta s)^2}{\alpha \, \Delta t} \tag{4.51}$$

the one-, two-, and three-dimensional explicit nodal equations for uniform and equal spatial increments are, respectively,

$$T_n^{t+1} = \frac{1}{M} (T_1^t + T_3^t) + \left(1 - \frac{2}{M} \right) T_n^t \tag{4.52}$$

$$T_n^{t+1} = \frac{1}{M} (T_1^t + T_2^t + T_3^t + T_4^t) + \left(1 - \frac{4}{M} \right) T_n^t \tag{4.53}$$

$$T_n^{t+1} = \frac{1}{M} (T_1^t + T_2^t + T_3^t + T_4^t + T_5^t + T_6^t) + \left(1 - \frac{6}{M} \right) T_n^t \tag{4.54}$$

Note that (4.52) implies that the surrounding nodes in the one-dimensional case are nodes 1 and 3, whereas (4.54) implies that the z-direction nodes in the three-dimensional case are nodes 5 and 6.

Stability of the numerical solution during repetitive computations for succeeding time steps is assured by requiring the coefficient of T_n^t in (4.52), (4.53), or (4.54) to be nonnegative. (A negative coefficient of T_n^t would mean that a larger value of T_n^t would result in a smaller value of T_n^{t+1}; this would in return result in a larger value of T_n^{t+2}, etc., yielding unstable fluctuations in the numerical solution.) To assure all coefficients being nonnegative we require

$$\begin{aligned} &M \geqslant 2 \quad \text{for one-dimensional problems} \\ &M \geqslant 4 \quad \text{for two-dimensional problems} \\ &M \geqslant 6 \quad \text{for three-dimensional problems} \end{aligned} \tag{4.55}$$

Equations (4.52) through (4.55) are appropriate for interior nodes of a body. For an exterior nodal point subjected to convective boundary conditions in a one-dimensional system, as shown in Fig. 4-17, an energy balance on the shaded area having unit depth is

$$\rho c \frac{(\Delta s)^2}{2} \left(\frac{T_n^{t+1} - T_n}{\Delta t} \right) = k \, \Delta s \left(\frac{T_1^t - T_n^t}{\Delta s} \right) + \bar{h} \, \Delta s \, (T_\infty - T_n^t)$$

or

$$T_n^{t+1} = \frac{2}{M} \left(T_1^t + \frac{\bar{h} \, \Delta s}{k} T_\infty \right) + \left[1 - \frac{2}{M} \left(\frac{\bar{h} \, \Delta s}{k} + 1 \right) \right] T_n^t \tag{4.56}$$

For an exterior nodal point subjected to a convective boundary condition in a two-dimensional system, as shown in Fig. 4-18, a similar approach yields

$$T_n^{t+1} = \frac{1}{M} \left(T_1^t + 2T_2^t + T_3^t + 2 \frac{\bar{h} \, \Delta s}{k} T_\infty \right) + \left[1 - \frac{2}{M} \left(\frac{\bar{h} \, \Delta s}{k} + 2 \right) \right] T_n^t \tag{4.57}$$

Sufficient stability requirements for convective boundary conditions are

$$M \geqslant 2 \left(\frac{\bar{h} \, \Delta s}{k} + 1 \right) \quad \text{for one-dimensional problems}$$

$$\tag{4.58}$$

$$M \geqslant 2 \left(\frac{\bar{h} \, \Delta s}{k} + 2 \right) \quad \text{for two-dimensional problems}$$

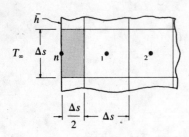

Fig. 4-17

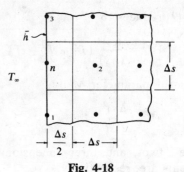

Fig. 4-18

Additional nodal equations for cases involving interior or exterior corners and/or surface heat flux such as that due to radiation can be derived in similar fashion. The stability requirement for each such situation can be obtained by requiring the grouped coefficient of T_n^t to be non-negative. The most stringent of the stability requirements applicable in a given problem dictates, through (4.51), the maximum time increment that can be used in the numerical solution.

The numerical solution proceeds by writing a nodal equation for each interior and each exterior nodal point not having fixed (constant) temperature, determining the maximum time increment Δt for the selected grid size and specified material thermal diffusivity, and then using the nodal equations to advance the solution in steps of Δt until the desired total time is reached.

Implicit Finite-Difference Conduction Equation

In the explicit formulation, the nodal equations were obtained by replacing (4.45) by

$$\left(\frac{\partial^2 T}{\partial x^2}\right)_{n,t,\text{cent}} + \left(\frac{\partial^2 T}{\partial y^2}\right)_{n,t,\text{cent}} = \frac{1}{\alpha}\left(\frac{\partial T}{\partial t}\right)_{n,t,\text{fwd}}$$

If, instead, (4.45) is replaced by

$$\left(\frac{\partial^2 T}{\partial x^2}\right)_{n,t+\Delta t,\text{cent}} + \left(\frac{\partial^2 T}{\partial y^2}\right)_{n,t+\Delta t,\text{cent}} = \frac{1}{\alpha}\left(\frac{\partial T}{\partial t}\right)_{n,t+\Delta t,\text{bkwd}}$$

one obtains the implicit form of the nodal equations:

$$\frac{1}{(\Delta x)^2}(T_3^{t+1} - 2T_n^{t+1} + T_1^{t+1}) + \frac{1}{(\Delta y)^2}(T_4^{t+1} - 2T_n^{t+1} + T_2^{t+1}) = \frac{1}{\alpha\,\Delta t}(T_n^{t+1} - T_n^t) \qquad (4.59)$$

or, when $\Delta x = \Delta y = \Delta s$,

$$\left(1 + \frac{4}{M}\right)T_n^{t+1} = \frac{1}{M}(T_1^{t+1} + T_2^{t+1} + T_3^{t+1} + T_4^{t+1}) + T_n^t \qquad (4.60)$$

In (4.59) or (4.60) the *future* temperature of node n is seen to depend upon the *future* temperatures of the adjacent nodes in addition to its *present* temperature. Consequently, a set of simultaneous algebraic equations must be solved for each time step, there being the same number of equations as the number of nodes. For more than three or four nodes, the implicit formulation normally requires computer solution using either matrix inversion or Gaussian elimination for each time step. Because all coefficients are positive in (4.60) there is no stability restriction on the size of M.

In order to treat boundary conditions in the implicit approach, we multiply (4.59) by $k(\Delta x)(\Delta y)$ and rearrange, to obtain

$$\frac{T_1^{t+1} - T_n^{t+1}}{R_{1n}} + \frac{T_2^{t+1} - T_n^{t+1}}{R_{2n}} + \frac{T_3^{t+1} - T_n^{t+1}}{R_{3n}} + \frac{T_4^{t+1} - T_n^{t+1}}{R_{4n}} = C_n\left(\frac{T_n^{t+1} - T_n^t}{\Delta t}\right) \qquad (4.61)$$

where, per unit depth,

$$R_{1n} = \frac{\Delta x}{k \, \Delta y}, \quad \text{etc.} \tag{4.62}$$

and

$$C_n = \rho c (\Delta x)(\Delta y) \tag{4.63}$$

Generalizing (4.61) through (4.63) to any number of spatial dimensions and to a grid spacing Δs that may vary with position and direction, we have:

$$\text{implicit:} \quad T_n^{t+1} = T_n^t + \Delta t \left[\sum_m \frac{T_m^{t+1} - T_n^{t+1}}{R_{mn} C_n} \right] \tag{4.64}$$

where m represents the nodes adjacent to node n;

$$R_{mn} = \frac{\Delta s}{k A_{k,mn}} \quad \text{for conduction}$$

$$\tag{4.65}$$

$$R_{mn} = \frac{1}{h_{mn} A_{c,mn}} \quad \text{for convection}$$

$$C_n = \rho c V_n \tag{4.66}$$

Here, $A_{k,mn}$ represents the area for conductive heat transfer between nodes m and n; $A_{c,mn}$ is the area for convective heat transfer between nodes m and n; and V_n is the volume element determined by the value(s) of Δs at node n. Boundary nodes are simply included by proper formulation of the R_{mn} and C_n.

The corresponding generalization of the explicit formulation is

$$\text{explicit:} \quad T_n^{t+1} = T_n^t + \Delta t \left[\sum_m \frac{T_m^t - T_n^t}{R_{mn} C_n} \right] \tag{4.67}$$

The advantage of the implicit formulation is that relatively large time steps can be taken even though the nodal spacing (or the thermal capacitance of a node) is small. This advantage is accompanied by the disadvantage of requiring a simultaneous solution of the complete set of nodal equations for each time step. Thus one is usually faced with a choice between large computer core storage (required for implicit solution by matrix inversion or Gaussian elimination) and large computational time (required for explicit solution with small nodal thermal capacity).

Solved Problems

4.1. Determine the Biot modulus for a 1.0-inch-diameter, mild steel sphere at 212 °F subjected to a convective air flow resulting in $\bar{h} = 10$ Btu/hr-ft²-°F.

From Appendix B, $k = 26$ Btu/hr-ft-°F. The characteristic linear dimension is

$$L = \frac{V}{A_s} = \frac{4\pi R^3/3}{4\pi R^2} = \frac{R}{3}$$

Thus

$$\mathbf{Bi} = \frac{\bar{h}(R/3)}{k} = \frac{(10 \text{ Btu/hr-ft}^2\text{-°F})[0.5/(3)(12)]\text{ft}}{26 \text{ Btu/hr-ft-°F}} = 0.0053$$

Clearly, internal temperature gradients are small, and a lumped thermal analysis would be quite accurate.

4.2. An iron (k = 64 W/m-K) billet measuring $20 \times 16 \times 80$ cm is subjected to free convective heat transfer resulting in $\bar{h}$ = 2 Btu/hr-ft²-°F. Determine the Biot number and the suitability of a lumped analysis to represent the cooling rate if the billet is initially hotter than the environment.

The characteristic length is $L = V/A_s$.

$$A_s = 2(20 \times 16) + 2(16 \times 80) + 2(20 \times 80) = 6400 \text{ cm}^2$$

so

$$L = \frac{(20 \times 16 \times 80) \text{ cm}^3}{6400 \text{ cm}^2} = 4.0 \text{ cm}$$

Using conversion factors of Appendix A,

$$\bar{h} = \frac{2 \text{ Btu}}{\text{hr-ft}^2\text{-°F}} \times \frac{3.1525 \text{ W/m}^2}{\text{Btu/hr-ft}^2} \times \frac{9 \text{ °F}}{5 \text{ K}} = 11.35 \text{ W/m}^2\text{-K}$$

Thus

$$\mathbf{Bi} = \frac{\bar{h}L}{k} = \frac{(11.35 \text{ W/m}^2\text{-K})(0.04 \text{ m})}{64 \text{ W/m-K}} = 0.0071$$

and a lumped analysis will represent the transient temperature quite well.

4.3. Determine the time required for a 0.5-inch-diameter mild steel sphere to cool from 1000 °F to 200 °F, if exposed to a cooling air flow at 80 °F resulting in $\bar{h}$ = 20 Btu/hr-ft²-°F.

The characteristic linear dimension is $L = R/3$ (Problem 4.1). From Appendix B, the thermal conductivity of mild steel at the average temperature, $(1000 + 200)/2 = 600$ °F, is approximately 25 Btu/hr-ft-°F. Thus,

$$\mathbf{Bi} = \frac{\bar{h}L}{k} = \frac{(20 \text{ Btu/hr-ft}^2\text{-°F})[0.25/(3)(12)]\text{ft}}{25 \text{ Btu/hr-ft-°F}} = 0.0056$$

and a lumped analysis is suitable. By (*4.9*) the time (**Fo**) for $T(t)$ to reach 200 °F satisfies

$$\frac{200 - 80}{1000 - 80} = 0.1304 = e^{-(\mathbf{Bi})(\mathbf{Fo})}$$

Solving,

$$(\mathbf{Bi})(\mathbf{Fo}) = 2.0369$$

$$\mathbf{Fo} = \frac{2.0369}{0.0056} = 363.73 = \frac{\alpha t}{L^2}$$

$$t = \frac{(363.73)(R/3)^2}{\alpha}$$

From Appendix B,

$$\alpha \approx \frac{25 \text{ Btu/hr-ft-°F}}{\left(490 \dfrac{\text{lbm}}{\text{ft}^3}\right)\left(\dfrac{0.11 \text{ Btu}}{\text{lbm-°F}}\right)} = 0.46 \text{ ft}^2/\text{hr}$$

Thus

$$t = \frac{(363.73)[0.25/(3)(12)]^2 \text{ ft}^2}{0.46 \text{ ft}^2/\text{hr}} = 0.0381 \text{ hr} = 2.29 \text{ min}$$

4.4. Determine the time constant for a spherically shaped, copper-constantan thermocouple at an average temperature of 32 °F, exposed to a convective environment where $\bar{h} = 8$ Btu/hr-ft^2-°F, for (a) bead diameter 0.005 in. and (b) bead diameter 0.010 in.

For a spherical object of radius a, $L = V/A_s = a/3$. From Appendix B the properties of the two metals are

$$k_{cu} = 224 \text{ Btu/hr-ft-°F} \qquad k_{con} = 12.4 \text{ Btu/hr-ft-°F}$$
$$c_{cu} = 0.091 \text{ Btu/lbm-°F} \qquad c_{con} = 0.10 \text{ Btu/lbm-°F}$$
$$\rho_{cu} = 558 \text{ lbm/ft}^3 \qquad \rho_{con} = 557 \text{ lbm/ft}^3$$

Assuming linear averaging to be valid, the thermocouple bead properties are

$$k = (224 + 12.4)/2 = 118.2 \text{ Btu/hr-ft-°F}$$
$$c = (0.091 + 0.10)/2 = 0.096 \text{ Btu/lbm-°F}$$
$$\rho = (558 + 557)/2 = 557.5 \text{ lbm/ft}^3$$

(a) From (4.8)

$$(RC)_{th} = \frac{\rho c V}{\bar{h} A_s} = \frac{\rho c a/3}{\bar{h}} = \frac{\left(557.5 \frac{\text{lbm}}{\text{ft}^3}\right)\left(0.096 \frac{\text{Btu}}{\text{lbm-°F}}\right)\left(\frac{0.0025}{3 \times 12} \text{ ft}\right)}{8 \text{ Btu/hr-ft}^2\text{-°F}}$$

$$= 4.646 \times 10^{-4} \text{ hr} = 1.6725 \text{ sec}$$

(b) The only difference from part (a) is that the radius is twice the previous value. Since $(RC)_{th}$ is linear in a,

$$(RC)_{th} = 2(1.6725 \text{ sec}) = 3.345 \text{ sec}$$

Before leaving this problem, we should verify that the Biot modulus is sufficiently small to insure the validity of a lumped analysis. Checking for the larger thermocouple,

$$\mathbf{Bi} = \frac{\bar{h} L}{k} = \frac{(8 \text{ Btu/hr-ft}^2\text{-°F})[0.005/(3)(12)]\text{ft}}{118.2 \text{ Btu/hr-ft-°F}} = 9.4 \times 10^{-6}$$

and the lumped analysis is certainly suitable.

4.5. If the initial temperature is 32 °F and the surrounding air temperature is 45 °F, how long will it take the 0.005-inch-diameter thermocouple of Problem 4.4 to reach (a) 44 °F, (b) 44.5 °F, (c) 44.9 °F and (d) 44.99 °F?

In general, the time required for the thermocouple junction to reach a given temperature is obtained by use of (4.9). Thus,

$$\ln\left[\frac{T(t) - T_\infty}{T_i - T_\infty}\right] = -\frac{t}{(RC)_{th}}$$

or

$$t = -(1.6725 \text{ sec}) \ln\left[\frac{T(t) - T_\infty}{T_i - T_\infty}\right]$$

where the numerical value of $(RC)_{th}$ was taken from Problem 4.4(a).

(a) For $T_i = 32$ °F, $T_\infty = 45$ °F, and $T(t) = 44$ °F,

$$t = -1.6725 \ln\left(\frac{44 - 45}{32 - 45}\right) = 4.29 \text{ sec}$$

(b) $$t = -1.6725 \ln\left(\frac{44.5 - 45}{32 - 45}\right) = 5.45 \text{ sec}$$

(c) $t = -1.6725 \ln \left(\dfrac{44.9 - 45}{32 - 45} \right) = 8.14$ sec

(d) $t = -1.6725 \ln \left(\dfrac{44.99 - 45}{32 - 45} \right) = 11.99$ sec

Notice that (4.9) indicates that a body subjected to a warming or cooling convective fluid can never reach the surrounding fluid temperature.   While this is mathematically correct, it is important to realize that the difference in temperature between the body and the convective fluid can be reduced to an insignificantly small value by allowing a sufficient time period, as illustrated by the present problem.

4.6.   A three-inch-diameter orange is subjected to a cold-air environment.   Assuming that the orange has properties similar to those of water at 68 °F and that $\bar{h} = 2$ Btu/hr-ft^2-°F, determine the suitability of a lumped analysis for predicting the temperature of the orange during cooling.

From Appendix B, the thermal conductivity of water at 68 °F is 0.345 Btu/hr-ft-°F.   Also, for a sphere

$$L = \frac{V}{A_s} = \frac{R}{3}$$

and

$$\mathbf{Bi} = \frac{\bar{h}L}{k} = \frac{(2 \text{ Btu/hr-ft}^2\text{-°F})[1.5/(3)(12)]\text{ft}}{0.345 \text{ Btu/hr-ft-°F}} = 0.24$$

Consequently a lumped analysis would not be highly accurate.

4.7.   What is the maximum edge dimension of a solid aluminum cube at 100 °C subjected to a convective heat transfer with $\bar{h} = 25$ W/m^2-K for a lumped analysis to be accurate to within 5 percent?

For an object of this shape, the lumped approach is accurate to ±5 percent if the Biot modulus is less than 0.1.   From Appendix B, $k = 205.82$ W/m-K.   Thus,

$$\mathbf{Bi} = 0.1 = \frac{\bar{h}L}{k} = \frac{(25 \text{ W/m}^2\text{-K})L}{205.82 \text{ W/m-K}} \quad \text{or} \quad L = 0.823 \text{ m}$$

Now

$$L = \frac{V}{A_s} = \frac{l^3}{6l^2}$$

$$l = 6L = 6(0.823 \text{ m}) = 4.94 \text{ m}$$

4.8.   For the situation of Problem 4.3, determine (a) the instantaneous heat-transfer rate 2 minutes after the start of cooling and (b) the total energy transferred from the sphere during the first 2 minutes.

Either we could calculate the numerical value of the temperature of the sphere at the instant specified and use this in Newton's law of cooling to determine q and Δu, or we could use the more general approach of substituting (4.9) into Newton's law of cooling to obtain an analytical expression for q.   Choosing the latter,

(a) $q = \bar{h}A_s(T - T_\infty) = \bar{h}A_s(T_i - T_\infty)e^{-t/(RC)_{\text{th}}}$

From Problem 4.3, at $t = 2.29$ min, $\mathbf{Bi} \cdot \mathbf{Fo} = 2.0369$ and $T_i - T_\infty = 920$ °F.   Thus at $t = 2.0$ min,

$$\frac{t_2}{(RC)_{\text{th}}} = (\mathbf{Bi} \cdot \mathbf{Fo})_2 = \frac{t_2}{t_{2.29}} (\mathbf{Bi} \cdot \mathbf{Fo})_{2.29} = \frac{2}{2.29}(2.0369) = 1.779$$

and

$$q = 20[4\pi(0.25/12)^2](920)e^{-1.779} = 16.94 \text{ Btu/hr}$$

(b) The total energy transferred is

$$Q = \int_0^{t_2} q\, dt = \int_0^{t_2} \bar{h}A_s(T_i - T_\infty)e^{-t/(RC)_{\text{th}}}\, dt = C_{\text{th}}(T_i - T_\infty)[1 - e^{-t_2/(RC)_{\text{th}}}]$$

where $C_{\text{th}} = \rho c V$ is the thermal capacitance [see (4.8)]. Using property data of Appendix B,

$$C_{\text{th}} = \rho c V = \left(490\,\frac{\text{lbm}}{\text{ft}^3}\right)\left(\frac{0.11\text{ Btu}}{\text{lbm-}°\text{F}}\right)\left(\frac{4\pi}{3}\right)\left(\frac{0.25}{12}\text{ ft}\right)^3 = 0.00204 \text{ Btu/}°\text{F}$$

and

$$Q = (0.00204 \text{ Btu/}°\text{F})(920\ °\text{F})(1 - e^{-1.779}) = 1.560 \text{ Btu}$$

4.9. Consider the container of fluid shown in Fig. 4-19. The system is brought to a uniform initial temperature $T_i > T_\infty$ by wrapping it in an insulating blanket (which corresponds to opening switch S_2), running the electric heater (close switch S_1), and waiting until both T_1 and T_2 reach T_i (at which point open S_1 and close S_2). Give a lumped analysis of the ensuing process, which is one of convective heat transfer from container to environment and from inner fluid to container.

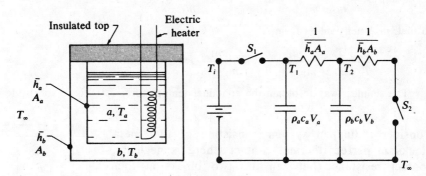

Fig. 4-19

The rate at which thermal energy (heat) leaves each body is equal to the negative of the rate of change of its stored thermal energy. Hence, the energy balances on the two lumps are

lump a: $\quad \bar{h}_aA_a(T_a - T_b) = -\rho_ac_aV_a\dfrac{dT_a}{dt}$ (1)

lump b: $\quad \bar{h}_aA_a(T_b - T_a) + \bar{h}_bA_b(T_b - T_\infty) = -\rho_bc_bV_b\dfrac{dT_b}{dt}$ (2)

These constitute a pair of simultaneous linear differential equations for $T_a(t)$ and $T_b(t)$. The initial conditions are

$$T_a(0) = T_b(0) = T_i$$ (3)

To solve (1) and (2) subject to (3), we first use (1) to obtain T_b in terms of T_a and dT_a/dt. Substituting this expression into (2) gives the following second-order equation for T_a:

$$\frac{d^2T_a}{dt^2} + C_1\frac{dT_a}{dt} + C_2T_a = C_2T_\infty$$ (4)

where

$$C_1 = \frac{\bar{h}_a A_a}{\rho_a c_a V_a} + \frac{\bar{h}_a A_a}{\rho_b c_b V_b} + \frac{\bar{h}_b A_b}{\rho_b c_b V_b}$$

$$C_2 = \left(\frac{\bar{h}_a A_a}{\rho_a c_a V_a}\right)\left(\frac{\bar{h}_b A_b}{\rho_b c_b V_b}\right)$$

The steady-state solution (particular integral) of (4) is clearly $T_a = T_\infty$, while the transient solution (general solution of the homogeneous equation) is

$$T_a = Ae^{m_1 t} + Be^{m_2 t}$$

where A and B are arbitrary constants and

$$m_1 = \frac{-C_1 + [C_1^2 - 4C_2]^{1/2}}{2} \qquad m_2 = \frac{-C_1 - [C_1^2 - 4C_2]^{1/2}}{2}$$

Thus,

$$T_a = Ae^{m_1 t} + Be^{m_2 t} + T_\infty \tag{5}$$

Finally, A and B are determined by applying to (5) the initial conditions $T_a(0) = T_i$ and

$$\left.\frac{dT_a}{dt}\right|_{t=0} = 0$$

which is implied by (3) and (1). The results are

$$A = \frac{m_2}{m_2 - m_1}(T_i - T_\infty) \qquad B = -\frac{m_1}{m_2 - m_1}(T_i - T_\infty)$$

so that, in dimensionless form,

$$\frac{T_a - T_\infty}{T_i - T_\infty} = \frac{m_2}{m_2 - m_1}e^{m_1 t} - \frac{m_1}{m_2 - m_1}e^{m_2 t} \tag{6}$$

The simplest way to obtain the container temperature, $T_b(t)$, is to substitute (6) into (1).

4.10. Consider a lumped system consisting of two metal blocks in perfect thermal contact (there is negligible contact resistance between the two blocks at their interface). Only one block is exposed to a convective environment at T_∞, as shown in Fig. 4-20. Both blocks are initially at T_i. Determine an analytical expression for the temperature of block a.

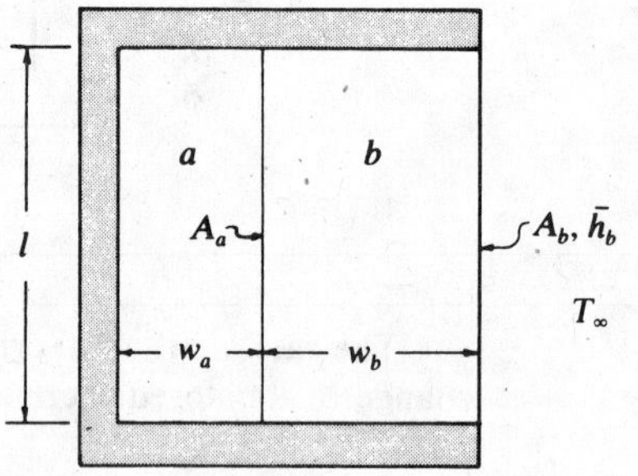

Fig. 4-20

The heat transfer rate from block a and to block b is

$$\frac{q}{A} = -k_a\left(\frac{T_{\text{int}} - T_a}{w_a/2}\right) = -k_b\left(\frac{T_b - T_{\text{int}}}{w_b/2}\right) \quad (1)$$

Eliminating T_{int}, the interfacial temperature, between these two equations for q/A gives

$$\frac{q}{A} = \frac{q_a}{A_a} = K(T_a - T_b) \tag{2}$$

where

$$K = \frac{2(k_a/w_a)(k_b/w_b)}{(k_a/w_a) + (k_b/w_b)} \tag{3}$$

Using (2) to write the energy balances on the two blocks, we find them to be identical to (1) and (2) of Problem 4.9 with $\bar{h}_a$ replaced by K. Furthermore, the initial conditions are identical to (3) of

Problem 4.9. It follows that all the results of Problem 4.9—in particular, (6)—apply to the present problem if $\bar{h}_a$ is everywhere replaced by K.

4.11. For a system consisting of two metallic blocks as shown in Fig. 4-20, with only block b exposed to the convective environment ($\bar{h}_b = 5$ Btu/hr-ft^2-°F), the following data apply:

Material a	Material b
Brass	Type 304 stainless steel
$c_a = 0.092$ Btu/lbm-°F	$c_b = 0.11$ Btu/lbm-°F
$k_a = 60.0$ Btu/hr-ft-°F	$k_b = 9.4$ Btu/hr-ft-°F
$w_a = 3.0$ in.	$w_b = 2.0$ in.
$\rho_a = 532$ lbm/ft^3	$\rho_b = 488$ lbm/ft^3

$$T_i = 300\ °F \qquad T_\infty = 100\ °F$$

(*a*) Is a lumped analysis suitable? (*b*) If the answer to (*a*) is yes, plot the temperature–time history of lump a from 300 °F down to 150 °F.

(*a*) An appropriate Biot modulus for material b is

$$\mathbf{Bi}_b = \frac{\bar{h}_b L_b}{k_b}$$

where

$$L_b = \frac{V_b}{A_s} = \frac{(l)(w_b)(1)}{(l)(1)} = w_b$$

if we neglect the heat transfer to body a. So

$$\mathbf{Bi}_b = \frac{(5)(2/12)}{9.4} = 0.09$$

As for $\mathbf{Bi}_a$, the resistance for heat transfer from lump a to the environment at T_∞ is clearly greater than that for heat transfer from lump b, hence a suitable $\bar{h}_a < \bar{h}_b$. Also,

$$\frac{w_a}{k_a} = \frac{3/12}{60} < \frac{2/12}{9.4} = \frac{w_b}{k_b}$$

so that $\mathbf{Bi}_a < \mathbf{Bi}_b$. As both moduli are smaller than 0.1, a lumped analysis of the system is suitable.

(*b*) Determining the constants to use in the solution given in Problem 4.10,

$$K = \frac{2(k_a/w_a)(k_b/w_b)}{(k_a/w_a) + (k_b/w_b)} = \frac{2[60(12)/3][9.4(12)/2]}{[60(12)/3] + [9.4(12)/2]} = 91.34\ \text{Btu/hr-ft}^2\text{-°F}$$

$$C_1 = \frac{KA_a}{\rho_a c_a V_a} + \frac{KA_a}{\rho_b c_b V_b} + \frac{\bar{h}_b A_b}{\rho_b c_b V_b} = \frac{K}{\rho_a c_a w_a} + \frac{K}{\rho_b c_b w_b} + \frac{\bar{h}_b}{\rho_b c_b w_b}$$

$$= \frac{(91.34)(12)}{532(0.092)(3)} + \frac{(91.34)(12)}{488(0.11)(2)} + \frac{5(12)}{488(0.11)(2)}$$

$$= 7.46 + 10.21 + 0.56 = 18.23\ \text{hr}^{-1}$$

$$C_2 = (7.46)(0.56) = 4.18\ \text{hr}^{-2}$$

$$m_1 = \frac{-18.23 + [(18.23)^2 - 4(4.18)]^{1/2}}{2} = -0.23\ \text{hr}^{-1}$$

$$m_2 = \frac{-18.23 - [(18.23)^2 - 4(4.18)]^{1/2}}{2} = -18.00\ \text{hr}^{-1}$$

Then, by (6) of Problem 4.9,

$$\frac{T_a - 100}{300 - 100} = \frac{-18.0}{-18.0 - (-0.23)} e^{-0.23\,t} - \frac{(-0.23)}{-18.0 - (-0.23)} e^{-18.0\,t}$$

$$T_a = 100 + 202.59 e^{-0.23\,t} - 2.59 e^{-18.0\,t}$$

where t is in hours and T_a has units of °F.   Plotting, we obtain Table 4-1 and Fig. 4-21.

Table 4-1

t, hr	T_a, °F
0	300.00
0.5	280.58
1.0	260.96
1.5	243.48
2.0	227.89
2.5	214.00
3.0	201.61
3.5	190.58
4.0	180.74
5.0	164.15
6.0	150.97
6.1	149.81
7.0	140.50

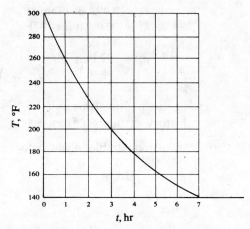

Fig. 4-21

4.12. At what depth should a water pipe be buried in wet soil ($\alpha = 0.03$ ft²/hr) initially at 40 °F for the surrounding soil temperature to remain above 33 °F, if the soil surface temperature suddenly drops to -5 °F and remains at this value for 10 hours?

Treating the soil as a semi-infinite body, (4.13) applies.   At the critical depth the temperature will just reach 33 °F after 10 hr.   Thus,

$$\frac{33 - (-5)}{40 - (-5)} = 0.84 = \text{erf}\left(\frac{x}{\sqrt{4\alpha t}}\right)$$

From Fig. 4-3, $x/\sqrt{4\alpha t} \approx 1.0$, so

$$x \approx 1.0[4(0.03 \text{ ft}^2/\text{hr})(10 \text{ hr})]^{1/2} = 1.1 \text{ ft}$$

4.13. A mild steel slab 5 cm thick, very wide and very long, is initially at 50 °C.   One surface is exposed to a fluid which suddenly causes the surface temperature to increase to and remain at 100 °C.   (a) What is the maximum time that the slab may be treated as a semi-infinite body?   (b) Determine the temperature at the center of the slab ($x = 2.5$ cm) one minute after the surface temperature change.

(a)   From Appendix B, $\alpha \approx 1.26 \times 10^{-5}$ m²/s.   By (4.17),

$$\sqrt{4\alpha t_{\max}} = 4L$$

$$t_{\max} = \frac{(2L)^2}{\alpha} = \frac{(5 \text{ cm})^2(10^{-2} \text{ m/cm})^2}{1.26 \times 10^{-5} \text{ m}^2/\text{s}} = 198.4 \text{ s} = 3.307 \text{ min}$$

(b)   By (4.13) at $x = 2.5$ cm and $t = 60$ s,

$$\frac{T - 100 \text{ °C}}{50 \text{ °C} - 100 \text{ °C}} = \text{erf}\left(\frac{2.5 \times 10^{-2} \text{ m}}{\sqrt{4(1.26 \times 10^{-5} \text{ m}^2/\text{s})(60 \text{ s})}}\right)$$

$$T = 100 - 50 \text{ erf } 0.45 \approx 100 - 50(0.48) = 76 \text{ °C}$$

4.14. A water pipe is buried 1.1 ft (cf. Problem 4.12) below ground in wet soil (α = 0.03 ft^2/hr and k = 1.5 Btu/hr-ft-°F). The soil is initially at a uniform temperature of 40 °F. For sudden application of a convective surface condition due to wind with $\bar{h}$ = 10 Btu/hr-ft^2-°F and T_∞ = −5 °F, will the pipe be exposed to freezing temperature in a 10-hr period?

We may treat this as a convective boundary condition applied to a semi-infinite body initially at T_i; thus (4.19) applies. The arguments of the erf terms in (4.19) are:

$$\frac{\bar{h}\sqrt{\alpha t}}{k} = \frac{10\sqrt{0.03 \times 10}}{1.5} = 3.65 \qquad \xi = \frac{x}{\sqrt{4\alpha t}} = \frac{1.1}{\sqrt{4(0.03)(10)}} = 1.00$$

Using Fig. 4-3,

$$\text{erf } \xi = \text{erf } 1.0 \approx 0.84$$

$$\text{erf}\left(\xi + \frac{\bar{h}\sqrt{\alpha t}}{k}\right) = \text{erf}(1.00 + 3.65) \approx 1.0$$

Thus, we need only keep the first two terms on the right of (4.19):

$$\frac{T - T_i}{T_\infty - T_i} \approx 1 - \text{erf } \xi$$

$$T \approx T_i + (1 - \text{erf } \xi)(T_\infty - T_i) = 40 + (1 - 0.84)(-5 - 40) = 32.8 \text{ °F}$$

Freezing will not occur.

4.15. Using the temperature expression (4.28), develop (4.29).

The energy transferred to or from the slab in the time interval from 0 to t_1 is

$$Q = \int_0^{t_1} q \, dt$$

where

$$q = -kA \frac{\partial T}{\partial x}\bigg|_{x=L} = -kA \frac{\partial \theta}{\partial x}\bigg|_{x=L}$$

Thus

$$Q = -kA \int_0^{t_1} \frac{\partial \theta}{\partial x}\bigg|_{x=L} dt \tag{1}$$

From (4.28),

$$\frac{\partial \theta}{\partial x} = 2\theta_i \sum_{n=1}^{\infty} e^{-\lambda_n^2 \alpha t} \left[\frac{\sin \lambda_n L}{\lambda_n L + (\sin \lambda_n L)(\cos \lambda_n L)}\right](-\sin \lambda_n x)\lambda_n$$

and consequently

$$\frac{\partial \theta}{\partial x}\bigg|_{x=L} = -2\theta_i \sum_{n=1}^{\infty} \frac{\lambda_n \sin^2 \lambda_n L}{\lambda_n L + (\sin \lambda_n L)(\cos \lambda_n L)} e^{-\lambda_n^2 \alpha t} \tag{2}$$

Substituting (2) into (1),

$$Q = 2kA\theta_i \sum_{n=1}^{\infty} \frac{\lambda_n \sin^2 \lambda_n L}{\lambda_n L + (\sin \lambda_n L)(\cos \lambda_n L)} \int_0^{t_1} e^{-\lambda_n^2 \alpha t} dt \tag{3}$$

The integral in (3) has the value $(1 - e^{-\lambda_n^2 \alpha t_1})/\lambda_n^2 \alpha$. Hence, noting that $\alpha = k/\rho c$,

$$Q = \frac{2kA\theta_i}{k/\rho c}\left(\frac{L}{L}\right) \sum_{n=1}^{\infty} \frac{1}{\lambda_n^2}\left[\frac{\lambda_n \sin^2 \lambda_n L}{\lambda_n L + (\sin \lambda_n L)(\cos \lambda_n L)}\right](1 - e^{-\lambda_n^2 \alpha t_1})$$

Noting that $Q_i = AL\rho c\theta_i$ and replacing t_1 by t,

$$\frac{Q}{Q_i} = 2\sum_{n=1}^{\infty} \frac{1}{\lambda_n L}\left[\frac{\sin^2 \lambda_n L}{\lambda_n L + (\sin \lambda_n L)(\cos \lambda_n L)}\right](1 - e^{-\lambda_n^2 \alpha t})$$

which is (4.29).

4.16. A large steel plate 2 in. thick is initially at a uniform temperature of 800 °F. It is suddenly exposed on both sides to a convective environment with $\bar{h} = 50$ Btu/hr-ft^2-°F and $T_\infty = 150$ °F. Determine the centerline temperature and the temperature inside the body 0.5 in. from the surface after 3 minutes.

From Appendix B, assuming an average temperature of 572 °F,

$$k = 25 \text{ Btu/hr-ft-°F} \qquad \alpha = 0.49 \times \frac{25}{26.5} = 0.46 \text{ ft}^2/\text{hr}$$

where the thermal conductivity at 32 °F is 26.5 Btu/hr-ft-°F. Then

$$\textbf{Fo} = \frac{\alpha t}{L^2} = \frac{(0.46 \text{ ft}^2/\text{hr})(3/60)\text{hr}}{(1/12)^2 \text{ ft}^2} = 3.31$$

$$\frac{1}{\textbf{Bi}} = \frac{k}{\bar{h}L} = \frac{25 \text{ Btu/hr-ft-°F}}{(50 \text{ Btu/hr-ft}^2\text{-°F})(1/12)\text{ft}} = 6.0$$

Note that $\textbf{Bi} = 0.17$; this is too large for a lumped analysis, although one would not expect large temperature differences at a given time within the body. From Fig. 4-6, using the above values of $\textbf{Fo}$ and $1/\textbf{Bi}$,

$$\frac{T_c - T_\infty}{T_i - T_\infty} \approx 0.6 \qquad \text{or} \qquad T_c \approx 150 + (0.6)(800 - 150) = 540 \text{ °F}$$

From Fig. 4-7 at $x/L = (1 - 0.5)/1 = 0.5$,

$$\frac{T - T_\infty}{T_c - T_\infty} \approx 0.98 \qquad \text{or} \qquad T \approx 150 + (0.98)(540 - 150) = 532 \text{ °F}$$

The closeness of the two temperatures is in keeping with the relatively small Biot number.

4.17. A large wall made of 4-in.-thick common brick is originally at a uniform temperature of 80 °F. It is well insulated on one side, as shown in Fig. 4-22. The uninsulated surface is suddenly exposed to free convective air flow at 30 °F, resulting in a heat transfer coefficient $\bar{h} = 2.0$ Btu/hr-ft^2-°F. Determine (a) the temperature of both surfaces ($x = 0$, $x = 4$) after 5.5 hr and (b) the total thermal energy removed per unit area of wall during this period.

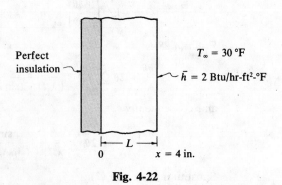

Fig. 4-22

This problem may be treated as 1/2 of the similar problem of an 8-in.-thick wall (with no insulation) exposed to the same convective environment on each surface. Clearly the Heisler chart for a slab of thickness $2L$ is suitable. From Appendix B,

$$k = 0.4 \text{ Btu/hr-ft-°F} \qquad \alpha = 0.02 \text{ ft}^2/\text{hr}$$
$$c = 0.2 \text{ Btu/lbm-°F} \qquad \rho = 100 \text{ lbm/ft}^3$$

The dimensionless parameters are

$$\mathbf{Fo} = \frac{\alpha t}{L^2} = \frac{(0.02 \text{ ft}^2/\text{hr})(5.5 \text{ hr})}{(4/12)^2 \text{ ft}^2} = 0.99$$

$$\frac{1}{\mathbf{Bi}} = \frac{k}{\bar{h}L} = \frac{0.4 \text{ Btu/hr-ft-}°\text{F}}{(2 \text{ Btu/hr-ft}^2\text{-}°\text{F})(4/12)\text{ft}} = 0.6$$

Note that $\mathbf{Bi} = 1.67$; clearly a lumped analysis is inappropriate.

(a) From Fig. 4-6, using the above $\mathbf{Fo}$ and $1/\mathbf{Bi}$,

$$\frac{T_c - T_\infty}{T_i - T_\infty} = \frac{T_0 - T_\infty}{T_i - T_\infty} \approx 0.41 \qquad \text{or} \qquad T_0 \approx 30 + (0.41)(80 - 30) = 50.5 \text{ °F}$$

where T_0 is the surface temperature adjacent to the insulation.   From Fig. 4-7 at $x/L = 1.0$ and $1/\mathbf{Bi} = 0.6$,

$$\frac{T - T_\infty}{T_c - T_\infty} = \frac{T_4 - T_\infty}{T_0 - T_\infty} \approx 0.52 \qquad \text{or} \qquad T_4 \approx 30 + (0.52)(50.5 - 30) = 40.7 \text{ °F}$$

and this is the exposed surface temperature.

(b) The thermal energy removed per square foot of wall surface is found with the aid of Fig. 4-8. At $\mathbf{Fo} = 0.99$ and $\mathbf{Bi} = 1.67$, $Q/Q_i \approx 0.64$.   Now

$$\frac{Q_i}{A} = \rho c L \theta_i = \left(100 \frac{\text{lbm}}{\text{ft}^3}\right) \left(\frac{0.2 \text{ Btu}}{\text{lbm-}°\text{F}}\right) \left(\frac{4}{12} \text{ ft}\right) [(80 - 30) \text{ °F}] = 333.33 \text{ Btu/ft}^2$$

and

$$\frac{Q}{A} \approx (0.64)(333.33) = 213.3 \text{ Btu/ft}^2$$

As a very crude check, the linear average temperature of the brick is

$$T_{\text{avg}} = \frac{50.5 + 40.7}{2} = 45.6 \text{ °F}$$

and

$$\frac{Q}{A} \approx \rho c L (T_i - T_{\text{avg}}) = (100)(0.2) \left(\frac{4}{12}\right) (80 - 45.6) = 229.3 \text{ Btu/ft}^2$$

The linear average temperature is, of course, too low during transient cooling (see Problem 4.26); it would likewise be too high during transient heating.

4.18. A long, 2.5-inch-diameter solid cylinder made of type 304 stainless steel is initially at a uniform temperature of 150 °F. It is suddenly exposed to a convective environment at $T_\infty = 70$ °F and the surface convective heat-transfer coefficient $\bar{h}$ is 50 Btu/hr-ft^2-°F. Calculate the temperature at (a) the axis of the cylinder and (b) a 1.0-inch radial distance, after 5 minutes of exposure to the cooling flow. (c) Determine the total heat transferred from the cylinder per foot of length during the first 5 minutes.

First check the Biot number based upon

$$L = \frac{V}{A_s} = \frac{\pi R^2 l}{2\pi R l} = \frac{R}{2}$$

With $k = 9$ Btu/hr-ft-°F (from Appendix B),

$$\mathbf{Bi} = \frac{50(1.25/24)}{9} = 0.289$$

Clearly a lumped analysis is inappropriate. The problem can be solved with the aid of the transient conduction charts.   The dimensionless terms for use with Fig. 4-9 are

$$\mathbf{Fo^*} = \frac{\alpha t}{R^2} = \frac{(0.15 \text{ ft}^2/\text{hr})(5/60)\text{hr}}{(1.25/12)^2 \text{ ft}^2} = 1.15$$

$$\frac{1}{\mathbf{Bi^*}} = \frac{k}{\bar{h}R} = \frac{9 \text{ Btu/hr-ft-}^\circ\text{F}}{(50 \text{ Btu/hr-ft}^2\text{-}^\circ\text{F})(1.25/12)\text{ft}} = 1.73$$

(a)   From Fig. 4-9,

$$\frac{T_c - T_\infty}{T_i - T_\infty} \approx 0.36 \qquad \text{or} \qquad T_c \approx 70 + (0.36)(150 - 70) = 98.8 \text{ }^\circ\text{F}$$

(b)   From Fig. 4-11 at $1/\mathbf{Bi^*} = 1.73$ and $r/R = 1.0/1.25 = 0.8$,

$$\frac{T - T_\infty}{T_c - T_\infty} \approx 0.84 \qquad \text{or} \qquad T \approx 70 + (0.84)(98.8 - 70) = 94.2 \text{ }^\circ\text{F}$$

(c)   From Fig. 4-10 at $\mathbf{Fo^*} = 1.15$ and $\mathbf{Bi^*} = 0.58$, $Q/Q_i \approx 0.69$.   Now, with ρ and c from Appendix B,

$$\frac{Q_i}{L} = \pi R^2 \rho c \theta_i$$

$$= \pi \left(\frac{1.25}{12} \text{ ft}\right)^2 (488 \text{ lbm/ft}^3)(0.11 \text{ Btu/lbm-}^\circ\text{F})[(150 - 70) \text{ }^\circ\text{F}] = 146.4 \text{ Btu/ft}$$

so

$$\frac{Q}{L} \approx (0.69)\frac{Q_i}{L} = 101 \text{ Btu/ft}$$

4.19.   A 3-inch-diameter orange originally at 80 °F is placed in a refrigerator where the air temperature is 35 °F and the average convective heat-transfer coefficient over the surface of the orange is $\bar{h} = 10$ Btu/hr-ft²-°F.   Estimate the time required for the center temperature of the orange to reach 40 °F.

Since an orange is mainly water, we will take the properties to be those of water at $T_{\text{av}} = (40 + 80)/2 = 60$ °F.   Thus, from Appendix B, by linear interpolation,

$$\alpha \approx 5.44 \times 10^{-3} \text{ ft}^2/\text{hr} \qquad k \approx 0.339 \text{ Btu/hr-ft-}^\circ\text{F}$$

The dimensionless parameters for the Heisler chart are

$$\frac{1}{\mathbf{Bi^*}} = \frac{k}{\bar{h}R} \approx \frac{0.339}{(10)(1.5/12)} = 0.27 \qquad \frac{T_c - T_\infty}{T_i - T_\infty} = \frac{40 - 35}{80 - 35} = 0.11$$

From Fig. 4-12, $\mathbf{Fo^*} \approx 0.5$ and thus

$$t = \frac{\mathbf{Fo^*} R^2}{\alpha} \approx \frac{(0.5)(1.5/12)^2 \text{ ft}^2}{5.44 \times 10^{-3} \text{ ft}^2/\text{hr}} = 1.44 \text{ hr}$$

Although we intuitively believe that the above solution is the correct approach, we should reexamine the Biot number for the lumped-capacity approach.   Since

$$L = \frac{V}{A_s} = \frac{4\pi R^3/3}{4\pi R^2} = \frac{R}{3}$$

$$\mathbf{Bi} = \frac{\bar{h}(R/3)}{k} = \frac{10(1.5)/(3)(12)}{0.339} = 1.23$$

and clearly a lumped-capacity approach is not suitable.

4.20. A solid, mild steel, 2-in.-diameter by 2.5-in.-long cylinder, initially at 1200 °F, is quenched during heat treatment in a fluid at 200 °F. The surface heat-transfer coefficient is 150 Btu/hr-ft²-°F. Determine the centerline temperature at the midpoint of length 2.7 min after immersion in the fluid.

Checking for suitability of a lumped analysis (length = $2a$),

$$L = \frac{V}{A_s} = \frac{\pi R^2(2a)}{2\pi R(2a) + 2\pi R^2} = \frac{R(2a)}{2(2a) + 2R}$$

$$= \frac{1(2.5)}{2(2.5) + 2} = 0.36 \text{ in.}$$

and, using k for mild steel at $T_{avg} = 572$ °F,

$$\mathbf{Bi} = \frac{\bar{h}L}{k} = \frac{(150 \text{ Btu/hr-ft}^2\text{-°F})(0.36/12)\text{ft}}{25 \text{ Btu/hr-ft-°F}} = 0.18$$

A lumped thermal-capacity analysis is not suitable.

An appropriate visualization of this problem is shown in Fig. 4-23. The axial conduction is treated by assuming the cylinder to be a slab of thickness $2a$ (the length of the cylinder), but infinite in the y- and z-directions. The radial conduction is treated by assuming the cylinder to be of finite radius R and infinite in length. From the solutions of the slab and cylinder subproblems, the dimensionless temperature is given by the product equation (*4.38*). The material properties at 572 °F are, from Appendix B,

$$k = 25 \text{ Btu/hr-ft-°F} \qquad k_{32} = 26.9 \text{ Btu/hr-ft-°F}$$

$$\alpha = \alpha_{32}\frac{k}{k_{32}} = (0.49)\frac{25}{26.9} = 0.462 \text{ ft}^2/\text{hr}$$

Cylindrical Subproblem
_____

$$\frac{1}{\mathbf{Bi^*}} = \frac{k}{\bar{h}R} = \frac{25}{(150)(1/12)} = 2.0 \qquad \mathbf{Fo^*} = \frac{\alpha t}{R^2} = \frac{(0.462)(2.7/60)}{(1/12)^2} = 2.99$$

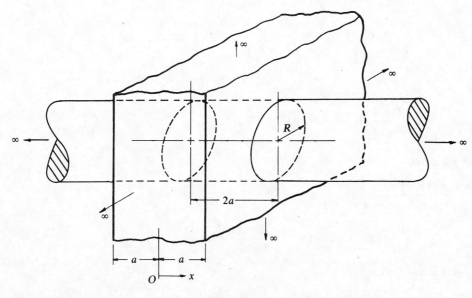

Fig. 4-23

From Fig. 4-9, $\left(\dfrac{T_c - T_\infty}{T_i - T_\infty}\right)_{\substack{\text{inf} \\ \text{cyl}}} \approx 0.079$.

Slab Subproblem

$$\frac{1}{\mathbf{Bi}} = \frac{k}{\bar{h}a} = \frac{25}{(150)(1.25/12)} = 1.6 \qquad \mathbf{Fo} = \frac{\alpha t}{a^2} = \frac{(0.462)(2.7/60)}{(1.25/12)^2} = 1.92$$

From Fig. 4-6, $\left(\dfrac{T_c - T_\infty}{T_i - T_\infty}\right)_{\substack{2a\text{-} \\ \text{slab}}} \approx 0.4$.

Complete Solution

$$\left(\frac{T_c - T_\infty}{T_i - T_\infty}\right)_{\substack{\text{short} \\ \text{cyl}}} \approx (0.079)(0.4) = 0.032$$

and

$$T_c = 200 + (0.032)(1200 - 200) = 232 \text{ °F}$$

which is the temperature at the axial and radial center of the cylinder.

4.21. For the conditions of Problem 4.20, determine the temperature at a location within the cylinder 0.5 in. from one flat face and at $r = 0.5$ in., 2.7 min after immersion.

Cylindrical Subproblem

From Problem 4.20,

$$\frac{1}{\mathbf{Bi}^*} = 2.0 \qquad \mathbf{Fo}^* = 2.99 \qquad \frac{T_c - T_\infty}{T_i - T_\infty} = 0.079$$

At $r/R = 0.5/1.0 = 0.5$, from Fig. 4-11,

$$\left(\frac{T - T_\infty}{T_c - T_\infty}\right)_{\substack{\text{inf} \\ \text{cyl}}} \approx 0.94$$

Now

$$\frac{T - T_\infty}{T_i - T_\infty} = \left(\frac{T - T_\infty}{T_c - T_\infty}\right)\left(\frac{T_c - T_\infty}{T_i - T_\infty}\right)$$

so

$$\left(\frac{T - T_\infty}{T_i - T_\infty}\right)_{\substack{\text{inf} \\ \text{cyl}}} \approx (0.94)(0.079) = 0.074$$

at $r = 0.5$ in.

Slab Subproblem

From Problem 4.20,

$$\frac{1}{\mathbf{Bi}} = 1.6 \qquad \mathbf{Fo} = 1.92 \qquad \frac{T_c - T_\infty}{T_i - T_\infty} \approx 0.4$$

Using Fig. 4-7 with

$$\frac{x}{a} = \frac{1.25 - 0.5}{1.25} = 0.6$$

yields

$$\left(\frac{T - T_\infty}{T_c - T_\infty}\right)_{\substack{2a- \\ slab}} \approx 0.905$$

Thus

$$\left(\frac{T - T_\infty}{T_i - T_\infty}\right)_{\substack{2a- \\ slab}} \approx (0.905)(0.4) = 0.36$$

at 0.5 in. from an end.

Complete Solution:

By (4.38),

$$\frac{T - T_\infty}{T_i - T_\infty} \approx (0.074)(0.36) = 0.0266$$

$$T \approx 200 + (0.0266)(1200 - 200) = 226.6 \text{ °F}$$

An important consideration in applying (4.38), or any of the multidimensional equations, is that the dimensionless temperatures used to form the product must be at the correct locations in the one-dimensional subproblems. In the present problem, the location is the circular curve formed by the intersection of the thin cylinder $r = 0.5$ in. with the plane $x = 0.75$ in., the plane being normal to the cylinder.

4.22. For the configuration and conditions of Problem 4.20, determine the time required for the temperature at the center (radially and axially) of the solid cylinder to reach 205 °F.

To find the time required to attain a given temperature in a transient multidimensional problem, a trial-and-error solution is required. From Problem 4.20, we know that the time is greater than 2.7 minutes. A logical approach is to use this as a beginning point and to calculate T_c for several larger values of time. A graph of the results will yield the required answer.

Try $t = 3.25$ min

Cyl. Solution: $\dfrac{1}{\textbf{Bi*}} = 2.0$; $\textbf{Fo*} = \dfrac{(0.462)(3.25/60)}{(1/12)^2} = 3.6$

From Fig. 4-9, $\left(\dfrac{T_c - T_\infty}{T_i - T_\infty}\right)_{\substack{inf \\ cyl}} \approx 0.047$

Slab Solution: $\dfrac{1}{\textbf{Bi}} = 1.6$; $\textbf{Fo} = \dfrac{(0.462)(3.25/60)}{(1.25/12)^2} = 2.3$

From Fig. 4-6, $\left(\dfrac{T_c - T_\infty}{T_i - T_\infty}\right)_{\substack{2a- \\ slab}} \approx 0.33$

Complete Solution: $\dfrac{T - T_\infty}{T_i - T_\infty} \approx (0.047)(0.33) = 0.016$

$$T \approx 200 + (0.016)(1200 - 200) = 215.5 \text{ °F}$$

Try $t = 3.9$ min

Cyl. Solution: $\dfrac{1}{\textbf{Bi*}} = 2.0$; $\textbf{Fo*} = \textbf{Fo*}_{3.25}\left(\dfrac{3.9}{3.25}\right) = 4.32$

From Fig. 4-9, $\left(\dfrac{T_c - T_\infty}{T_i - T_\infty}\right)_{\substack{\text{inf} \\ \text{cyl}}} \approx 0.025$

Slab Solution: $\dfrac{1}{\mathbf{Bi}} = 1.6;$ $\mathbf{Fo} = \mathbf{Fo}_{3.25}\left(\dfrac{3.9}{3.25}\right) = 2.76$

From Fig. 4-6, $\left(\dfrac{T_c - T_\infty}{T_i - T_\infty}\right)_{\substack{2a- \\ \text{slab}}} \approx 0.26$

Complete Solution: $\dfrac{T - T_\infty}{T_i - T_\infty} \approx (0.025)(0.26) = 0.0065$

$$T \approx 200 + (0.0065)(1200 - 200) = 206.5 \ ^\circ\text{F}$$

Try $t = 4.3$ min

Cyl. Solution: $\dfrac{1}{\mathbf{Bi}^*} = 2.0;$ $\mathbf{Fo}^* = \mathbf{Fo}^*_{3.25}\left(\dfrac{4.3}{3.25}\right) = 4.76$

From Fig. 4-9, $\left(\dfrac{T_c - T_\infty}{T_i - T_\infty}\right)_{\substack{\text{inf} \\ \text{cyl}}} \approx 0.017$

Slab Solution: $\dfrac{1}{\mathbf{Bi}} = 1.6;$ $\mathbf{Fo} = \mathbf{Fo}_{3.25}\left(\dfrac{4.3}{3.25}\right) = 3.04$

From Fig. 4-6, $\left(\dfrac{T_c - T_\infty}{T_i - T_\infty}\right)_{\substack{2a- \\ \text{slab}}} \approx 0.22$

Complete Solution: $\dfrac{T - T_\infty}{T_i - T_\infty} \approx (0.017)0.22) = 0.0037$

$$T \approx 200 + (0.0037)(1200 - 200) = 203.7 \ ^\circ\text{F}$$

Plotting these results in Fig. 4-24, we find $t = 4.07$ min.

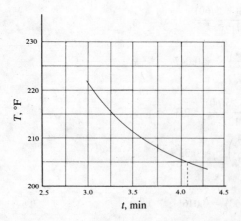

Fig. 4-24

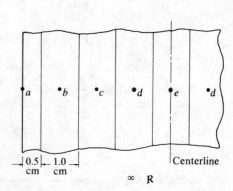

Fig. 4-25

4.23.   An 8.0-cm-thick concrete slab having very large dimensions in the plane normal to the thickness is initially at a uniform temperature of 20 °C. Both surfaces of the slab are suddenly raised to and held at 100 °C. The material properties are $k = 1.40$ W/m-K and

$\alpha = 0.0694 \times 10^{-5}$ m^2/s. Using a nodal spacing of 1 cm, numerically determine by the explicit method the temperature history in the slab during a 1/4-hr period.

The nodal designations are given in Fig. 4-25; since the problem is symmetrical about the centerline, only one-half of the slab is treated. The stability requirement for the explicit method, one-dimensional case, is

$$M \equiv \frac{(\Delta s)^2}{\alpha \, \Delta t} \geqslant 2$$

from which

$$(\Delta t)_{max} = \frac{(\Delta s)^2}{2\alpha} = \frac{(0.01 \text{ m})^2}{2(0.0694 \times 10^{-5}) \text{ m}^2/\text{s}} = 72.05 \text{ s} = 0.020 \text{ hr}$$

For this time increment, $M = 2$ and (4.52) simplifies to

$$T_n^{t+1} = \frac{T_1^t + T_3^t}{2}$$

where the subscripts 1 and 3 are for the two nodes adjacent to node n. Thus, for the present problem the nodal equations are

$$T_b^{t+1} = (T_a^t + T_c^t)/2$$
$$T_c^{t+1} = (T_b^t + T_d^t)/2$$
$$T_d^{t+1} = (T_c^t + T_e^t)/2$$
$$T_e^{t+1} = (T_d^t + T_d^t)/2 = T_d^t$$

No equation is required for node a since it has constant temperature. Since T_a is initially 20 °C and is suddenly changed to 100 °C, an appropriate value for T_a during the first time increment is the average of these two extremes. The solution is carried out in Table 4-2, where the units are hr and °C. In this solution we chose the maximum time step for simplicity of the resulting equations. In practice, especially when using a digital computer, a smaller time step is preferable since this would (i) reduce truncation (round-off) error and (ii) further ensure stability.

Table 4-2

Node Time	a	b	c	d	e
0.0	60	20	20	20	20
0.02	100	40	20	20	20
0.04	100	60	30	20	20
0.06	100	65	40	25	20
0.08	100	70	45	30	25
0.10	100	72.5	50	35	30
0.12	100	75	53.8	40	35
0.14	100	76.9	57.5	44.4	40
0.16	100	78.8	60.6	48.8	44.4
0.18	100	80.3	63.8	52.5	48.8
0.20	100	81.9	66.4	56.3	52.5
0.22	100	83.2	69.1	59.4	56.3
0.24	100	84.6	71.3	62.7	59.4
0.26	100	85.6	73.6	65.4	62.7

Supplementary Problems

4.24. A magnesium cube 20 cm along each edge is exposed to a convective flow resulting in $\bar{h} = 300$ W/m^2-K. The initial metal temperature is 100 °C. Determine the Biot number, using the conductivity evaluated at the initial temperature.   *Ans.* **Bi** = 0.0627

4.25. Repeat Problem 4-12 for dry soil ($\alpha = 0.01$ ft^2/hr) and $T_i = 50$ °F.   *Ans.* $x = 0.46$ ft

4.26. In the solution to Problem 4-17, it was stated that the linear average temperature was too low because this was a cooling problem. Show this by determining temperatures throughout the slab at $t = 5.5$ hr and plotting these together with the linear representation obtained with the two surface temperatures only.

Ans.

x, in.	0.8	1.6	2.0	2.4	3.2	3.6
T, °F	50.1	48.8	47.8	46.8	44.0	42.4

4.27. Determine the temperature at the center of a lead cube, having edge dimension 3.0 in., one minute after exposure to a convective fluid at 100 °F with $\bar{h} = 75$ Btu/hr-ft^2-°F. The initial uniform temperature of the cube is 300 °F. Is a lumped thermal-capacity analysis suitable? Use properties at $T_{avg} = 212$ °F, i.e. $k = 19$ Btu/hr-ft-°F, $\alpha = 0.9$ ft^2/hr.

Ans. **Bi**, based upon $L = V/A_s$, is 0.16; **Fo**, based upon $L_{slab} = 1.5$ in., is 0.96; **Bi**, based upon $L = 1.5$ in., is 0.49;

$$\left(\frac{T_c - T_\infty}{T_i - T_\infty} \right)_{\substack{2a- \\ slab}} \approx 0.7; \quad T_c \approx 168.6 \text{ °F at } t = 1 \text{ min}$$

Chapter 5

Fluid Mechanics

5.1 FLUID STATICS

In the study of convective heat transfer (Chapters 6, 7, and 8) we shall be primarily concerned with the behavior of fluids in motion. However, we must also be able to analyze static fluids—fluids at rest or moving at constant velocity—to understand many flow situations.

Pascal's Law

The pressure at a point in a static or uniformly moving fluid is equal in all directions. Pascal's law is also valid for an accelerating frictionless ($\mu = 0$) fluid.

Hydrostatics

The pressure differential, $p_2 - p_1$, between two points in a static or uniformly moving fluid is proportional to the difference in elevation, $y_2 - y_1$, between the points, the fluid density ρ, and the local acceleration of gravity g.

$$p_2 - p_1 = -\frac{\rho g}{g_c}(y_2 - y_1) \tag{5.1}$$

If elevation y_2 is taken as a datum at zero pressure, (5.1) simplifies to

$$p = \gamma h \tag{5.2}$$

where $\gamma = \rho g/g_c$; (5.2) is known as the *hydrostatic equation.*

It is often more convenient to use *gauge pressure*, the pressure above that of the atmosphere, $p - p_{atm}$, since many pressure-measuring devices indicate pressure with respect to the surroundings. Figure 5-1 shows the relationship among absolute, gauge, and vacuum pressures.

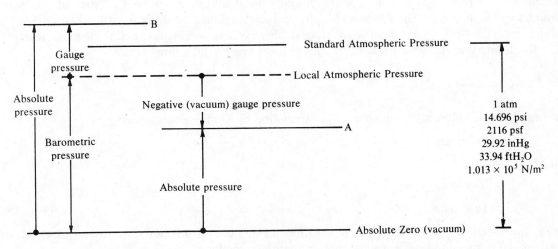

Fig. 5-1

107

Manometry

Problems 5-7 and 5-8 illustrate the use of the hydrostatic equation in manometry.

Buoyancy

A body seems to weigh less when partially or wholly immersed in a fluid. This apparent loss of weight is the *buoyant force*, $\mathbf{F}_B$, which is the vertical resultant of the pressure distribution exerted by the fluid on the body.

Buoyancy, which is central in natural convection (parcels of lesser density are buoyed upward), is governed by *Archimedes' principle. The buoyant force is equal to the weight of fluid displaced.* The line of action of the buoyant force is through the centroid of the volume of fluid displaced.

5.2 FLUID DYNAMICS

Analogous to the heat flow lines of the preceding chapters are *streamlines* in the flow of fluids. Shown in Fig. 5-2, a streamline is an imaginary line, taken at an instant of time in a flow field, such that the fluid velocity at every point of the line is tangent to it. Since movement occurs only in the direction of the velocity vector, no mass crosses a streamline.

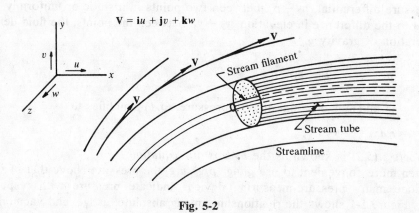

Fig. 5-2

A family of streamlines which forms a cylinder of infinitesimal cross section is a *stream filament*. A *stream tube* is a finite surface, made up of an infinite number of streamlines, across which there is no flow. The concept of a stream tube simplifies the analysis of fluid flow since fluid which enters a stream tube must leave it, assuming no creation or destruction of mass within.

By noting that the velocity components in the x- and y-directions are

$$u = \frac{dx}{dt} \qquad v = \frac{dy}{dt} \tag{5.3}$$

we can get the differential equations of a streamline by eliminating dt, giving

$$u\,dy = v\,dx \tag{5.4}$$

Similarly,

$$v\,dz = w\,dy \tag{5.5}$$

$$w\,dx = u\,dz \tag{5.6}$$

If, at the specified instant, u, v, and w are known functions of position, any two of (5.4) through (5.6) may be integrated to give the equation of the streamlines.

Substantial Derivative

When concentrating attention on a fixed region in space without regard to the identities of the fluid particles within it at a given time—known as the *Eulerian approach*, as contrasted with the *Lagrangian method*, which focuses on the motion of individual particles—the velocity field is given in cartesian coordinates by

$$\mathbf{V} = \mathbf{i}u + \mathbf{j}v + \mathbf{k}w \qquad (5.7)$$

where the velocity components are functions of space and time, i.e.

$$u = u(x, y, z, t)$$
$$v = v(x, y, z, t) \qquad (5.8)$$
$$w = w(x, y, z, t)$$

Using the chain rule for partial differentiation, the rate of change of the velocity $\mathbf{V}$ is given by

$$\mathbf{a} = \frac{d\mathbf{V}}{dt} = \frac{\partial \mathbf{V}}{\partial x}\frac{dx}{dt} + \frac{\partial \mathbf{V}}{\partial y}\frac{dy}{dt} + \frac{\partial \mathbf{V}}{\partial z}\frac{dz}{dt} + \frac{\partial \mathbf{V}}{\partial t} \qquad (5.9)$$

Since, for a moving particle, $u = dx/dt$, $v = dy/dt$, and $w = dz/dt$, (5.9) may be written

$$\blacksquare \qquad \mathbf{a} = \frac{d\mathbf{V}}{dt} = \left(u\frac{\partial \mathbf{V}}{\partial x} + v\frac{\partial \mathbf{V}}{\partial y} + w\frac{\partial \mathbf{V}}{\partial z} \right) + \frac{\partial \mathbf{V}}{\partial t} \qquad (5.10)$$

which is known as the *substantial, total, or fluid derivative*, designated $D\mathbf{V}/Dt$. The influence of time on a particle's behavior is given by the *local acceleration*, $\partial \mathbf{V}/\partial t$; space dependence is given by the *convective acceleration*, the terms in parentheses.

Types of Motion

When the local acceleration is zero, $\partial \mathbf{V}/\partial t = 0$, the motion is *steady*. Even though the velocity may change with respect to space, it does not change with respect to time. Streamlines are fixed in steady flow. A flow which is time dependent, $\partial \mathbf{V}/\partial t \neq 0$, is *unsteady*.

Uniform flow occurs when the convective acceleration is zero. The velocity vector is identical at every point in the flow field. The flow may be unsteady, but the velocity must change identically at every point. Streamlines are straight. An example is a frictionless fluid flowing through a long straight pipe. *Nonuniform* flow is space dependent. A frictionless fluid would flow nonuniformly in a pipe elbow.

In *laminar* flow, fluid particles move very smoothly parallel to each other. A dye stream injected in a laminar flow field would move in a thin line. Low velocities in smooth channels can produce laminar flow. At high velocities, however, *turbulent* flow, characterized by random motion of fluid particles, occurs. A dye injected in the stream would break up and diffuse throughout the flow field. Turbulent flow is always unsteady in the strict sense. We sometimes, however, think in terms of steady and unsteady turbulent flow, illustrated in Fig. 5-3, which also shows steady and unsteady laminar flow.

In discovering the difference between laminar and turbulent flow in 1883, Osborne Reynolds noted that the type of flow depended upon the dimensionless parameter VD/ν, where V is the average fluid velocity in a pipe of diameter D and $\nu \equiv \mu_m/\rho$ is the kinematic viscosity of the flowing fluid. More generally, the *Reynolds number* is defined as

$$\mathbf{Re} \equiv \frac{Vl}{\nu} \qquad (5.11)$$

where l is a characteristic length. In pipe flow the motion usually is turbulent for $\mathbf{Re} > 2000$. In flow over a flat plate, when the plate length is taken as the characteristic length, the transition from laminar to turbulent flow commonly occurs at $300{,}000 < \mathbf{Re} < 600{,}000$.

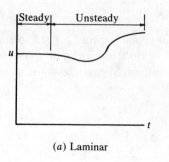 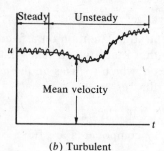

(a) Laminar   (b) Turbulent

Fig. 5-3

The velocity in turbulent flow consists of the average value $\bar{V}$ and a fluctuating part V':

$$V = \bar{V} + V' \tag{5.12}$$

Taking the time average over a long period of time, we get

$$\bar{V} = \lim_{\Delta t \to \infty} \frac{1}{\Delta t} \int_{t_0}^{t} V \, dt \tag{5.13}$$

i.e. the fluctuations cancel out in the long term. Any fluid property, say viscosity μ, may be similarly time-averaged: $\mu = \bar{\mu} + \mu'$.

The simple equation for shear stress in laminar flow

$$\text{laminar:} \qquad \tau = \mu_f \frac{du}{dy} \tag{5.14}$$

is not valid in turbulent flow. The relation is complicated by the *eddy viscosity* ϵ, which is a function of the fluid motion as well as its density.

$$\text{turbulent:} \qquad \tau = \left(\mu_f + \frac{\rho}{g_c} \epsilon \right) \frac{du}{dy} \tag{5.15}$$

Eddy viscosity ϵ is not a fluid property as the absolute viscosity μ is.

A *perfect fluid* has zero viscosity or negligible viscosity; a *viscous fluid* is a real fluid.

Similitude

Because of the impossibility of individually varying the several parameters in fluid mechanics and heat transfer studies, it is often desirable to group variables in order to compare one system (*model*) with another system (*prototype*). Three techniques of methodically grouping variables are in common use:

1. An algebraic method known as the *Rayleigh* or *Buckingham Pi theorem*.

2. Use of the governing differential equations. This technique will be illustrated in Section 7.2.

3. Similitude. Requiring geometric, kinematic and dynamic similarity, this method is the most commonly practiced technique.

For *geometric similarity* between model and prototype the fields and boundaries of both systems must be in the same geometric proportions and have the same orientation. *Kinematic similarity* requires that the velocity fields of the corresponding systems be proportional in magnitude and be identical in orientation, producing streamlines, or heat flow lines, of the same pattern for model and prototype. These two modes of similarity are adequate for describing flows which involve nearly perfect fluids ($\mu \approx 0$).

Real fluids, however, additionally require *dynamic similarity*. In dynamically similar systems each of several force ratios is respectively equal at corresponding points in the model and prototype fields. Some ratios which occur often enough in dimensionless form to warrant special designation are:

Reynolds number:
$$\mathbf{Re} \equiv \frac{Vl}{\nu} = \frac{\text{inertial force, } [\rho V^2 A]}{\text{viscous force, } [lV\mu_f]} \tag{5.16}$$

Mach number:
$$\mathbf{M} \equiv \frac{V}{c} = \sqrt{\frac{\text{inertial force, } [\rho V^2 A]}{\text{elastic force, } [EA]}} \tag{5.17}$$

Froude number:
$$\mathbf{Fr} \equiv \frac{V}{\sqrt{gl}} = \sqrt{\frac{\text{inertial force, } [\rho V^2 A]}{\text{weight, } [\rho (lA)g]}} \tag{5.18}$$

Euler number:
$$\mathbf{Eu} \equiv \frac{p}{\rho V^2} = \frac{\text{pressure force, } [pA]}{\text{inertial force, } [\rho V^2 A]} \tag{5.19}$$

Weber number:
$$\mathbf{We} \equiv \frac{V^2 l \rho}{\sigma} = \frac{\text{inertial force, } [\rho V^2 A]}{\text{surface force, } [\sigma l]} \tag{5.20}$$

Here, l is a characteristic length, $A = l^2$, c is the local speed of sound, E is the bulk modulus, σ is the surface tension, and square brackets denote dimensions.

Often two of the dimensionless numbers suffice to assure similarity. In wind tunnel testing, for example, flows are similar when the Reynolds and Mach numbers are each identical in model and prototype. Equality of Reynolds numbers is sufficient in geometrically similar, incompressible pipe flow.

5.3 CONSERVATION OF MASS

For steady flow the mass entering a stream tube is equal to the mass leaving. Thus, if $\dot{m}$ denotes the rate of mass transport through a cross section,

$$\dot{m} = \rho_1 A_1 V_1 = \rho_2 A_2 V_2 = \text{constant} \tag{5.21}$$

where V is the average velocity taken normal to the cross-sectional area A, and ρ is the density, assumed uniform over a cross section. This equation is known as the *continuity equation*. The *mass velocity*, $G \equiv \rho V$, is often used in heat transfer calculations, giving

$$\dot{m} = AG = \text{constant} \tag{5.22}$$

If, in addition to being steady, the flow is *incompressible* ($\rho = $ constant), the continuity equation reduces to

$$Q = A_1 V_1 = A_2 V_2 = \text{constant} \tag{5.23}$$

where Q is termed the *volumetric flow rate*.

The differential form of the continuity equation, which holds for steady or unsteady flow, is derived in Problem 5.20. In cartesian coordinates, it is

$$\frac{\partial \rho}{\partial t} + \frac{\partial (\rho u)}{\partial x} + \frac{\partial (\rho v)}{\partial y} + \frac{\partial (\rho w)}{\partial z} = 0 \tag{5.24}$$

and in general vector form,

$$\frac{\partial \rho}{\partial t} + \nabla \cdot (\rho \mathbf{V}) = 0 \tag{5.25}$$

where $\nabla \cdot$ is the *divergence*, which may be conveniently expressed in any orthogonal coordinate system. If the flow is steady and incompressible, the continuity equation in the cartesian coordinate system reduces to

$$\nabla \cdot \mathbf{V} = \frac{\partial u}{\partial x} + \frac{\partial v}{\partial y} + \frac{\partial w}{\partial z} = 0 \tag{5.26}$$

5.4   EQUATION OF MOTION ALONG A STREAMLINE

In Problem 5.23 Newton's second law is used to derive an equation for the motion of any fluid along a streamline. It is:

$$\frac{1}{\rho}\frac{\partial p}{\partial s} + \frac{g}{g_c}\frac{\partial z}{\partial s} + \frac{\tau}{\rho R_h} + \frac{1}{g_c}\frac{\partial}{\partial s}\left(\frac{V^2}{2}\right) + \frac{1}{g_c}\frac{\partial V}{\partial t} = 0 \tag{5.27}$$

Here, s is the arc length along the streamline and z is the vertical coordinate. Equation (5.27) is valid for viscous or frictionless fluid and for steady or unsteady flow.

The frictional term in the equation of motion is referred to as the *head loss*, i.e.

$$\frac{\partial h_L}{\partial s} \equiv \frac{\tau}{\rho R_h}$$

In this book we shall suppose it is given in its integrated form, h_L, and use empirical data to evaluate it in engineering calculations.

Assuming a uniform gravitational field (g = constant) and steady, incompressible flow ($\partial V/\partial t = 0$, ρ = constant), we can integrate (5.27) along the streamline from $s = s_1$ to $s = s_2$. The result is:

steady,
incompressible:
$$\frac{p_2 - p_1}{\rho} + \frac{g}{g_c}(z_2 - z_1) + \frac{V_2^2 - V_1^2}{2g_c} + h_L = 0 \tag{5.28}$$

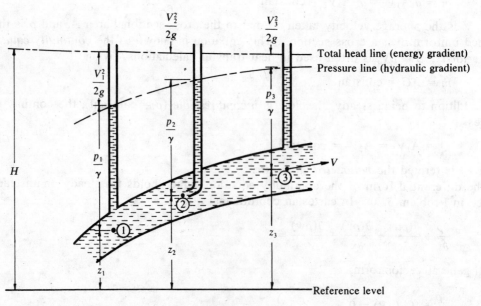

Fig. 5-4

in which head loss is measured relative to station 1. If the flow is also frictionless ($h_L = 0$), (5.28) reduces to *Bernoulli's equation*:

■
$$\frac{p_1}{\gamma} + z_1 + \frac{V_1^2}{2g} = \frac{p_2}{\gamma} + z_2 + \frac{V_2^2}{2g} = \text{constant} \qquad (5.29)$$

where $\gamma = \rho g / g_c$. The "constant" in (5.29) will, in general, vary from streamline to streamline.

Figure 5-4 depicts the physical significance of the terms in Bernoulli's equation: the *pressure head*, p/γ; the *elevation head*, z; and the *velocity head*, $V^2/2g$. The sum of these three heads is called the *total head*, H. The hydraulic and energy grade lines are also defined in the figure. The hydraulic and energy gradients are parallel for sections of pipe of equal cross-sectional area where frictional effects are negligible.

5.5 CONSERVATION OF ENERGY

The terms of Bernoulli's equation represent mechanical energy possessed by the flowing fluid due to pressure, position, and velocity. The concept of the energy gradient reflects this. In (5.28) the head loss term results from energy lost (dissipated into internal energy) in the flow process. If, in addition to these energy terms, we permit energy to be added to or extracted from a given flowing mass, an energy balance requires that

$$\begin{bmatrix} \text{energy at} \\ \text{station 1} \end{bmatrix} + \begin{bmatrix} \text{energy} \\ \text{added} \end{bmatrix} - \begin{bmatrix} \text{energy} \\ \text{lost} \end{bmatrix} - \begin{bmatrix} \text{useful} \\ \text{energy} \\ \text{extracted} \end{bmatrix} = \begin{bmatrix} \text{energy at} \\ \text{station 2} \end{bmatrix} \qquad (5.30)$$

where the "energy added" and "energy extracted" terms have meaning only when energy is transmitted across the boundaries of the system. The heat generation term q''' in Chapter 2 is an example of energy addition.

In algebraic form (5.30) may be expressed as

$$\left(\frac{p_1}{\rho} + \frac{gz_1}{g_c} + \frac{V_1^2}{2g_c} \right) + q - h_L - w_s = \left(\frac{p_2}{\rho} + \frac{gz_2}{g_c} + \frac{V_2^2}{2g_c} \right) \qquad (5.31)$$

where q is the heat transfer, positive when added to the system, and w_s is the work transfer, positive when done by the system. Each term in (5.31) is an energy per unit mass. The work term is often called *shaft work* to distinguish it from the *flow work*, p/ρ, which is the work required to maintain the flow.

Identifying the head loss with the gain in internal energy between stations 1 and 2, i.e. $h_L = u_2 - u_1$, puts (5.31) into the more convenient form

$$\left(\frac{p_1}{\rho} + u_1 + \frac{gz_1}{g_c} + \frac{V_1^2}{2g_c} \right) + q - w_s = \left(\frac{p_2}{\rho} + u_2 + \frac{gz_2}{g_c} + \frac{V_2^2}{2g_c} \right) \qquad (5.32)$$

Both the internal energy u and the *enthalpy*

$$h \equiv u + \frac{p}{\rho}$$

are tabulated for common fluids.

Solved Problems

5.1. Demonstrate Pascal's law by taking a force balance on the triangular fluid element shown in Fig. 5-5.

Assume the pressures are uniformly distributed over the faces of the element. The force on each face is the product of the pressure on that face and its area. The forces on the vertical xy-faces are equal and opposite, and the weight W is the product of specific weight, $\gamma = \rho g/g_c$, and volume.

Summing forces in the x-direction,

$$p_x\,\Delta y\,(1) = \left[p(1)\frac{\Delta x}{\cos\alpha}\right]\sin\alpha$$

and, since

$$W = \gamma\frac{\Delta x\,\Delta y\,(1)}{2}$$

the y-direction force balance yields

$$p_y\,\Delta x\,(1) = \left[p(1)\frac{\Delta x}{\cos\alpha}\right]\cos\alpha + \gamma\frac{\Delta x\,\Delta y\,(1)}{z}$$

Noting that $(\sin\alpha)/(\cos\alpha) = \tan\alpha = \Delta y/\Delta x$, we have

$$p_x = p \qquad p_y = p + \gamma\frac{\Delta y}{2}$$

and the latter equation reduces to $p_y = p$ as Δx and Δy approach zero, i.e., as the element approaches a point in the limit. Therefore, the pressure at a point in a static fluid is equal in all directions, which is Pascal's law.

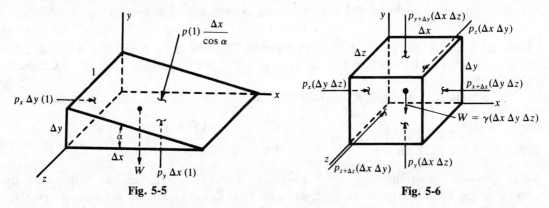

Fig. 5-5   Fig. 5-6

5.2. Determine the pressure variations in the x-, y- (vertical) and z-directions at an arbitrary point in an unaccelerated fluid.

Figure 5-6 shows a small fluid parallelepiped centered on the given point. The forces acting on this element consist of *surface forces* (pressure times area) and *body forces* (weight).

The conditions for equilibrium in the x-, y- and z-directions are

$$p_x(\Delta y\,\Delta z) - p_{x+\Delta x}(\Delta y\,\Delta z) = 0$$
$$p_y(\Delta x\,\Delta z) - p_{y+\Delta y}(\Delta x\,\Delta z) - \gamma(\Delta x\,\Delta y\,\Delta z) = 0$$
$$p_z(\Delta x\,\Delta y) - p_{z+\Delta z}(\Delta x\,\Delta y) = 0$$

Dividing each equation by the volume, $\Delta x\,\Delta y\,\Delta z$, and taking the limit as Δx, Δy and Δz approach zero, we get

$$\frac{\partial p}{\partial x} = 0 \qquad \frac{\partial p}{\partial y} = -\gamma \qquad \frac{\partial p}{\partial z} = 0$$

Thus, the pressure is independent of x and z.   To determine its dependence on y it is necessary to know the variation of γ, i.e. the variation of ρ and g.

5.3. How does the pressure vary with depth in an unaccelerated incompressible fluid in a constant gravitational field (with gravity acting in the negative y-direction)?

By Problem 5.2, $p = p(y)$ and

$$\frac{dp}{dy} = -\gamma = \text{constant}$$

since ρ and g are both constant.   Integrating,

$$\int_{p_0}^{p} dp = -\gamma \int_{y_0}^{y} dy$$
$$p - p_0 = -\gamma(y - y_0)$$

which is the form (5.1) of the hydrostatic equation.

5.4. How deep can a diver descend in ocean water ($\gamma = 64$ lbf/ft^3) without damaging his watch which will withstand an absolute pressure of 80 lbf/in^2?

Assuming a standard atmospheric pressure of 14.7 psia, the hydrostatic equation,

$$p - p_{\text{atm}} = \gamma h$$

gives

$$h = \frac{p - p_{\text{atm}}}{\gamma} = \frac{[(80 - 14.7)\ \text{lbf/in}^2](144\ \text{in}^2/\text{ft}^2)}{64\ \text{lbf/ft}^3} = 146.9\ \text{ft}$$

5.5. How might an ordinary garden hose be used to establish whether the corners of a building under construction are at the same level?

With its ends by a pair of corners, fill the hose with water and expose the ends to the atmosphere. Because the pressures at the water surfaces will be equal, the surfaces will be at the same height ($p = \gamma h$). This principle holds regardless of the length of the hose or its shape.

5.6. A mercury ($\gamma = 13.6\ \gamma_{\text{water}}$) barometer, depicted schematically in Fig. 5-7, has a column height h of 28.67 in.   What is the barometric pressure (a) in psi? (b) in feet of water?

(a)   A force balance on the fluid column gives

$$p_b = p_{\text{atm}} = p_v + \gamma h$$

But the vapor pressure of mercury is negligible; therefore,

$$p_b = \gamma h = 13.6\ \gamma_{\text{water}} h$$
$$= 13.6 \left(62.4\ \frac{\text{lbf}}{\text{ft}^3} \right) \left(\frac{28.67}{12}\ \text{ft} \right) \left(\frac{1\ \text{ft}^2}{144\ \text{in}^2} \right) = 14.08\ \text{psi}$$

Alternately, using the pressure equivalences from Fig. 5-1,

$$p_b = (28.67\ \text{inHg}) \left(\frac{14.696\ \text{psi}}{29.92\ \text{inHg}} \right) = 14.08\ \text{psi}$$

(b) $$p_b = (28.67\ \text{inHg}) \left(\frac{33.94\ \text{ftH}_2\text{O}}{29.92\ \text{inHg}} \right) = 32.52\ \text{ftH}_2\text{O}$$

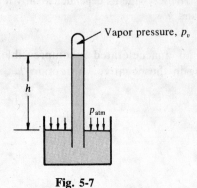

Fig. 5-7

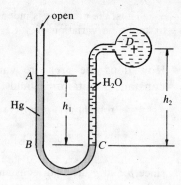

Fig. 5-8

5.7. For the manometer shown in Fig. 5-8, what is the pressure at D? Express in terms of the column heights given and the specific gravity $S \equiv \gamma/\gamma_w$, where γ_w is the specific weight of water.

For this class of problems it is convenient to begin at some point (any point, usually an interface) and add pressures resulting from fluid columns when going downward; subtract when going upward. Beginning at D, we get

$$p_D + \gamma_w h_2 - \gamma_{Hg} h_1 = p_{atm}$$

or

$$p_D - p_{atm} = \gamma_w (S_{Hg} h_1 - h_2)$$

or

$$p_D|_{gauge} = \gamma_w (S_{Hg} h_1 - h_2)$$

since the point B and the interface C are at the same elevation and joined by a common fluid, making their pressures identical.

5.8. A *differential manometer* (Fig. 5-9) is used to measure the pressure drop across a porous plug in a horizontal oil ($S_1 = 0.8$) line. For a deflection of 30 cm in the mercury ($S_2 = 13.6$) column, what is the pressure drop?

Beginning at A and proceeding through the manometer to point E, we have

$$p_A - \gamma_1 h_1 - \gamma_2 h_2 + \gamma_1 (h_1 + h_2) = p_E$$

or

$$p_A - p_E = h_2 (\gamma_2 - \gamma_1) = \gamma_w h_2 (S_2 - S_1)$$

$$= \left[\left(\frac{10^3 \text{ kg}}{\text{m}^3} \right) \left(9.8 \frac{\text{m}}{\text{s}^2} \right) \right] (0.30 \text{ m})(13.6 - 0.8) = 0.376 \times 10^5 \text{ N/m}^2$$

(See Table 1-1 for the definition of the newton.) It should be noted that it was unnecessary to know the height h_1.

5.9. The hydrometer shown in Fig. 5-10 has a stem diameter of 0.5 cm and a mass of 0.040 kg. What is the distance on the scale between specific gravity (S) markings 1.0 and 1.04?

Let $\mathcal{V}_w$ be the volume displaced when the hydrometer floats in water ($S = 1$). The buoyant force is then $\gamma_w \mathcal{V}_w$, which is equal to the weight of the hydrometer. In a fluid of unknown specific

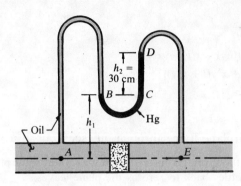

Fig. 5-9

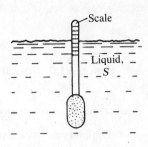

Fig. 5-10

gravity S, the buoyant force is $\gamma(\mathcal{V}_w - A\,\Delta y)$, where A is the stem cross-sectional area and Δy is the decrease in depth. Therefore,

$$\gamma_w \mathcal{V}_w = \gamma(\mathcal{V}_w - A\,\Delta y)$$

or

$$\Delta y = \frac{\mathcal{V}_w}{A}\frac{S-1}{S}$$

Substituting

$$A = \frac{\pi d^2}{4} = \frac{\pi(0.5\ \text{cm})^2}{4} = 0.1963\ \text{cm}^2$$

$$\mathcal{V}_w = \frac{W}{\gamma_w} = \frac{M}{\rho_w} = \frac{0.040\ \text{kg}}{10^{-3}\ \text{kg/cm}^3} = 40\ \text{cm}^3$$

(M is mass) and $S = 1.04$ yields

$$\Delta y = \frac{40\ \text{cm}^3}{0.1963\ \text{cm}^2}\frac{1.04-1.0}{1.04} = 7.837\ \text{cm}$$

5.10. For the flow field described by $\mathbf{V} = 2xy\mathbf{i} + x\mathbf{j}$ determine (*a*) the equation of a streamline, (*b*) the equation of the streamline which passes through the point $(1, 2)$.

(*a*) From (*5.4*)

$$\frac{dy}{dx} = \frac{v}{u} = \frac{x}{2xy} \qquad \text{or} \qquad 2\int y\,dy = \int dx$$

or $y^2 = x + C$.

(*b*) $C(1, 2) = 4 - 1 = 3$ whence $y^2 = x + 3$

5.11. Find (*a*) the velocity and (*b*) the acceleration of a particle at $x = 1$, $y = 2$, $t = 1$ in a field described by $\mathbf{V} = 3y^2\mathbf{i} + x^3yt\mathbf{j}$.

(*a*) $\mathbf{V}(1, 2, 1) = 12\mathbf{i} + 2\mathbf{j}$

(*b*) The acceleration is given by (*5.10*).

$$\mathbf{a}(x, y, t) = u\frac{\partial \mathbf{V}}{\partial x} + v\frac{\partial \mathbf{V}}{\partial y} + \frac{\partial \mathbf{V}}{\partial t}$$

$$= 3y^2(3x^2yt\mathbf{j}) + x^3yt(6y\mathbf{i} + x^3t\mathbf{j}) + x^3y\mathbf{j}$$

$$= 6x^3y^2t\mathbf{i} + (9x^2y^3t + x^6yt^2 + x^3y)\mathbf{j}$$

Hence, $\mathbf{a}(1, 2, 1) = 24\mathbf{i} + 76\mathbf{j}$.

5.12. Describe the motion produced by an airplane flying through quiescent air at a velocity $\mathbf{V}$.

The fluid velocity at a point in the path of the aircraft is unsteady. It is zero before the plane reaches it, varies widely as the plane passes, and settles back to zero as the plane leaves.

If the same airplane is fixed in a large wind tunnel with the air blowing past it at velocity $\mathbf{V}_\infty = -\mathbf{V}$, the same effects are produced on the aircraft. But the flow is now steady, since the local acceleration is zero at each point.

5.13. What is the maximum discharge (in gallons per minute) of fuel oil at 100 °F and standard atmospheric pressure ($\nu = 3 \times 10^{-3}$ ft^2/sec) from a 3/4-in.-diameter tube if laminar flow is maintained?

To assure laminar flow the Reynolds number should not exceed 2000; therefore,

$$\mathbf{Re} = \frac{VD}{\nu} = 2000 \quad \text{or} \quad V = \frac{2000(3 \times 10^{-3}\text{ ft}^2/\text{sec})}{0.75/12\text{ ft}} = 96\text{ fps}$$

and the discharge is

$$Q = VA = \left(96\,\frac{\text{ft}}{\text{sec}}\right)\left[\frac{\pi(0.75)^2}{4(144)}\text{ ft}^2\right]\left(\frac{7.48\text{ gal}}{\text{ft}^3}\right)\left(\frac{60\text{ sec}}{\text{min}}\right) = 132.18\text{ gpm}$$

5.14. Atmospheric air at 150 °F ($\nu = 2 \times 10^{-4}$ ft^2/sec) flows over a long flat plate at 40 fps. At what length will the flow cease to be laminar?

Laminar flow can be expected at Reynolds numbers up to 300,000; hence,

$$\mathbf{Re} = \frac{Vl}{\nu} = 300{,}000 \quad \text{or} \quad l = (300{,}000)\frac{2 \times 10^{-4}\text{ ft}^2/\text{sec}}{40\text{ ft/sec}} = 1.5\text{ ft}$$

5.15. Water having a dynamic viscosity $\mu_f = 1 \times 10^{-5}$ lbf-sec/ft^2 flows through a long, 2-in.-diameter tube at a rate of 25 gpm. Is the flow laminar or turbulent?

Since Reynolds number is a measure of whether the flow is laminar or turbulent, determining it will suffice.

$$V = \frac{Q}{A} = \frac{\left(25\,\dfrac{\text{gal}}{\text{min}}\right)\left(\dfrac{\text{ft}^3}{7.48\text{ gal}}\right)\left(\dfrac{\text{min}}{60\text{ sec}}\right)}{\dfrac{\pi(2)^2}{4(144)}\text{ ft}^2} = 2.55\,\frac{\text{ft}}{\text{sec}}$$

$$\mathbf{Re} = \frac{VD\rho}{\mu_m} = \frac{\left(2.55\,\dfrac{\text{ft}}{\text{sec}}\right)\left(\dfrac{2}{12}\text{ ft}\right)\left(62.4\,\dfrac{\text{lbm}}{\text{ft}^3}\right)}{\left(1 \times 10^{-5}\,\dfrac{\text{lbf-sec}}{\text{ft}^2}\right)\left(\dfrac{32.2\text{ lbm-ft}}{\text{lbf-sec}^2}\right)} = 82{,}466$$

The flow is turbulent.

5.16.   A model is used in the design of a spillway for a dam.   For a model-to-prototype ratio of 1:12, what is the velocity at a point in the prototype when the velocity at the corresponding point in the model is 4.6 fps?

Assuming geometric similarity of model and prototype, equal Froude numbers assure dynamic similarity; i.e., from (*5.18*),

$$\frac{V_m}{\sqrt{g_m l_m}} = \frac{V_p}{\sqrt{g_p l_p}}$$

Since $g_m = g_p$,

$$V_p = V_m \sqrt{\frac{l_p}{l_m}} = (4.6 \text{ fps})\sqrt{12} = 15.93 \text{ fps}$$

5.17.   A flow meter model is 1/5 the size of its prototype.   The model is tested with 70 °F water ($\nu = 10^{-5}$ ft²/sec), while the prototype operates at 175 °F ($\nu = 4 \times 10^{-6}$ ft²/sec).   For a velocity of 10 fps in the 12-in.-diameter throat of the prototype, what discharge is required in the model for similitude?

For incompressible flow, identical Reynolds numbers will assure similarity, assuming geometrical similarity; therefore,

$$\mathbf{Re}_m = \mathbf{Re}_p$$
$$\frac{V_m D_m}{\nu_m} = \frac{V_p D_p}{\nu_p}$$

or

$$V_m = \frac{D_p}{D_m}\frac{\nu_m}{\nu_p} V_p = 5 \frac{10^{-5}}{4 \times 10^{-6}}(10 \text{ fps}) = 125 \text{ fps}$$

This gives a discharge of

$$Q_m = V_m A_m = \left(125 \frac{\text{ft}}{\text{sec}}\right)\frac{\pi \left[\frac{12}{5(12)}\right]^2 \text{ft}^2}{4} = 3.93 \text{ cfs}$$

5.18.   Derive the continuity equation for steady compressible flow, considering flow through the stream tube of Fig. 5-11.

Taking sections 1 and 2 normal to the streamlines forming the stream tube, the mass passing section 1 per unit time is $\rho_s A_s V_s$, or $(\rho A V)_s$; similarly, $(\rho A V)_{s+\Delta s}$ passes section 2.   For steady flow

$$\dot{m} = (\rho A V)_{s+\Delta s} = (\rho A V)_s$$

Dividing by Δs and taking the limit as Δs approaches zero, we get

$$\lim_{\Delta s \to 0} \frac{(\rho A V)_{s+\Delta s} - (\rho A V)_s}{\Delta s} = 0$$

But this is

$$\frac{d}{ds}(\rho A V) = 0$$

from the definition of a derivative.   Therefore,

$$\rho A V = \text{constant}$$

which is (*5.21*).   By taking the natural logarithm,

$$\ln \rho + \ln A + \ln V = \ln C$$

and differentiating, we get the useful form

$$\frac{d\rho}{\rho} + \frac{dA}{A} + \frac{dV}{V} = 0$$

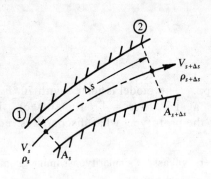

Fig. 5-11

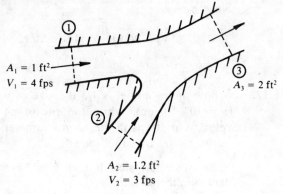

Fig. 5-12

5.19.   Water flows through the Y-section shown in Fig. 5-12.   What is the velocity at section 3 for one-dimensional flow (i.e. flow in which the fluid properties may be expressed in terms of one space coordinate and time)?

Equation (*5.21*) modified for multiple inlets and outlets is

$$\sum (\rho A V)_{\text{in}} = \sum (\rho A V)_{\text{out}}$$

Therefore, $A_1 V_1 + A_2 V_2 = A_3 V_3$, where the constant density ρ has been canceled out.   Using the given values, we get

$$(1)(4) + (1.2)(3) = 2V_3 \qquad \text{or} \qquad V_3 = 3.8 \text{ fps}$$

5.20.   Develop the differential form of the continuity equation by considering the unsteady flow of a fluid through an element having mutually orthogonal dimensions Δx, Δy, and Δz (Fig. 5-13).

The mass accumulated within the parallelepiped, due to the unsteady flow, is equal to the mass flowing into the element minus the mass flowing out of it.   Using an average value of density over the volume element, the mass accumulation per unit time is

$$\text{accumulation} = \frac{\partial \rho_{\text{av}}}{\partial t} \Delta x \, \Delta y \, \Delta z$$

and the respective fluxes are

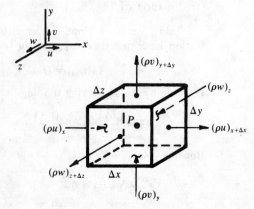

Fig. 5-13

$$\text{influx} = (\rho u)_x \, \Delta y \, \Delta z + (\rho v)_y \, \Delta x \, \Delta z + (\rho w)_z \, \Delta x \, \Delta y$$
$$\text{efflux} = (\rho u)_{x+\Delta x} \, \Delta y \, \Delta z + (\rho v)_{y+\Delta y} \, \Delta x \, \Delta z + (\rho w)_{z+\Delta z} \, \Delta x \, \Delta y$$

Thus,

$$\frac{\partial \rho_{av}}{\partial t} \Delta x \, \Delta y \, \Delta z = [(\rho u)_x \, \Delta y \, \Delta z + (\rho v)_y \, \Delta x \, \Delta z + (\rho w)_z \, \Delta x \, \Delta y]$$

$$- [(\rho u)_{x+\Delta x} \, \Delta y \, \Delta z + (\rho v)_{y+\Delta y} \, \Delta x \, \Delta z + (\rho w)_{z+\Delta z} \, \Delta x \, \Delta y]$$

Dividing by the volume $\Delta x \, \Delta y \, \Delta z$ and rearranging, we have

$$\frac{\partial \rho_{av}}{\partial t} + \frac{(\rho u)_{x+\Delta x} - (\rho u)_x}{\Delta x} + \frac{(\rho v)_{y+\Delta y} - (\rho v)_y}{\Delta y} + \frac{(\rho w)_{z+\Delta z} - (\rho w)_z}{\Delta z} = 0$$

By letting Δx, Δy, and Δz approach zero, the volume element approaches point P, and the average value of density approaches the local value. From the definition of the derivative the flux terms become gradients in their respective directions, giving (5.24):

$$\frac{\partial \rho}{\partial t} + \frac{\partial (\rho u)}{\partial x} + \frac{\partial (\rho v)}{\partial y} + \frac{\partial (\rho w)}{\partial z} = 0$$

5.21. Does the velocity field $\mathbf{V}(x, y, z) = (2x + \cos y)\mathbf{i} + (\sin x - 2y)\mathbf{j} - 4\mathbf{k}$ represent a possible incompressible flow?

The incompressible continuity equation,

$$\frac{\partial u}{\partial x} + \frac{\partial v}{\partial y} + \frac{\partial w}{\partial z} = 0$$

must be satisfied for the flow to be possible. In this case,

$$u = 2x + \cos y \qquad v = \sin x - 2y \qquad w = -4$$

$$\frac{\partial u}{\partial x} = 2 \qquad\qquad \frac{\partial v}{\partial y} = -2 \qquad\qquad \frac{\partial w}{\partial z} = 0$$

so the flow is possible.

5.22. (a) If the velocity field satisfies the condition

$$\nabla \times \mathbf{V} \equiv \operatorname{curl} \mathbf{V} = \begin{vmatrix} \mathbf{i} & \mathbf{j} & \mathbf{k} \\ \dfrac{\partial}{\partial x} & \dfrac{\partial}{\partial y} & \dfrac{\partial}{\partial z} \\ u & v & w \end{vmatrix} = 0$$

the flow is *irrotational*; the fluid does not rotate as it translates. (For example, a toothpick placed on the surface of an irrotational stream will have the same orientation at every point downstream.) Show that the flow field described by

$$\mathbf{V}(x, y) = (x^2 - y^2)\mathbf{i} - 2xy\mathbf{j}$$

is irrotational. (b) In two-dimensional irrotational flow a *velocity potential*, $\phi(x, y)$, exists such that

$$u = \frac{\partial \phi}{\partial x} \qquad v = \frac{\partial \phi}{\partial y}$$

Determine the velocity potential for the flow of part (a).

(a) $$\operatorname{curl} \mathbf{V} = \begin{vmatrix} \mathbf{i} & \mathbf{j} & \mathbf{k} \\ \dfrac{\partial}{\partial x} & \dfrac{\partial}{\partial y} & \dfrac{\partial}{\partial z} \\ x^2 - y^2 & -2xy & 0 \end{vmatrix} = \mathbf{k} \begin{vmatrix} \dfrac{\partial}{\partial x} & \dfrac{\partial}{\partial y} \\ x^2 - y^2 & -2xy \end{vmatrix} = \mathbf{k}(-2y + 2y) = 0$$

so the flow is irrotational.

(b) The velocity potential must satisfy

$$\frac{\partial \phi}{\partial x} = x^2 - y^2 \qquad \frac{\partial \phi}{\partial y} = -2xy$$

Integrating the first equation with respect to x, and the second with respect to y, gives

$$\phi = \frac{x^3}{3} - y^2 x + P(y) \qquad \phi = -xy^2 + Q(x)$$

These two expressions for ϕ will be equal only if

$$\frac{x^3}{3} - Q(x) = -P(y) = \text{constant}$$

Therefore,

$$\phi = \frac{x^3}{3} - y^2 x - \text{constant}$$

and the constant may be equated to zero, since only the derivatives of ϕ have significance.

5.23. Develop (5.27), the equation of motion along a streamline.

Figure 5-14 is the free-body diagram for an element of fluid, of average length Δs, within a stream tube. The pressure forces act normally on the ends of the element. The frictional force, which acts on the circumferential surface to retard motion, is the product of shearing stress τ, the mean circumference c, and the average length Δs. The weight is given by

$$W = \gamma \, \Delta \mathcal{V} = \gamma \, \Delta s \left[\frac{A_{s+\Delta s} + A_s}{2} \right] \qquad (1)$$

where $\gamma = \rho g / g_c$; the component of the weight in the $+s$-direction is

$$-W \cos \theta = -W \frac{\Delta z}{\Delta s}$$

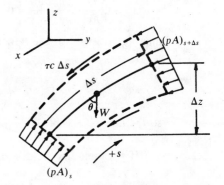

Fig. 5-14

Writing Newton's second law for a fixed mass, (1.19), in the $+s$-direction,

$$(pA)_s - (pA)_{s+\Delta s} - W \frac{\Delta z}{\Delta s} - \tau c \, \Delta s = \frac{W}{g} \frac{dV}{dt} \qquad (2)$$

or, substituting (1) into (2),

$$(pA)_{s+\Delta s} - (pA)_s + \gamma \, \Delta s \left[\frac{A_{s+\Delta s} + A_s}{2} \right] \frac{\Delta z}{\Delta s} + \tau c \, \Delta s = -\frac{\gamma}{g} \Delta s \left[\frac{A_{s+\Delta s} + A_s}{2} \right] \frac{dV}{dt} \qquad (3)$$

Using the *hydraulic radius*, $R_h \equiv A/c$, defined as the cross-sectional area divided by the wetted perimeter, replacing γ by $\rho g / g_c$, and dividing by $\rho \, \Delta s$, we get

$$\frac{1}{\rho} \frac{(pA)_{s+\Delta s} - (pA)_s}{\Delta s} + \frac{g}{g_c} \left[\frac{A_{s+\Delta s} + A_s}{2} \right] \frac{\Delta z}{\Delta s} + \frac{\tau A}{\rho R_h} = -\frac{1}{g_c} \left[\frac{A_{s+\Delta s} + A_s}{2} \right] \frac{dV}{dt} \qquad (4)$$

Taking the limit as $\Delta s \to 0$, and noting that $A_{s+\Delta s} \to A_s$ so that in the limit the area divides out, we have

$$\frac{1}{\rho} \frac{\partial p}{\partial s} + \frac{g}{g_c} \frac{\partial z}{\partial s} + \frac{\tau}{\rho R_h} + \frac{1}{g_c} \frac{dV}{dt} = 0 \qquad (5)$$

In (5) the independent variables are arc length s and time t. However, by definition, $ds/dt = V$ along the streamline. Therefore,

$$\frac{dV}{dt} = \frac{\partial V}{\partial t} + \frac{\partial V}{\partial s}\frac{ds}{dt} = \frac{\partial V}{\partial t} + V\frac{\partial V}{\partial s} = \frac{\partial V}{\partial t} + \frac{\partial}{\partial s}\left(\frac{V^2}{2}\right)$$

and (5) becomes

$$\frac{1}{\rho}\frac{\partial p}{\partial s} + \frac{g}{g_c}\frac{\partial z}{\partial s} + \frac{\tau}{\rho R_h} + \frac{1}{g_c}\frac{\partial}{\partial s}\left(\frac{V^2}{2}\right) + \frac{1}{g_c}\frac{\partial V}{\partial t} = 0$$

which is (5.27).

5.24. A Pitot-static tube (Fig. 5-15) is used with a mercury manometer to measure the flow of an inviscid fluid with $S = 1$ in a 4-in.-diameter pipe. What is the flow rate through the duct?

Assume that the flow is steady, incompressible and frictionless, making Bernoulli's equation valid.

$$\frac{p_1}{\gamma} + z_1 + \frac{V_1^2}{2g} = \frac{p_2}{\gamma} + z_2 + \frac{V_2^2}{2g} \qquad (1)$$

Station 2 is a stagnation point, where the flow is completely stopped ($V_2 = 0$), and the two stations are at the same elevation ($z_1 = z_2$); therefore,

$$\frac{V_1^2}{2g} = \frac{p_2 - p_1}{\gamma} \qquad (2)$$

The pressure differential can be determined by using the hydrostatic equation, $p = \gamma h$, along the path $1 \to 3 \to 4 \to 2$ through the manometer.

$$p_1 + \gamma h + \gamma_{Hg}h_0 - \gamma(h + h_0) = p_2$$

or

$$\frac{p_2 - p_1}{\gamma} = \frac{\gamma_{Hg}}{\gamma}h_0 - h_0 = h_0(S_{Hg} - 1) \qquad (3)$$

Substituting (3) into (2),

$$V_1 = \sqrt{2gh_0(S_{Hg} - 1)} = \sqrt{2\left(32.2\,\frac{ft}{sec^2}\right)\left(\frac{2}{12}\,ft\right)(13.6 - 1)} = 11.63\ \text{fps}$$

and

$$Q = AV = \left[\frac{\pi\left(\frac{4}{12}\right)^2 ft^2}{4}\right]\left(11.63\,\frac{ft}{sec}\right)\left(\frac{7.48\ \text{gal}}{ft^3}\right)\left(\frac{60\ \text{sec}}{\text{min}}\right) = 455.49\ \text{gpm}$$

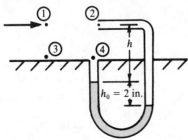

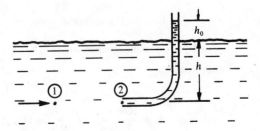

Fig. 5-15   **Fig. 5-16**

5.25. A simple Pitot tube, sometimes called an *impact tube*, may be used to measure the velocity in a flowing stream as shown in Fig. 5-16. Derive an expression for the velocity in terms of the given quantities.

Writing Bernoulli's equation along the streamline between 1 and 2, we get

$$\frac{V_1^2}{2g} + \frac{p_1}{\gamma} = \frac{p_2}{\gamma}$$

But, from the hydrostatic equation, $p_1 = p_{atm} + \gamma h$, $p_2 = p_{atm} + \gamma(h_0 + h)$, whence

$$\frac{p_2 - p_1}{\gamma} = h_0$$

Therefore, $V_1 = \sqrt{2gh_0}$.

5.26. A liquid discharges through a small orifice in a large tank (Fig. 5-17) which is kept full. What is the discharge velocity?

For flow along a streamline from the surface to the orifice, Bernoulli's equation gives

$$\frac{p_1}{\gamma} + z_1 + \frac{V_1^2}{2g} = \frac{p_2}{\gamma} + z_2 + \frac{V_2^2}{2g}$$

But $V_1 \approx 0$, $p_1 = p_2 = p_{atm}$, and $z_1 - z_2 = h$; hence

$$V_2 = \sqrt{2gh}$$

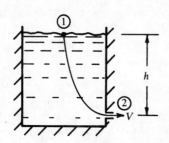

Fig. 5-17

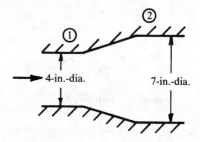

Fig. 5-18

5.27. Gasoline ($S = 0.82$) flows with negligible loss through a divergent section (Fig. 5-18) at a rate of 1.0 ft³/sec. What is the pressure at station 2 if the pressure at 1 is 26 psig?

Assuming steady, one-dimensional flow with no change in elevation,

Bernoulli's equation: $\dfrac{p_1}{\gamma} + \dfrac{V_1^2}{2g} = \dfrac{p_2}{\gamma} + \dfrac{V_2^2}{2g}$

continuity equation: $Q = V_1 A_1 = V_2 A_2$

Therefore,

$$V_1 = \frac{Q}{A_1} = \frac{1.0 \dfrac{\text{ft}^3}{\text{sec}}}{\dfrac{\pi}{4}\left(\dfrac{4}{12}\right)^2 \text{ft}^2} = 11.46 \text{ fps} \qquad V_2 = \frac{Q}{A_2} = \frac{1.0}{\dfrac{\pi}{4}\left(\dfrac{7}{12}\right)^2} = 3.74 \text{ fps}$$

and

$$p_2 = p_1 + \gamma\left(\frac{V_1^2 - V_2^2}{2g}\right)$$

$$= 26\,\frac{\text{lbf}}{\text{in}^2} + (0.82)\left(62.4\,\frac{\text{lbf}}{\text{ft}^3}\right)\left[\frac{(11.46)^2 - (3.74)^2}{2(32.2\,\text{ft/sec}^2)}\right]\frac{\text{ft}^2}{\text{sec}^2}\left(\frac{\text{ft}^2}{144\,\text{in}^2}\right)$$

$$= 26 + 0.65 = 26.65\ \text{psig}$$

5.28. A gas turbine operates with an average inlet velocity of 400 fps and an inlet temperature of 2200 °F.   The exit velocity and temperature are 900 fps and 1400 °F.   For a mass-flow rate of 40 lbm/min and a heat loss of 200 Btu/hr, what horsepower does it develop?

Writing (*5.32*) in terms of enthalpy and as a rate equation, we have

$$\dot{Q} - \dot{W}_s = \dot{m}\left[(h_2 - h_1) + \frac{g}{g_c}(z_2 - z_1) + \frac{V_2^2 - V_1^2}{2g_c}\right]$$

where Q and W_s are used for the total heat and work transfer terms.   Neglecting the change in elevation, which is reasonable for a gas, and assuming the gas has properties approximating those of air, $z_2 - z_1 = 0$ and

$$h_2 - h_1 = c_p(T_2 - T_1) = \left(0.24\,\frac{\text{Btu}}{\text{lbm-°R}}\right)(T_2 - T_1)$$

Therefore,

$$\dot{W}_s = \dot{Q} - \dot{m}\left[c_p(T_2 - T_1) + \frac{V_2^2 - V_1^2}{2g_c}\right]$$

$$= -200\,\frac{\text{Btu}}{\text{hr}} - \left(40\,\frac{\text{lbm}}{\text{min}}\right)\left[\left(0.24\,\frac{\text{Btu}}{\text{lbm-°R}}\right)(1400 - 2200)°\text{R}\right.$$

$$\left. + \frac{(900)^2 - (400)^2}{2}\,\frac{\text{ft}^2}{\text{sec}^2}\left(\frac{\text{lbf-sec}^2}{32.2\,\text{lbm-ft}}\right)\right]$$

$$= -200\,\frac{\text{Btu}}{\text{hr}} + \left(7680\,\frac{\text{Btu}}{\text{min}}\right)\left(\frac{60\,\text{min}}{\text{hr}}\right) - \left(403{,}727\,\frac{\text{ft-lbf}}{\text{min}}\right)\left(\frac{\text{Btu}}{778\,\text{ft-lbf}}\right)\left(\frac{60\,\text{min}}{\text{hr}}\right)$$

$$= \left(429{,}464\,\frac{\text{Btu}}{\text{hr}}\right)\left(\frac{\text{hp-hr}}{2544\,\text{Btu}}\right) = 168.8\ \text{hp}$$

5.29. A centrifugal air compressor receives ambient air at 14.6 psia and 80 °F.   At a discharge pressure of 60 psia the temperature is 400 °F and the velocity is 300 fps.   What power is required to drive the compressor for a mass-flow rate of 200 lbm/min?

Assume the inlet velocity to be so low that its effect is negligible.   Take the flow as steady, assume the process is adiabatic (no heat transfer), and neglect elevation changes.   The energy equation (*5.32*) reduces to

$$w_s = h_1 - h_2 - \frac{V_2^2}{2g_c}$$

The corresponding rate equation is, assuming constant specific heat,

$$\dot{W}_s = \dot{m}\left[c_p(T_1 - T_2) - \frac{V_2^2}{2g_c}\right]$$

$$= \left(200\,\frac{\text{lbm}}{\text{min}}\right)\left[\left(0.24\,\frac{\text{Btu}}{\text{lbm-°R}}\right)(80 - 400)\,°\text{R} - \frac{(300)^2}{2}\,\frac{\text{ft}^2}{\text{sec}^2}\left(\frac{\text{lbf-sec}^2}{32.2\,\text{lbm-ft}}\right)\left(\frac{\text{Btu}}{778\,\text{ft-lbf}}\right)\right]$$

$$= \left(-15{,}719\,\frac{\text{Btu}}{\text{min}}\right)\left(\frac{\text{hp-hr}}{2544\,\text{Btu}}\right)\left(\frac{60\,\text{min}}{\text{hr}}\right) = -370.7\ \text{hp}$$

The negative sign indicates that work must be done on the system.

Supplementary Problems

5.30. A 3-ft by 3-ft tank contains hydraulic fluid ($S = 0.84$). With the deflection of mercury shown in the manometer (Fig. 5-19), what pressure exists at A in the tank?   *Ans.* 0.9675 N/cm^2 gauge

5.31. Neglecting friction, what pressure is required to pump water to the top of the Empire State building, 1250 ft high?   *Ans.* 541.67 psi

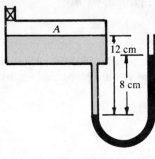

Fig. 5-19

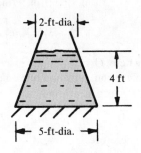

Fig. 5-20

5.32. An inverted conical tank contains water as shown in Fig. 5-20. What force is exerted on the tank bottom by the water?   *Ans.* 4900.88 lbf

5.33. What is the maximum depth from which water may be pumped by a suction-type (shallow well) pump?   *Ans.* 33.94 ft

5.34. In a flood two people jump onto an empty 55-gal oil drum, which then floats barely, but totally, submerged. What is the combined weight of the two people if the drum weighs 50 lbf?   *Ans.* 408.82 lbf

5.35. What is the acceleration of a particle at the point $(2, 1)$ in the two-dimensional field described by $\mathbf{V} = 2xy^2\mathbf{i} + x^3y\mathbf{j}$?   *Ans.* $72\mathbf{i} + 112\mathbf{j}$

5.36. Determine the maximum flow rate of gasoline ($\mu_f = 7 \times 10^{-6}$ lbf-sec/ft^2; $S = 0.68$) for laminar flow in a 1-in.-dia. tube.   *Ans.* 6.95×10^{-4} cfs

5.37. Air at 100 °F ($\nu = 1.7 \times 10^{-4}$ ft^2/sec) flows parallel with a 20-ft-long flat plate at 20 fps. Assume the flow to remain laminar until the Reynolds number, based on the distance from the leading edge, reaches 300,000. Over what fraction of the plate is the flow laminar?   *Ans.* $l_{lam}/l = 0.128$

5.38. A model of a river is one-hundredth scale. Predict the surface velocity in the river if the model surface velocity is 0.3 m/s.   *Ans.* 3.0 m/sec

5.39. A jet engine burns 4000 lbm of fuel per hour while inducting air at 150 lbm/sec in flight at 500 mph. The combustion products exhaust at 2000 fps relative to the engine. Determine the engine thrust, assuming $p_1A_1 = p_2A_2$.   *Ans.* $T = (\dot{m}_a + \dot{m}_f)V_2 - \dot{m}_aV_1 = 5969.6$ lbf

5.40. What gauge pressure is required in a 3-in.-i.d. fire hose just upstream of a horizontal nozzle to result in a 1-in.-dia. water jet having a velocity of 60 ft/sec at the nozzle exit plane? Assume standard gravitational acceleration and $\gamma = 62.4$ lbf/ft^3.   *Ans.*   23.92 psig

5.41. Water flows through a turbine at the rate of 10 cfs. Inlet and outlet pressures are 25 psia and -4 psig, respectively.   The 12-in.-dia. inlet is 6 ft above the 24-in.-dia. outlet.   What horsepower is delivered by the water to the turbine?   *Ans.*   46.93 hp

5.42. What power is available from a 3-in.-dia. jet of water discharging into the atmosphere at 100 fps?   *Ans.*   173.1 hp

Chapter 6

Forced Convection: Laminar Flow

The primary resistance to heat transfer by convection is normally controlled within in a thin layer of the fluid, adjacent to the immersed body, in which viscous effects are important. The quantity of heat transferred is highly dependent upon the fluid motion within this *boundary layer*, being determined chiefly by the thickness of the layer. While greatly affecting heat transfer, the boundary layer and the general velocity field can be treated independently of it, provided fluid properties do not vary with temperature. Otherwise, the heat-transfer and fluid-flow processes are interlinked.

6.1 HYDRODYNAMIC (ISOTHERMAL) BOUNDARY LAYER: FLAT PLATE

Fundamentals of boundary layer flow can best be understood from laminar flow along a flat plate, since the fluid motion can be visualized and since an exact solution exists for the fluid's behavior.

Consider a very thin, flat plate with an unbounded, incompressible, viscous fluid flowing parallel to it with an approach (free-stream) velocity V_∞. This is shown in Fig. 6-1, with the y-dimensions greatly exaggerated with respect to the x-dimensions. Shown also are velocity profiles at two stations, indicating how the velocity varies from zero at the surface (*no-slip condition*) to 0.99 V_∞ at what, by convention, is taken as the edge of the boundary layer. A streamline is shown for reference in the inviscid region, where Bernoulli's equation is valid. Continuity requires that streamlines diverge as the fluid is retarded more and more in moving along the plate. The divergence of streamlines suggests motion in the y-direction, normal to the plate. (At this point it should be noted that a mirror profile exists underneath our hypothetically thin plate.)

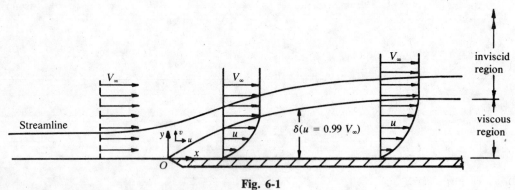

Fig. 6-1

Prandtl's Boundary Layer Equations

By applying Newton's second law and the continuity equation to an infinitesimal, two-dimensional control volume within the boundary layer, and assuming that

1. fluid viscosity is constant
2. shear in the y-direction is negligible
3. the flow is steady, and the fluid is incompressible
4. the vertical pressure gradient is negligible ($\partial p/\partial y = 0$)

128

we obtain (see Problem 6.1):

x-momentum: $\qquad u\dfrac{\partial u}{\partial x} + v\dfrac{\partial u}{\partial y} = -\dfrac{g_c}{\rho}\dfrac{dp}{dx} + v\dfrac{\partial^2 u}{\partial y^2}$ $\qquad\qquad$ (6.1)

y-momentum: $\qquad \dfrac{\partial p}{\partial y} = 0$ $\qquad\qquad\qquad\qquad\qquad\qquad$ (6.2)

continuity: $\qquad \dfrac{\partial u}{\partial x} + \dfrac{\partial v}{\partial y} = 0$ $\qquad\qquad\qquad\qquad\qquad$ (6.3)

Since the pressure can be determined in the inviscid region from Bernoulli's equation,

$$\frac{p}{\rho} + \frac{V^2}{2g_c} = \text{constant}$$

the pressure at a given lengthwise location within the boundary layer is known from (6.2). Equations (6.1) and (6.3) can then be solved simultaneously to give the velocity distribution.

Blasius' Solution (Exact)

Problem 6.2 establishes the order-of-magnitude relationship

$$\frac{\delta}{x} \sim \sqrt{\frac{v}{V_\infty x}} \qquad\qquad (6.4)$$

Experiments have shown that velocity profiles at different locations along the plate are geometrically similar, i.e. they differ only by a *stretching factor* in the y-direction. (Refer again to Fig. 6-1.) This means that the dimensionless velocity u/V_∞ can be expressed at any location x as a function of the dimensionless distance from the wall, y/δ:

$$\frac{u}{V_\infty} = \phi\left(\frac{y}{\delta}\right) \qquad\qquad (6.5)$$

or, using (6.4),

$$\frac{u}{V_\infty} = g(\eta) \qquad\qquad (6.6)$$

where $\eta \equiv y\sqrt{V_\infty/vx}$ denotes the stretching factor.

We seek the function $g(\eta)$ which will satisfy Prandtl's equations. To this end let us define a continuous *stream function* ψ, having continuous first partial derivatives, such that

$$u = \frac{\partial \psi}{\partial y} \qquad v = -\frac{\partial \psi}{\partial x} \qquad\qquad (6.7)$$

Then (6.3) is automatically satisfied. Holding x constant,

$$\psi = \int u\, dy + C(x) = V_\infty \int g(\eta)\frac{dy}{d\eta}\, d\eta + C(x) = \sqrt{V_\infty vx}\int g(\eta)\, d\eta + C(x)$$
$$= \sqrt{V_\infty vx}\, f(\eta) + C(x) \qquad\qquad (6.8)$$

where we have dropped the integration "constant" on dimensional grounds (ψ and $\sqrt{V_\infty vx}$ have the same units, e.g. m²/s). We now use the chain rule with (6.8) to express u and v and their derivatives in terms of f and its derivatives. Substituting the results in (6.1), and further assuming constant pressure ($dP/dx = 0$), we obtain

■ $\qquad f\dfrac{d^2 f}{d\eta^2} + 2\dfrac{d^3 f}{d\eta^3} = 0$ $\qquad\qquad\qquad\qquad\qquad$ (6.9)

an ordinary (but nonlinear) differential equation for f. The boundary conditions are:

Physical System	Transformed System
$u = 0$ at $y = 0$	$\dfrac{df}{d\eta} = 0$ at $\eta = 0$
$v = 0$ at $y = 0$	$f = 0$ at $\eta = 0$
$\dfrac{\partial u}{\partial y} \to 0$ as $y \to \infty$	$\dfrac{df}{d\eta} \to 1.0$ as $\eta \to \infty$

Blasius' numerical solution of (6.9), with the corresponding values of u and v, is plotted in Fig. 6-2. It is seen that the boundary layer edge, $u/V_\infty = 0.99$, corresponds to $\eta \approx 5.0$; hence, since the edge is also given by $y = \delta$,

$$\delta \sqrt{\frac{V_\infty}{\nu x}} \approx 5.0$$

or

$$\frac{\delta}{x} \approx \frac{5.0}{\sqrt{\dfrac{V_\infty x}{\nu}}} = \frac{5.0}{\sqrt{Re_x}} \qquad (6.10)$$

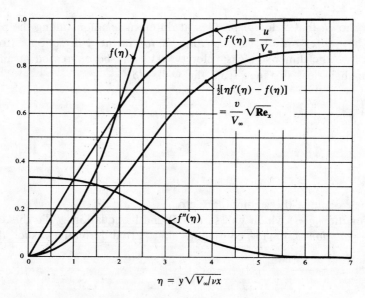

$$\eta = y\sqrt{V_\infty/\nu x}$$

Fig. 6-2

Using the fundamental equation for shear stress, $\tau = \mu_f(\partial u/\partial y)$, we can get the drag on the plate by evaluating the local shear stress at the surface, $\tau_s(x)$:

$$\tau_s(x) = \mu_f \frac{\partial u}{\partial y}\bigg|_{y=0} = \mu_f V_\infty \sqrt{\frac{V_\infty}{\nu x}} f''(0) = \mu_f V_\infty \sqrt{\frac{V_\infty}{\nu x}} (0.332)$$

in which Fig. 6-2 has been used to evaluate $f''(0)$, equal to the slope of the f' curve at $\eta = 0$. The *local skin-friction coefficient* is defined by

$$c_f \equiv \frac{\tau_s}{\rho_\infty V_\infty^2/2g_c} \qquad (6.11)$$

Substituting for τ_s, and using $\nu = \mu_m/\rho = g_c\mu_f/\rho$ and $\rho = \rho_\infty$ (incompressible flow),

$$c_f = \frac{0.664}{\sqrt{\dfrac{V_\infty x}{\nu}}} = \frac{0.644}{\sqrt{\mathbf{Re}_x}} \tag{6.12}$$

The *average skin-friction coefficient* for $0 < x < L$ is then

■ $$C_f \equiv \frac{1}{L}\int_0^L c_f\,dx = \frac{1.328}{\sqrt{\mathbf{Re}_L}} \tag{6.13}$$

where $\mathbf{Re}_L = V_\infty L/\nu$.

Integral Momentum Equation (Approximate)

The technique of the Blasius solution is limited by one's ingenuity in discovering transformations which satisfy the boundary conditions. And, of course, the Blasius solution was for the simplest of cases. Although approximate, there is a simpler, more general method which yields good engineering results.

The *von Kármán integral technique* involves the application of Newton's second law to a finite control volume, as opposed to the infinitesimal element of Blasius. Problem 6.7 obtains the result for the flat-plate case, cited here.

$$\tau_s = \frac{\rho}{g_c}\frac{d}{dx}\left[\int_0^\delta (V_\infty - u)u\,dy\right] + \frac{\rho}{g_c}\frac{dV_\infty}{dx}\int_0^\delta (V_\infty - u)\,dy \tag{6.14}$$

If, as in the Blasius solution, a constant pressure is assumed, then V_∞ is also constant and the

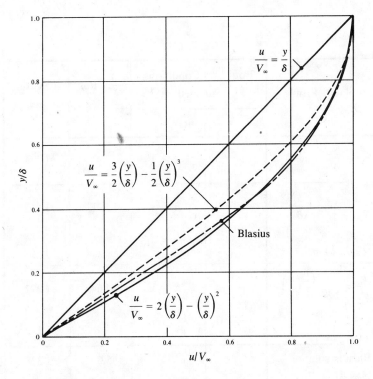

Fig. 6-3

second term in (6.14) vanishes.   In that case, equating τ_s to

$$\mu_f \frac{\partial u}{\partial y}\bigg|_{y=0}$$

gives the following integral equation for the boundary layer thickness:

$$\blacksquare \qquad \nu \frac{\partial u}{\partial y}\bigg|_{y=0} = \frac{d}{dx}\left[\int_0^\delta (V_\infty - u)u\,dy\right] \tag{6.15}$$

To solve (6.15), or the more general (6.14), requires knowing, or assuming, a velocity distribution within the boundary layer that satisfies the boundary conditions

$$
\begin{aligned}
\text{at } y = 0: \quad & (1)\ u = 0; \quad (2)\ \frac{\partial^2 u}{\partial y^2} = 0 \\
&\\
\text{at } y = \delta: \quad & (3)\ u = V_\infty; \quad (4)\ \frac{\partial u}{\partial y} = 0
\end{aligned}
\tag{6.16}
$$

Boundary conditions (1), (3) and (4) arise from the physics of the velocity profile; boundary condition (2) then comes from (6.1) for a constant-pressure condition.

At first glance the mere suggestion of assuming a velocity profile seems ridiculous.   In actuality, however, the thinness of the boundary layer helps to lessen the significance of errors inherent in assumptions.   Figure 6-3 shows three possible assumptions, ranging from a simple linear to a cubic profile; the Blasius exact profile is shown for reference.   The results yielded by these three profiles for the boundary layer thickness and the average skin-friction coefficient are displayed in Table 6-1.   Problems 6.8 and 6.9 show how these results are obtained.

Table 6-1

Velocity Profile	Boundary Conditions Satisfied		$\frac{\delta}{x}\sqrt{\mathbf{Re}_x}$	$C_f\sqrt{\mathbf{Re}_L}$
	at $y = 0$	at $y = \delta$		
$\dfrac{u}{V_\infty} = \dfrac{y}{\delta}$	$u = 0$	$u = V_\infty$	3.46	1.156
$\dfrac{u}{V_\infty} = 2\dfrac{y}{\delta} - \left(\dfrac{y}{\delta}\right)^2$	$u = 0$	$u = V_\infty$ $\dfrac{\partial u}{\partial y} = 0$	5.47	1.462
$\dfrac{u}{V_\infty} = \dfrac{3}{2}\dfrac{y}{\delta} - \dfrac{1}{2}\left(\dfrac{y}{\delta}\right)^3$	$u = 0$ $\dfrac{\partial^2 u}{\partial y^2} = 0$	$u = V_\infty$ $\dfrac{\partial u}{\partial y} = 0$	4.64	1.292
Blasius solution (exact)			5.0	1.328

6.2 THERMAL BOUNDARY LAYER: FLAT PLATE

When a fluid at one temperature flows along a surface which is at another temperature, the behavior of the fluid cannot be described by the hydrodynamic equations alone.   In addition to the hydrodynamic boundary layer, a thermal boundary layer develops.

Figure 6-4 shows temperature distributions within the thermal boundary layer, having a gradient which is infinite at the leading edge and approaches zero as the layer develops downstream.   Shown also is a heat balance at the plate's surface, where the heat conducted from the plate must equal the heat convected into the fluid; thus,

$$-k\frac{\partial T}{\partial y}\bigg|_s = h_x(T_s - T_\infty)$$

or

$$h_x = -\frac{k}{T_s - T_\infty}\frac{\partial T}{\partial y}\bigg|_s \qquad\qquad (6.17)$$

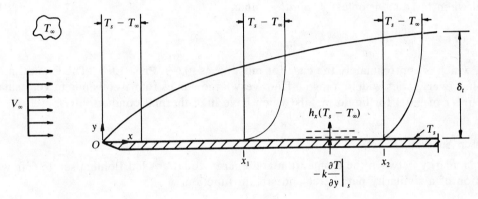

Fig. 6-4

Our problem in this section is to obtain an expression for the convective heat-transfer coefficient h_x, which obviously reduces to finding the temperature distribution.   To that end, in Problem 6.12 an energy balance is made on an infinitesimal control volume within the boundary layer, giving

$$u\frac{\partial T}{\partial x} + v\frac{\partial T}{\partial y} = \alpha\frac{\partial^2 T}{\partial y^2} \qquad\qquad (6.18)$$

with boundary conditions, analogous to (6.16),

at $y = 0$:   (1) $T = T_s$;   (2) $\dfrac{\partial^2 T}{\partial y^2} = 0$

at $y = \delta_t$:   (3) $T = T_\infty$;   (4) $\dfrac{\partial T}{\partial y} = 0$

(6.19)

Analogous to the hydrodynamic case, the thermal boundary layer thickness δ_t is defined as the distance required for the temperature T to reach 99 percent of its free-stream value T_∞.   The assumptions underlying (6.18) are:

1. steady, incompressible flow

2. constant fluid properties evaluated at the *film temperature,*

$$T_f \equiv \frac{T_s + T_\infty}{2}$$

3. negligible body forces, viscous heating (low velocity), and conduction in the flow direction

Recalling that the x-momentum equation for a constant-pressure field is

$$u\frac{\partial u}{\partial x} + v\frac{\partial u}{\partial y} = \nu\frac{\partial^2 u}{\partial y^2} \qquad (6.20)$$

we note the similarity between this and the energy equation (*6.18*). The temperature and velocity variations are identical when the thermal diffusivity α is equal to the kinematic viscosity ν. Complete analogy exists when the temperature is nondimensionalized by

$$\theta \equiv \frac{T - T_s}{T_\infty - T_s}$$

and the velocities are nondimensionalized by dividing by V_∞, making the thermal boundary condition $\theta = 0$ at $y = 0$ analogous to $u/V_\infty = 0$ at $y = 0$. Temperature and velocity profiles are identical when the dimensionless *Prandtl number,*

$$\mathbf{Pr} \equiv \frac{\nu}{\alpha} \qquad (6.21)$$

is unity, which is approximately the case for most gases ($0.6 < \mathbf{Pr} < 1.0$). The Prandtl number for liquids, however, varies widely, ranging from very large values for viscous oils to very small values (on the order of 0.01) for liquid metals which have high thermal conductivities.

Pohlhausen Solution (Exact)

The similarity between the momentum and energy equations led Pohlhausen to follow Blasius' assumption of a similarity parameter and stream function,

$$\eta \equiv y\sqrt{\frac{V_\infty}{\nu x}} \qquad \psi = \sqrt{\nu x V_\infty}\, f(\eta)$$

giving the ordinary linear differential equation

$$\frac{d^2\theta}{d\eta^2} + \frac{\mathbf{Pr}}{2} f \frac{d\theta}{d\eta} = 0 \qquad (6.22)$$

with boundary conditions $\theta(0) = 0$, $\theta(\infty) = 1$. The solution is

$$\theta(\eta) = \frac{\displaystyle\int_0^\eta \exp\left(-\frac{\mathbf{Pr}}{2}\int_0^\beta f(\alpha)\,d\alpha\right) d\beta}{\displaystyle\int_0^\infty \exp\left(-\frac{\mathbf{Pr}}{2}\int_0^\beta f(\alpha)\,d\alpha\right) d\beta} \qquad (6.23)$$

where $f(\alpha)$ is known (in numerical form at least) from the Blasius solution. The temperature distribution (*6.23*) is plotted in Fig. 6-5 for several Prandtl numbers; the curve for $\mathbf{Pr} = 0.7$ is typical for air and several other gases.

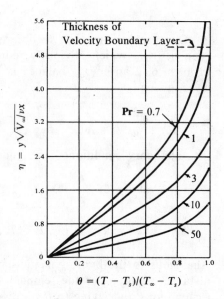

$$\theta = (T - T_s)/(T_\infty - T_s)$$

Fig. 6-5

The slope of the temperature profile at the surface, $y = 0$, is well represented by

$$\left(\frac{\partial T}{\partial y}\right)_{y=0} = (T_\infty - T_s)\sqrt{\frac{V_\infty}{\nu x}}\left(\frac{d\theta}{d\eta}\right)_{\eta=0} \approx (T_\infty - T_s)\sqrt{\frac{V_\infty}{\nu x}}(0.332)\mathbf{Pr}^{1/3} \tag{6.24}$$

for $0.6 < \mathbf{Pr} < 15$.   Substitution of this expression into (6.17) yields

$$h_x = (0.332)k\sqrt{\frac{V_\infty}{\nu x}}\,\mathbf{Pr}^{1/3} \tag{6.25}$$

Multiplying through by x/k, we get the local dimensionless *Nusselt number*,

$$\mathbf{Nu}_x \equiv \frac{h_x x}{k} = (0.332)\mathbf{Re}_x^{1/2}\mathbf{Pr}^{1/3} \tag{6.26}$$

Taking averages over the interval $0 < x < L$, we find the average heat-transfer coefficient and Nusselt number to be

$$\bar{h} = 2h_x|_{x=L} \tag{6.27}$$

$$\overline{\mathbf{Nu}} \equiv \frac{\bar{h}L}{k} = (0.664)\mathbf{Re}_L^{1/2}\mathbf{Pr}^{1/3} \tag{6.28}$$

Integral Energy Equation (Approximate)

The von Kármán integral technique can be applied to the thermal-hydrodynamic boundary layer just as it was to the hydrodynamic layer alone.   As shown in Problem 6.14, the result for the flat plate is

$$\blacksquare \qquad \alpha\frac{\partial T}{\partial y}\bigg|_{y=0} = \frac{d}{dx}\left[\int_0^{\delta_t}(T_\infty - T)u\,dy\right] \tag{6.29}$$

which should be compared with (6.15).   To solve (6.29) for δ_t we need to know, or to assume, a velocity profile and a temperature distribution within the boundary layer that respectively satisfy the boundary conditions (6.16) and (6.19).   If third-degree polynomials are used for u/V_∞ and θ (see Problem 6.15), the resultant local Nusselt number for convective heat-transfer from a flat plate is

$$\mathbf{Nu}_x \equiv \frac{h_x x}{k} = (0.332)\mathbf{Re}_x^{1/2}\,\mathbf{Pr}^{1/3}\left[1 - \left(\frac{x_i}{x}\right)^{3/4}\right]^{-1/3} \tag{6.30}$$

Here, x_i is the length of an unheated leading section of the plate; when $x_i = 0$, (6.30) is identical to the Pohlhausen solution, (6.26).

6.3 ISOTHERMAL PIPE FLOW

If we imagine the flat plate of the preceding sections to be rolled into a duct, we can apply the concepts of the boundary layer to flow in pipes.   There is one major difference, however. Whereas the boundary layer continues to increase in thickness as a fluid passes along a flat plate, the boundary layer thickness in a pipe is physically limited to the radius of the pipe.

Figure 6-6 shows the successive stages of development of the boundary layer of an incompressible viscous fluid in the entrance region of a circular tube.   At the tube's entrance *slug flow*, or uniform flow at the free-stream velocity, exists.   As the fluid moves down the tube, shear between the fluid and the wall, and between adjacent fluid particles, retards the motion, causing the boundary layer to grow until it is *fully developed* at station 3.   From this point on, the velocity profile remains unchanged.

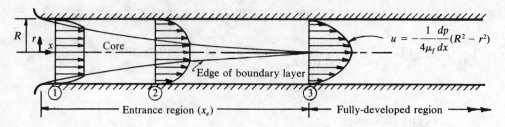

Fig. 6-6

Entrance Region

The length required for the velocity profile to become invariant with axial position is known as the *entry length*, x_e. It may be approximated by the simple *Langhaar equation*

$$x_e \approx (0.05)\mathbf{Re}_D \, D \tag{6.31}$$

In most cases the entry length of a pipe is negligible when compared with its total length. Most engineering calculations are, therefore, made assuming fully developed flow throughout.

Fully-Developed Region

For steady, fully developed, laminar flow in a tube, the velocity profile is parabolic:

$$u = -\frac{1}{4\mu_f}\frac{dp}{dx}(R^2 - r^2) \tag{6.32}$$

where the minus sign is required because the pressure decreases in the flow direction. This result is obtained from a force balance in Problem 6-17. The maximum velocity, which obviously occurs when $r = 0$, is

$$V_0 = -\frac{1}{4\mu_f}\frac{dp}{dx}R^2$$

The average velocity, V, may be obtained by equating the volumetric flow to the integrated paraboloidal flow, i.e.

$$V\pi R^2 = \int_0^R u(2\pi r)\,dr$$

giving

$$V = \frac{V_0}{2} = -\frac{1}{8\mu_f}\frac{dp}{dx}R^2 \tag{6.33}$$

In engineering practice it is customary to express the pressure gradient in terms of a *friction factor*, f, defined by

$$-\frac{dp}{dx} = \frac{f}{D}\frac{\rho V^2}{2g_c} \tag{6.34}$$

where $\rho V^2/2g_c$ is the dynamic pressure of the mean flow and D is the tube diameter. Integrating this expression, we get the *Darcy-Weisbach equation*,

$$\frac{\Delta p}{L} = \frac{f}{D}\frac{\rho V^2}{2g_c} \tag{6.35}$$

where $\Delta p = p_1 - p_2$ and $L = x_2 - x_1$. By (5.28) the head loss between stations 1 and 2 is

$h_L = (p_1 - p_2)/\rho$; thus,

■ $$h_L = f \frac{L}{D} \frac{V^2}{2g_c} \qquad (6.36)$$

Together, (6.33) and (6.34) give the friction factor as a simple function of Reynolds number, i.e.

$$f = \frac{64}{\mathbf{Re}_D} \qquad (6.37)$$

which is valid for laminar tube flow, **Re** < 2000.

In terms of volumetric flow rate, $Q = AV$, the head loss is given by

$$h_L = \frac{\Delta p}{\rho} = 128 \frac{QL\mu_f}{\pi D^4 \rho} \qquad (6.38)$$

Or, the volumetric flow rate may be conveniently expressed in terms of the pressure drop, i.e.

■ $$Q = \frac{\pi}{8\mu_f} \frac{R^4}{L} (p_1 - p_2) \qquad (6.39)$$

a result known as the *Hagen-Poiseuille equation.*

Noncircular Ducts

The Darcy-Weisbach equation, (6.35), is valid for noncircular ducts when the geometric diameter is replaced by the *hydraulic diameter* D_h defined by

$$D_h \equiv \frac{4A}{P} \qquad (6.40)$$

where A is the cross-sectional area and P is the wetted duct perimeter (the perimeter touched by the fluid). The friction factor must, or course, be evaluated for the particular duct configuration. Values of the product of friction factor and Reynolds number for the two most important industrial duct configurations were determined by Lundgren et al. and are given in Tables 6-2 and 6-3.

Table 6-2. Annular Ducts

Ratio of Radii	$f \cdot \mathbf{Re}$
0.001	74.68
0.01	80.11
0.05	86.27
0.10	89.37
0.20	92.35
0.40	94.71
0.60	95.59
0.80	95.92
1.00	96.00

Table 6-3. Rectangular Ducts

Ratio of Sides	$f \cdot \mathbf{Re}$
0.05	89.91
0.10	84.68
0.125	82.34
0.166	78.81
0.25	72.93
0.40	65.47
0.50	62.19
0.75	57.89
1.00	56.91

6.4 HEAT TRANSFER IN PIPE FLOW

A large class of heat transfer problems of engineering importance involves the flow of fluids through pipes, particularly in heat exchangers. The thermal boundary layer, which develops similarly to the hydrodynamic boundary layer shown in Fig. 6-6, is significant in the heat transfer process. Although heat transfer in laminar flow is not too common because the rate is lower than that encountered in turbulent flow, it is sometimes desirable due to the lower pumping power required in the laminar case.

In purely laminar flow the heat transfer mechanism is conduction, resulting in large heat-transfer coefficients for fluids with high thermal conductivities, such as liquid metals. Figure 6-7 shows the variation of local heat-transfer coefficient (Nusselt number) with axial distance along a tube in developing laminar flow for $\mathbf{Pr} = 0.7$. Three cases of boundary conditions are shown. A unique feature of these curves is the asymptotic values of Nusselt number as the flow becomes fully developed.

$$\text{uniform heat flux}$$
$$\text{or}$$
$$\text{constant temperature difference:} \quad \mathbf{Nu}_{D\infty} \equiv \frac{h_\infty D}{k} = 4.364 \tag{6.41}$$

$$\text{constant wall temperature:} \quad \mathbf{Nu}_{D\infty} \equiv \frac{h_\infty D}{k} = 3.656 \tag{6.42}$$

The result for uniform heat flux is derived in Problem 6.21. Two dimensionless numbers, the *Peclet number*, $\mathbf{Pe}$, and the *Graetz number*, $\mathbf{Gz}$, which are common in heat transfer literature, are introduced in Fig. 6-7.

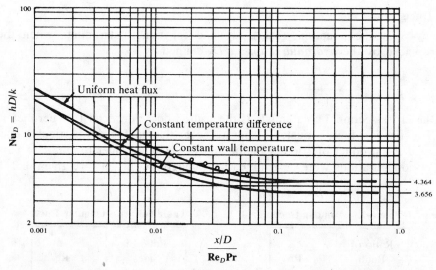

Fig. 6-7. (Adapted from Knudsen, J. G. and D. L. Katz, *Fluid Dynamics and Heat Transfer*, McGraw-Hill, New York, 1958. Used by permission.)

While the asymptotic Nusselt numbers furnish constant values for fully developed flows, many engineering applications involve tubes which are too short for the flow to fully develop. The length over which the flow develops, called the *thermal entry length* $x_{e,t}$, is given by

$$x_{e,t} \approx 0.05 \, \mathbf{Re}_D \, \mathbf{Pr} \, D \tag{6.43}$$

An analytical formula that supplements the results of Fig. 6-7 was developed by H. Hausen:

$$\text{uniform heat flux:} \quad \mathbf{Nu}_D = \mathbf{Nu}_{D\infty} + \frac{K_1[(D/x)\mathbf{Re}_D\mathbf{Pr}]}{1 + K_2[(D/x)\mathbf{Re}_D\mathbf{Pr}]^n} \tag{6.44}$$

$$\text{constant wall temperature:} \quad \overline{\mathbf{Nu}}_D = \mathbf{Nu}_{D\infty} + \frac{K_1[(D/L)\mathbf{Re}_D\mathbf{Pr}]}{1 + K_2[(D/L)\mathbf{Re}_D\mathbf{Pr}]^n}$$

(Note that $\mathbf{Nu}_D$ reflects the local heat-transfer coefficient, whereas $\overline{\mathbf{Nu}}_D$ reflects the coefficient averaged over the range $0 < x < L$.) Table 6-4 lists the parametric values to be used in (6.44) corresponding to four common sets of inlet conditions. For oils, or other fluids in which the viscosity varies considerably with temperature, the K_1-term in (6.44) should be multiplied by $(\mu/\mu_s)^{0.14}$.

Table 6-4

Wall Condition	Inlet Velocity	Pr	$Nu_{D\infty}$	K_1	K_2	n
Uniform q/A	parabolic	any	4.36	0.023	0.0012	1.0
Uniform q/A	developing	0.7	4.36	0.036	0.0011	1.0
Constant T	parabolic	any	3.66	0.0668	0.04	2/3
Constant T	developing	0.7	3.66	0.104	0.016	0.8

Except for μ_s, which is evaluated at the wall temperature, all fluid properties occurring in (6.41) through (6.44) are evaluated at the *bulk*, or *mixing-cup, temperature* T_b, which is the temperature that would be obtained if the fluid at a given cross section were directed into an insulated chamber and allowed to reach equilibrium (mix thoroughly). This definition makes the basic convection equation simply

$$q = hA(T_s - T_b) = \dot{m}c_p(T_{b,o} - T_{b,i}) \tag{6.45}$$

Mathematically, the bulk temperature is the enthalpy-average temperature of the bulk:

$$T_b \int_0^R \rho c_p u \, 2\pi r \, dr = \int_0^R \rho c_p u T \, 2\pi r \, dr$$

which, for an incompressible fluid having constant specific heat, reduces to

$$T_b = \frac{\displaystyle\int_0^R uTr \, dr}{\displaystyle\int_0^R ur \, dr} \tag{6.46}$$

In engineering practice, a simple approximate average value,

$$T_b = \frac{T_{\text{inlet}} + T_{\text{outlet}}}{2} \tag{6.47}$$

is used in the calculation of average heat-transfer coefficients.

Solved Problems

6.1. Determine the x-direction equation of motion for steady flow in a laminar boundary layer. Assume ρ and μ are constant.

For a fixed control volume and steady flow, Newton's second law, for the x-direction, states that the resultant applied x-force equals the net rate of x-momentum transfer out of the volume, i.e.

$$\sum F_x = \frac{1}{g_c}[(x\text{-momentum efflux}) - (x\text{-momentum influx})] \tag{1}$$

We apply this equation to an infinitesimal control volume (Δx by Δy by unit depth) from within

the boundary layer (see Fig. 6-8).   Mass enters the left
face at the rate $\rho u_x\,\Delta y$, producing an x-momentum influx of
$(\rho u_x\,\Delta y)u_x$.   Similarly, momentum

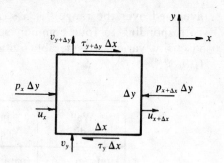

Fig. 6-8

efflux through right face: $(\rho u_{x+\Delta x}\,\Delta y)u_{x+\Delta x}$

influx through bottom face: $(\rho v_y\,\Delta x)u_y$

efflux through top face: $(\rho v_{y+\Delta y}\,\Delta x)u_{y+\Delta y}$

The acting pressure and shear forces are as shown in the
figure.   Substituting in (1), for constant ρ,

$$p_x\,\Delta y - p_{x+\Delta x}\,\Delta y + \tau_{y+\Delta y}\,\Delta x - \tau_y\,\Delta x$$

$$= \frac{\rho}{g_c}[(uu)_{x+\Delta x}\,\Delta y + (vu)_{y+\Delta y}\,\Delta x]$$

$$\qquad - \frac{\rho}{g_c}[(uu)_x\,\Delta y + (vu)_y\,\Delta x]$$

Rearranging, dividing by $\Delta x\,\Delta y$, and taking the limit as Δx and Δy approach zero, the result is

$$-\frac{\partial p}{\partial x} + \frac{\partial \tau}{\partial y} = \frac{\rho}{g_c}\left[\frac{\partial}{\partial x}(uu) + \frac{\partial}{\partial y}(vu)\right]$$

But

$$\frac{\partial}{\partial x}(uu) + \frac{\partial}{\partial y}(vu) = u\frac{\partial u}{\partial x} + v\frac{\partial u}{\partial y} + \left(\frac{\partial u}{\partial x} + \frac{\partial v}{\partial y}\right)u$$

and the term in the parentheses vanishes, by the continuity equation.   Moreover, $\tau = \mu_f \partial u/\partial y$;
therefore,

$$\frac{\rho}{g_c}\left(u\frac{\partial u}{\partial x} + v\frac{\partial u}{\partial y}\right) = -\frac{\partial p}{\partial x} + \mu_f\frac{\partial^2 u}{\partial y^2} \tag{2}$$

since μ_f is constant.   Assuming that $p = p(x)$ in agreement with Prandtl's boundary layer assumptions,
and recalling that $\nu \equiv \mu_m/\rho = g_c\mu_f/\rho$, we see that (2) is equivalent to (6.1).

6.2.   By use of an order-of-magnitude analysis find a functional relationship for the boundary
layer thickness δ, assuming constant fluid properties and zero pressure gradient.

The governing equations are:

continuity: $\dfrac{\partial u}{\partial x} + \dfrac{\partial v}{\partial y} = 0$

x-momentum: $u\dfrac{\partial u}{\partial x} + v\dfrac{\partial u}{\partial y} = \nu\dfrac{\partial^2 u}{\partial y^2}$

Except very close to the surface, the velocity u within the boundary layer is of the order of the
free-stream velocity V_∞, i.e. $u \sim V_\infty$.   And the y-dimension within the boundary layer is of the order
of the boundary layer thickness, $y \sim \delta$.   We can then approximate the continuity equation as

$$\frac{V_\infty}{x} + \frac{v}{\delta} \approx 0$$

giving

$$v \sim \frac{V_\infty\delta}{x}$$

Using the estimates of u, y, and v in the x-momentum equation gives

$$V_\infty \frac{V_\infty}{x} + \frac{V_\infty \delta}{x} \frac{V_\infty}{\delta} \approx \nu \frac{V_\infty}{\delta^2} \qquad \text{or} \qquad \delta^2 \sim \frac{\nu x}{V_\infty}$$

Dividing by x^2 to make dimensionless,

$$\frac{\delta}{x} \sim \sqrt{\frac{\nu}{V_\infty x}} = \sqrt{\frac{1}{\mathbf{Re}_x}}$$

which is (6.4).

6.3.   Air at 150 °F and 14.7 psia ($\nu \approx 2 \times 10^{-4}$ ft^2/sec) flows along a smooth flat plate at 40 fps.   For laminar flow, at what length from the leading edge does the boundary layer thickness reach 0.186 in.?

For laminar flow (6.10) can be solved for x:

$$\frac{\delta}{x} = \frac{5.0}{\sqrt{\dfrac{V_\infty x}{\nu}}}$$

$$x = \frac{\delta^2 V_\infty}{25\nu} = \frac{\left(\dfrac{0.186}{12}\, \text{ft}\right)^2 \left(40\, \dfrac{\text{ft}}{\text{sec}}\right)}{25\left(2 \times 10^{-4}\, \dfrac{\text{ft}^2}{\text{sec}}\right)} = 1.922\ \text{ft}$$

6.4.   A Pitot tube, located on the undercarriage of an airship 4 in. aft of its leading edge, is to be used to monitor airspeed which varies from 20 to 80 mph.   The undercarriage is approximately flat, making the pressure gradient negligible.   Air temperature is 40 °F and the pressure is 24.8 inHg.   To be outside of the boundary layer, at what distance should the Pitot tube be located from the undercarriage?

From Figure B-1 the dynamic viscosity is $\mu_f = 3.7 \times 10^{-7}$ lbf-sec/ft^2.   The density may be determined from the ideal gas relation,

$$\rho = \frac{p}{RT}$$

where, for air, $R = 53.3$ ft-lbf/lbm-°R and

$$p = (24.8\ \text{inHg}) \left(\frac{2116\ \text{lbf/ft}^2}{29.92\ \text{inHg}}\right) = 1753.9\ \text{lbf/ft}^2$$

Hence,

$$\rho = \frac{1753.9\ \text{lbf/ft}^2}{(53.3\ \text{ft-lbf/lbm-°R})(500\ \text{°R})} = 0.0658\ \text{lbm/ft}^3$$

and the kinematic viscosity is

$$\nu = \frac{\mu_f g_c}{\rho} = \frac{\left(3.7 \times 10^{-7}\, \dfrac{\text{lbf-sec}}{\text{ft}^2}\right)\left(32.2\, \dfrac{\text{lbm-ft}}{\text{lbf-sec}^2}\right)}{0.0658\ \text{lbm/ft}^3} = 1.81 \times 10^{-4}\, \frac{\text{ft}^2}{\text{sec}}$$

Assuming the flow is laminar, which must be checked, the boundary layer thickness at a given location varies inversely with the square root of Reynolds number, see (6.10).   The critical case occurs, therefore, at the minimum speed.

$$\mathbf{Re} = \frac{V_\infty x}{\nu}$$

$$\mathbf{Re}_{max} = \frac{(80 \text{ mph})\left(\dfrac{88 \text{ ft/sec}}{60 \text{ mph}}\right)\left(\dfrac{4}{12}\text{ ft}\right)}{1.81 \times 10^{-4} \text{ ft}^2/\text{sec}} = 216,083 < 300,000$$

and the flow is laminar;

$$\mathbf{Re}_{min} = \frac{20\left(\dfrac{88}{60}\right)\left(\dfrac{4}{12}\right)}{1.81 \times 10^{-4}} = 54,021$$

Therefore,

$$\delta_{max} = \frac{5.0\, x}{\sqrt{\mathbf{Re}_{min}}} = \frac{(5.0)(4 \text{ in.})}{\sqrt{54,021}} = 0.086 \text{ in.}$$

The Pitot tube must be located at some distance greater than this, which is easy to do since the boundary layer is so thin at this point.

6.5.   Approximate the skin-friction drag on a 3-ft-long by 2-ft-diameter cylinder, located axially in a wind tunnel, when the air speed is 15 fps. The pressure is atmospheric and the temperature is 140 °F.

From Figure B-1, $\nu = 2 \times 10^{-4}$ ft²/sec. Since the drag is required, the characteristic length for the calculation of Reynolds number is the length of the cylinder rather than its diameter; therefore,

$$\mathbf{Re}_L = \frac{V_\infty L}{\nu} = \frac{(15 \text{ ft/sec})(3 \text{ ft})}{2 \times 10^{-4} \text{ ft}^2/\text{sec}} = 225,000$$

and the laminar flow equations are valid.

The average skin-friction coefficient, from (6.13), is

$$C_f = \frac{1.328}{\sqrt{\mathbf{Re}_L}} = \frac{1.328}{\sqrt{225,000}} = 0.0028$$

The skin-friction drag force, F_f, is given by the product of average shear stress and total area, i.e.

$$F_f = C_f \left(\frac{\rho_\infty V_\infty^2}{2g_c}\right) A$$

where $\rho_\infty = 0.066$ lbm/ft³ from Table B-4. Hence,

$$F_f = 0.0028 \frac{\left(0.066\, \dfrac{\text{lbm}}{\text{ft}^3}\right)\left(15\, \dfrac{\text{ft}}{\text{sec}}\right)^2}{2\left(32.2\, \dfrac{\text{lbm-ft}}{\text{lbf-sec}^2}\right)} [\pi(2 \text{ ft})(3 \text{ ft})] = 0.0122 \text{ lbf}$$

It should be noted here that another type of drag—*form* or *profile drag*, resulting from the pile-up of air at the cylinder's end—is much more significant. It will be considered in Chapter 7.

6.6.   A thermal sensor is to be located 2 m from the leading edge of a flat plate along which 10 °C glycerine flows at 19 m/s. The pressure is atmospheric. In order to calibrate the sensor the velocity components, u and v, must be known. At $y = 4.5$ cm determine the velocity components.

From Figure B-2, $\nu = 2.79 \times 10^{-3}$ m²/s. The Reynolds number is

$$\mathbf{Re}_x = \frac{V_\infty x}{\nu} = \frac{(19 \text{ m/s})(2 \text{ m})}{2.79 \times 10^{-3} \text{ m}^2/\text{s}} = 13{,}620$$

and the flow is laminar. But is the sensor within the boundary layer? From (6.10), at $x = 2$ m,

$$\delta = \frac{5.0 \, x}{\sqrt{\mathbf{Re}_x}} = \frac{5(2 \text{ m})}{\sqrt{13{,}620}} = 8.57 \text{ cm} > 4.5 \text{ cm}$$

so that the sensor is within the boundary layer.

The velocity components are obtained from the Blasius solution (Fig. 6-2). At

$$\eta = y \sqrt{\frac{V_\infty}{\nu x}} = (0.045 \text{ m}) \sqrt{\frac{19 \text{ m/s}}{(2.79 \times 10^{-3} \text{ m}^2/\text{s})(2 \text{ m})}} = 2.63$$

we have

$$\frac{u}{V_\infty} \approx 0.775 \qquad \frac{v}{V_\infty} \sqrt{\mathbf{Re}_x} \approx 0.48$$

giving

$$u \approx (0.775) V_\infty = (0.775)(19 \text{ m/s}) = 14.73 \text{ m/s}$$

$$v \approx \frac{(0.48) V_\infty}{\sqrt{\mathbf{Re}_x}} = \frac{(0.48)(19)}{\sqrt{13{,}620}} = 0.078 \text{ m/s}$$

6.7. Derive (6.14) for steady, incompressible, laminar flow over a flat plate.

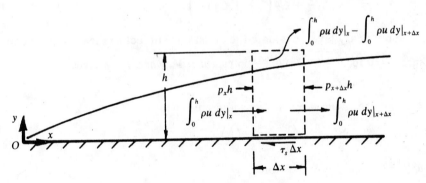

Fig. 6-9

Consider the dashed control volume in Fig. 6-9, which is infinitesimal in the x-direction but finite in the y-direction. The forces acting on the volume are shown (there is no shear at the upper face, which is outside the boundary layer), as well as the mass fluxes through the faces. Notice that the mass efflux through the upper face exactly cancels the net mass influx through the other two faces; this is required by the equation of continuity. The corresponding x-momentum fluxes are:

influx through left face: $\displaystyle \int_0^h \rho u^2 \, dy \bigg|_x$

efflux through right face: $\displaystyle \int_0^h \rho u^2 \, dy \bigg|_{x+\Delta x}$

efflux through upper face: $\displaystyle V_\infty \left[\int_0^h \rho u \, dy \bigg|_x - \int_0^h \rho u \, dy \bigg|_{x+\Delta x} \right]$

where, in the last expression, V_∞ is evaluated at some point between x and $x + \Delta x$. Newton's law,

(1) of Problem (6.1), then gives

$$p_x h - p_{x+\Delta x} h - \tau_s \, \Delta x = \frac{1}{g_c} \left\{ \left. \int_0^h \rho u^2 \, dy \right|_{x+\Delta x} + V_\infty \left[\left. \int_0^h \rho u \, dy \right|_x - \left. \int_0^h \rho u \, dy \right|_{x+\Delta x} \right] - \left. \int_0^h \rho u^2 \, dy \right|_x \right\}$$

Rearranging, dividing by Δx, and taking the limit as Δx approaches zero, we get ($\rho =$ constant):

$$h \frac{dp}{dx} + \tau_s = \frac{\rho}{g_c} \left\{ -\frac{d}{dx} \int_0^h u^2 \, dy + V_\infty \frac{d}{dx} \int_0^h u \, dy \right\} \tag{1}$$

But, outside the boundary layer, Bernoulli's equation gives

$$\frac{dp}{dx} = -\frac{\rho}{g_c} V_\infty \frac{dV_\infty}{dx} \tag{2}$$

Also,

$$V_\infty \frac{d}{dx} \int_0^h u \, dy = \frac{d}{dx} \int_0^h V_\infty u \, dy - \frac{dV_\infty}{dx} \int_0^h u \, dy \tag{3}$$

Substituting (2) and (3) into (1) and recombining terms, we get

$$\tau_s = \frac{\rho}{g_c} \frac{d}{dx} \left[\int_0^h (V_\infty - u) u \, dy \right] + \frac{\rho}{g_c} \frac{dV_\infty}{dx} \int_0^h (V_\infty - u) \, dy \tag{4}$$

The upper limit of integration in (4) may be replaced by δ, yielding (6.14), since $V_\infty - u$ is zero outside the boundary layer.

6.8. Assuming a velocity profile of the form

$$\frac{u}{V_\infty} = C + C_1 \frac{y}{\delta} + C_2 \left(\frac{y}{\delta}\right)^2 + C_3 \left(\frac{y}{\delta}\right)^3$$

within the boundary layer, evaluate the constants subject to the boundary conditions (6.16).

Since $u = 0$ for $y = 0$, $C = 0$. Applying $u = V_\infty$ when $y = \delta$ gives

$$1 = C_1 + C_2 + C_3 \tag{1}$$

Differentiating the assumed profile (V_∞ and δ are functions of x alone),

$$\frac{1}{V_\infty} \frac{\partial u}{\partial y} = \frac{C_1}{\delta} + \frac{2C_2}{\delta} \left(\frac{y}{\delta}\right) + \frac{3C_3}{\delta} \left(\frac{y}{\delta}\right)^2 \tag{2}$$

But $\partial u/\partial y = 0$ when $y = \delta$; therefore,

$$0 = C_1 + 2C_2 + 3C_3 \tag{3}$$

A second differentiation of the velocity profile gives

$$\frac{1}{V_\infty} \frac{\partial^2 u}{\partial y^2} = \frac{2C_2}{\delta^2} + \frac{6C_3}{\delta^2} \left(\frac{y}{\delta}\right) \tag{4}$$

from which (at $y = 0$, $\partial^2 u/\partial y^2 = 0$) we get

$$0 = C_2 \tag{5}$$

Solving (1), (3) and (5) simultaneously, the resulting constants are:

$$C_1 = \frac{3}{2} \qquad C_3 = -\frac{1}{2}$$

giving the cubic velocity profile cited in Fig. 6-3 and Table 6-1, viz.

$$\frac{u}{V_\infty} = \frac{3}{2} \frac{y}{\delta} - \frac{1}{2} \left(\frac{y}{\delta}\right)^3$$

6.9.   Using the cubic velocity profile from Problem 6.8 in the von Kármán integral equation
(*6.15*), determine the boundary layer thickness and the average skin-friction coefficient for
laminar flow over a flat plate.

From (*2*) of Problem 6.8,

$$\frac{1}{V_\infty}\frac{\partial u}{\partial y}\Big|_{y=0} = \frac{C_1}{\delta} = \frac{3}{2\delta} \tag{1}$$

where now, by hypothesis, V_∞ is constant.   Equation (*6.15*) becomes

$$\frac{3\nu}{2V_\infty\delta} = \frac{d}{dx}\int_0^\delta \left[1 - \frac{3}{2}\frac{y}{\delta} + \frac{1}{2}\left(\frac{y}{\delta}\right)^3\right]\left[\frac{3}{2}\frac{y}{\delta} - \frac{1}{2}\left(\frac{y}{\delta}\right)^3\right] dy$$

which integrates to give

$$\frac{3\nu}{2V_\infty\delta} = \frac{39}{280}\frac{d\delta}{dx}$$

which is a first-order differential equation for δ.   Separating the variables and integrating ($\delta = 0$ at
$x = 0$),

$$\frac{140\nu}{13V_\infty}\int_0^x dx = \int_0^\delta \delta\, d\delta$$

and this gives

$$\frac{\delta^2}{x^2} = \frac{280/13}{\dfrac{V_\infty x}{\nu}} \quad \text{or} \quad \frac{\delta}{x} = \frac{4.64}{\sqrt{\mathbf{Re}_x}} \tag{2}$$

as shown in Table 6-1.

From the definition of local skin-friction coefficient, (*6.11*),

$$c_f \equiv \frac{\tau_s}{\dfrac{\rho V_\infty^2}{2g_c}}$$

and Newton's law of viscosity,

$$\tau_s = \mu_f \frac{\partial u}{\partial y}\Big|_{y=0}$$

we get, using (*1*),

$$c_f = \frac{3\nu}{V_\infty\delta} = \frac{3}{(\mathbf{Re}_x)(\delta/x)}$$

Substituting for δ/x from (*2*), we get

$$c_f = \frac{0.646}{\sqrt{\mathbf{Re}_x}} \tag{3}$$

from which the average value may be determined as

$$C_f = \frac{1}{L}\int_0^L c_f\, dx = \frac{1.292}{\sqrt{\mathbf{Re}_L}} \tag{4}$$

6.10.   Air at 20 °C and one atmosphere flows parallel to a flat plate at a velocity of 3.5 m/s.   Compare the boundary layer thickness and the local skin-friction coefficient at $x = 1$ m from the
exact Blasius solution and from the approximate von Kármán integral technique assuming
the cubic velocity profile.

From Figure B-2, $\nu = 1.5 \times 10^{-5}$ m²/s and the Reynolds number is

$$\mathbf{Re}_x = \frac{V_\infty x}{\nu} = \frac{(3.5 \text{ m/s})(1 \text{ m})}{1.5 \times 10^{-5} \text{ m}^2/\text{s}} = 233{,}300$$

Therefore, from the exact solution, (6.10) and (6.12):

$$\delta = \frac{(5.0)x}{\sqrt{\mathbf{Re}_x}} = \frac{(5.0)(1 \text{ m})}{\sqrt{233{,}300}} = 0.01035 \text{ m} = 1.035 \text{ cm}$$

$$c_f = \frac{0.664}{\sqrt{\mathbf{Re}_x}} = \frac{0.664}{\sqrt{233{,}300}} = 1.37 \times 10^{-3}$$

From the approximate solution, (2) and (3) of Problem 6.9:

$$\delta = \frac{(4.64)x}{\sqrt{\mathbf{Re}_x}} = 0.961 \text{ cm} \qquad c_f = \frac{0.646}{\sqrt{\mathbf{Re}_x}} = 1.34 \times 10^{-3}$$

We note that the approximate solution deviates from the exact solution by 7.2 percent for the boundary layer thickness and 2.9 percent for the local skin-friction coefficient, deviations which are quite acceptable for the usual engineering accuracy.

6.11. What is the drag per unit width on one side of the plate of Problem 6.10 for a 1-m length beginning at the leading edge?

The drag force F_f is given by

$$F_f = C_f \left(\frac{\rho_\infty V_\infty^2}{2g_c} \right) A \qquad (1)$$

From Table B-4, $\rho_\infty = 1.210$ kg/m³, and in SI units

$$g_c = 1.00 \frac{\text{kg-m}}{\text{N-s}^2}$$

The average coefficient of friction (Blasius solution) is given by (6.13):

$$C_f = \frac{1.328}{\sqrt{\mathbf{Re}_L}} = \frac{1.328}{\sqrt{233{,}300}} = 2.75 \times 10^{-3}$$

where the Reynolds number based on $L = 1$ is the same as that based on $x = 1$. Therefore,

$$F_f = (2.75 \times 10^{-3}) \frac{\left(1.210 \dfrac{\text{kg}}{\text{m}^3} \right) \left(3.5 \dfrac{\text{m}}{\text{s}} \right)^2}{2 \left(1 \dfrac{\text{kg-m}}{\text{N-s}^2} \right)} (1 \text{ m}) = 0.02038 \text{ N/m}$$

6.12. Assuming constant fluid properties (k, c_p, μ), determine the energy equation for a steady, incompressible laminar boundary layer by making an energy balance on a small control volume from within the boundary layer. Neglect conduction in the flow direction.

Choosing an element, Δx by Δy by unit depth, from the boundary layer of Fig. 6-4, the respective energy terms are shown in the enlarged view of Fig. 6-10. The convective terms q_h may be written in terms of the specific heat, assumed constant, i.e.

$$q_h = \dot{m} c_p (T - T_{\text{ref}})$$

where $\dot{m} = \rho u$ in the x-direction and $\dot{m} = \rho v$ in the y-direction. The conductive terms are, from Fourier's law,

$$q_k = -k \frac{\partial T}{\partial y}$$

The remaining terms, designated $u\tau \, \Delta x$, account for the heat generated by fluid friction—a work rate in which $\tau \, \Delta x$ is the force, and u is the velocity at which the shear occurs. This term arises because fluid on the top face of the control volume moves faster than fluid on the bottom face.

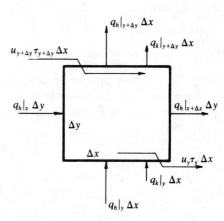

Fig. 6-10

Equating the rate of energy entering the control volume to the rate of energy leaving, we get

$$q_h|_x \, \Delta y + q_h|_y \, \Delta x + q_k|_y \, \Delta x + u_{y+\Delta y}\tau_{y+\Delta y} \, \Delta x$$

$$= q_h|_{x+\Delta x} \, \Delta y + q_h|_{y+\Delta y} \, \Delta x + q_k|_{y+\Delta y} \, \Delta x$$

$$+ u_y\tau_y \, \Delta x$$

Rearranging, dividing by $\Delta x \, \Delta y$, and taking the limit as Δx and Δy approach zero, the equation simplifies to

$$\frac{\partial q_h}{\partial x} + \frac{\partial q_h}{\partial y} + \frac{\partial q_k}{\partial y} = \frac{\partial (u\tau)}{\partial y}$$

or, in terms of measurable parameters, where $\tau = \mu_f \, \partial u / \partial y$,

$$\frac{\partial}{\partial x}(\rho u c_p T) + \frac{\partial}{\partial y}(\rho v c_p T) + \frac{\partial}{\partial y}\left(-k\frac{\partial T}{\partial y}\right) = \frac{\partial}{\partial y}\left(u\mu_f \frac{\partial u}{\partial y}\right)$$

Further simplification yields

$$\rho c_p\left[\left(u\frac{\partial T}{\partial x} + v\frac{\partial T}{\partial y}\right) + T\left(\frac{\partial u}{\partial x} + \frac{\partial v}{\partial y}\right)\right] = k\frac{\partial^2 T}{\partial y^2} + \frac{\partial}{\partial y}\left(u\mu_f \frac{\partial u}{\partial y}\right)$$

where the term multiplying T is zero from continuity and the last term is negligible except in high-speed flows; therefore,

$$u\frac{\partial T}{\partial x} + v\frac{\partial T}{\partial y} = \alpha\frac{\partial^2 T}{\partial y^2}$$

where $\alpha = k/\rho c_p$. This is *(6.18)*.

6.13. How do the thermal and hydrodynamic boundary layer thicknesses from the Pohlhausen and Blasius results compare?

From Fig. 6-2,

$$\frac{\partial(u/V_\infty)}{\partial \eta}\bigg|_{\eta=0} = 0.332$$

But this can be approximated as

$$\frac{\partial(u/V_\infty)}{\partial \eta}\bigg|_{\eta=0} \approx \frac{1}{\delta} = 0.332 \qquad (1)$$

From *(6.24)*,

$$\frac{d\theta}{d\eta}\bigg|_{\eta=0} = 0.332\,\mathbf{Pr}^{1/3}$$

which can be approximated as

$$\frac{d\theta}{d\eta}\bigg|_{\eta=0} \approx \frac{\theta_\infty - \theta_s}{\delta_t} = \frac{1}{T_\infty - T_s}\frac{T_\infty - T_s}{\delta_t} = \frac{1}{\delta_t} = 0.332\,\mathbf{Pr}^{1/3} \qquad (2)$$

Combining (1) and (2), we get

■ $$\delta_t = \frac{\delta}{\mathbf{Pr}^{1/3}}$$

6.14. By making an energy balance on the finite control volume shown in Fig. 6-11, find an expression for the heat conducted into the laminar, incompressible boundary layer at the surface of a flat plate, assuming ρ, c_p and k are constant.

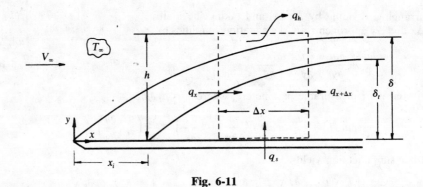

Fig. 6-11

Mass enters the left face of the control volume, of unit depth, at the rate

$$\int_0^h \rho u \, dy \bigg|_x$$

carrying with it (convecting) energy (enthalpy)

$$q_x = \int_0^h \rho u(c_p T) \, dy \bigg|_x$$

Similarly, at $x + \Delta x$ the enthalpy flux is

$$q_{x+\Delta x} = \int_0^h \rho u(c_p T) \, dy \bigg|_{x+\Delta x}$$

The mass leaving the top face of the volume is, by continuity,

$$\int_0^h \rho u \, dy \bigg|_x - \int_0^h \rho u \, dy \bigg|_{x+\Delta x}$$

and the temperature is constant at T_∞, since this face is outside the thermal boundary layer. Hence, the energy flux is

$$q_h = c_p T_\infty \left[\int_0^h \rho u \, dy \bigg|_x - \int_0^h \rho u \, dy \bigg|_{x+\Delta x} \right]$$

Energy is conducted into the lower face of the control volume at the rate

$$q_s = -k \frac{\partial T}{\partial y} \bigg|_{y=0} \Delta x$$

Balancing these individual energy terms for the element, we get

$$q_x + q_s = q_{x+\Delta x} + q_h$$

$$\rho c_p \int_0^h uT\,dy\bigg|_x - k\frac{\partial T}{\partial y}\bigg|_{y=0}\Delta x = \rho c_p \int_0^h uT\,dy\bigg|_{x+\Delta x} + \rho c_p T_\infty\left[\int_0^h u\,dy\bigg|_x - \int_0^h u\,dy\bigg|_{x+\Delta x}\right]$$

Transposing the first term to the right-hand side, dividing by $-\rho c_p\,\Delta x$, and letting $\Delta x \to 0$, we obtain (6.29):

$$\alpha\frac{\partial T}{\partial y}\bigg|_{y=0} = -\frac{d}{dx}\int_0^h uT\,dy + T_\infty\frac{d}{dx}\int_0^h u\,dy$$

$$= \frac{d}{dx}\left[\int_0^h (T_\infty - T)u\,dy\right] = \frac{d}{dx}\left[\int_0^{\delta_t} (T_\infty - T)u\,dy\right]$$

since $T_\infty - T = 0$ for $y > \delta_t$.

6.15. Using cubic velocity and temperature distributions in the integral boundary layer energy equation, (6.29), determine the convective heat-transfer coefficient for laminar flow over a flat plate that has an unheated starting length x_i.

The cubic velocity profile was obtained in Problem 6.8. Since the boundary conditions on

$$\theta \equiv \frac{T - T_s}{T_\infty - T_s}$$

[see (6.19)] have the same form as those on u/V_∞, the cubic temperature distribution is simply

$$\theta = \frac{3}{2}\frac{y}{\delta_t} - \frac{1}{2}\left(\frac{y}{\delta_t}\right)^3 \tag{1}$$

and (6.29) becomes

$$\alpha\frac{\partial T}{\partial y}\bigg|_{y=0} = (T_\infty - T_s)V_\infty\frac{d}{dx}\int_0^{\delta_t}\left[1 - \frac{3}{2}\frac{y}{\delta_t} + \frac{1}{2}\left(\frac{y}{\delta_t}\right)^3\right]\left[\frac{3}{2}\frac{y}{\delta} - \frac{1}{2}\left(\frac{y}{\delta}\right)^3\right]dy$$

Multiplying and simplifying, we get

$$\alpha\frac{\partial T}{\partial y}\bigg|_{y=0} = (T_\infty - T_s)V_\infty\frac{d}{dx}\int_0^{\delta_t}\left[\frac{3}{2\delta}y - \frac{9}{4\delta\delta_t}y^2 - \frac{1}{2\delta^3}y^3 + \left(\frac{3}{4\delta\delta_t^3} + \frac{3}{4\delta^3\delta_t}\right)y^4 - \frac{1}{4\delta^3\delta_t^3}y^6\right]dy$$

$$= (T_\infty - T_s)V_\infty\frac{d}{dx}\left[\frac{3}{4}\frac{\delta_t^2}{\delta} - \frac{3}{4}\frac{\delta_t^2}{\delta} - \frac{1}{8}\frac{\delta_t^4}{\delta^3} + \frac{3}{20}\left(\frac{\delta_t^2}{\delta} + \frac{\delta_t^4}{\delta^3}\right) - \frac{1}{28}\frac{\delta_t^4}{\delta^3}\right]$$

Differentiating (1) to get the temperature gradient for the left side of the equation, and letting $\zeta \equiv \delta_t/\delta$, the result simplifies to

$$\frac{3\alpha}{2\zeta\delta} = V_\infty\frac{d}{dx}\left[\delta\left(\frac{3}{20}\zeta^2 - \frac{3}{280}\zeta^4\right)\right] \tag{2}$$

From Problem 6.13, $\delta_t = \delta/\mathbf{Pr}^{1/3}$. Since **Pr** is close to unity for most gases and quite large for most viscous fluids (except for liquid metals), $\zeta \leqslant 1$, and $\zeta^4 \ll \zeta^2$; hence, we may neglect the ζ^4-term, yielding

$$\frac{10\alpha}{\zeta\delta} = V_\infty\frac{d}{dx}(\delta\zeta^2) = V_\infty\left(2\delta\zeta\frac{d\zeta}{dx} + \zeta^2\frac{d\delta}{dx}\right)$$

or

$$\frac{10\alpha}{V_\infty}dx = 2\zeta^2\delta^2\,d\zeta + \zeta^3\delta\,d\delta \tag{3}$$

From Problem 6.9, the hydrodynamic solution,

$$\delta^2 = \frac{280}{13}\frac{\nu x}{V_\infty}$$

Using this and $\delta\,d\delta = (140/13)(\nu/V_\infty)\,dx$ in (3) yields

$$\frac{\alpha}{V_\infty}\,dx = 2\zeta^2\frac{28}{13}\frac{\nu x}{V_\infty}\,d\zeta + \zeta^3\frac{14}{13}\frac{\nu}{V_\infty}\,dx$$

which simplifies to

$$\zeta^3 + 4x\zeta^2\frac{d\zeta}{dx} = \frac{13}{14}\frac{\alpha}{\nu}$$

This equation can be expressed as an ordinary differential equation in ζ^3:

$$\frac{4x}{3}\frac{d\zeta^3}{dx} + \zeta^3 = \frac{13}{14}\mathbf{Pr} \qquad (4)$$

with boundary condition

$$\text{at } x = x_i: \quad \zeta^3 = \left(\frac{\delta_t}{\delta}\right)^3 = 0 \qquad (5)$$

The general solution of (4) is

$$\zeta^3 = Cx^{-3/4} + \frac{13}{14}\mathbf{Pr}^{-1}$$

and (5) gives

$$C = -\frac{13}{14}\frac{x_i^{3/4}}{\mathbf{Pr}}$$

Therefore,

$$\zeta = \frac{\delta_t}{\delta} = \frac{0.976}{\mathbf{Pr}^{1/3}}\left[1 - \left(\frac{x_i}{x}\right)^{3/4}\right]^{1/3} \qquad (6)$$

The local heat-transfer coefficient h_x is defined by

$$q_s'' = h_x(T_s - T_\infty) = -k\frac{\partial T}{\partial y}\bigg|_{y=0}$$

or

$$h_x = -\frac{k}{(T_s - T_\infty)}\frac{3}{2\delta_t}(T_\infty - T_s) = \frac{3k}{2\delta_t}$$

Substituting the expression (6) for δ_t and (2) of Problem 6.9 for δ, the local heat-transfer coefficient is

$$h_x = \frac{3}{2}\frac{k}{x}\left[1 - \left(\frac{x_i}{x}\right)^{3/4}\right]^{-1/3}\frac{\mathbf{Pr}^{1/3}\,\mathbf{Re}_x^{1/2}}{(0.976)(4.46)}$$

which, upon multiplying both sides by x/k, gives (6.30):

$$\mathbf{Nu}_x \equiv \frac{h_x x}{k} = (0.332)\mathbf{Re}_x^{1/2}\,\mathbf{Pr}^{1/3}\left[1 - \left(\frac{x_i}{x}\right)^{3/4}\right]^{-1/3}$$

6.16. In a pharmaceutical process 100 °F castor oil flows over a wide, 20-ft-long, heated plate at 0.2 fps. For a surface temperature of 200 °F, determine the: (a) hydrodynamic boundary layer thickness δ at the end of the plate, (b) total drag on the surface per unit width, (c) thermal boundary layer thickness δ_t at the end of the plate, (d) local heat-transfer coefficient h_x at the end of the plate, (e) total heat flux from the surface per unit width. Assume the thermal diffusivity α to be 2.8×10^{-3} ft²/hr and the thermal conductivity to be 0.123 Btu/hr-ft-°F at the film temperature.

For moderate temperature differences between the free stream and surface, good results are obtained by assuming constant fluid properties evaluated at the film temperature T_f; therefore,

$$T_f = \frac{T_s + T_\infty}{2} = 150 \text{ °F}$$

At the end of the plate the Reynolds number is

$$\mathbf{Re}_L = \frac{V_\infty L}{\nu} = \frac{(0.2 \text{ ft/sec})(20 \text{ ft})}{7 \times 10^{-4} \text{ ft}^2/\text{sec}} = 5714$$

where the kinematic viscosity is taken from Figure B-2. The flow is laminar throughout. Using the dynamic viscosity from Figure B-1, at the film temperature, the density is

$$\rho = \frac{\mu_m}{\nu} = \frac{\left(1.3 \times 10^{-3} \dfrac{\text{lbf-sec}}{\text{ft}^2}\right)\left(32.2 \dfrac{\text{lbm-ft}}{\text{lbf-sec}^2}\right)}{7 \times 10^{-4} \dfrac{\text{ft}^2}{\text{sec}}} = 59.8 \text{ lbm/ft}^3$$

(a) From (6.10),

$$\delta = \frac{5.0 \, L}{\sqrt{\mathbf{Re}_L}} = \frac{5.0 \,(20 \text{ ft})}{\sqrt{5714}} = 1.323 \text{ ft}$$

(b) The average skin-friction coefficient is given by (6.13):

$$C_f = \frac{1.328}{\sqrt{\mathbf{Re}_L}} = \frac{1.328}{\sqrt{5714}} = 0.01757$$

and the drag F_f is given by

$$F_f = C_f A \frac{\rho_\infty V_\infty^2}{2g_c} = (0.01757)\left(20 \frac{\text{ft}^2}{\text{ft}}\right)\frac{\left(59.8 \dfrac{\text{lbm}}{\text{ft}^3}\right)\left(0.2 \dfrac{\text{ft}}{\text{sec}}\right)^2}{2\left(32.2 \dfrac{\text{lbm-ft}}{\text{lbf-sec}^2}\right)} = 0.013 \text{ lbf/ft}$$

(c) To get the thermal boundary layer thickness the Prandtl number is needed; it is

$$\mathbf{Pr} = \frac{\nu}{\alpha} = \frac{7 \times 10^{-4} \text{ ft}^2/\text{sec}}{2.8 \times 10^{-3} \text{ ft}^2/\text{hr}}\left(3600 \frac{\text{sec}}{\text{hr}}\right) = 900$$

From Problem 6.13,

$$\delta_t = \frac{\delta}{\mathbf{Pr}^{1/3}} = \frac{1.323}{900^{1/3}} = 0.137 \text{ ft}$$

and we see that the thermal boundary layer thickness is an order of magnitude less than the hydrodynamic boundary layer thickness.

(d) The local heat-transfer coefficient at $x = L$ is given by (6.25) as

$$h_L = (0.332)k\sqrt{\frac{V_\infty}{\nu L}}\,\mathbf{Pr}^{1/3}$$

$$= (0.332)\left(0.123 \frac{\text{Btu}}{\text{hr-ft-°F}}\right)\sqrt{\frac{0.2 \text{ ft/sec}}{(7 \times 10^{-4} \text{ ft}^2/\text{sec})(20 \text{ ft})}}\,900^{1/3}$$

$$= 1.49 \text{ Btu/hr-ft}^2\text{-°F}$$

(e)   From (6.27) we can get the average heat-transfer coefficient required in evaluating the total heat flux, i.e.

$$\bar{h} = 2h_L = 2(1.49) = 2.98 \text{ Btu/hr-ft}^2\text{-}°F$$

The total heat transfer is then given by

$$q_s = \bar{h}A(T_s - T_\infty)$$

$$= \left(2.98 \frac{\text{Btu}}{\text{hr-ft}^2\text{-}°F}\right)\left(20 \frac{\text{ft}^2}{\text{ft}}\right)[(200 - 100) °F] = 5960 \text{ Btu/hr-ft}$$

6.17.   Find the velocity distribution for fully developed, steady, laminar flow in a tube by considering the force equilibrium of a cylindrical element of fluid.

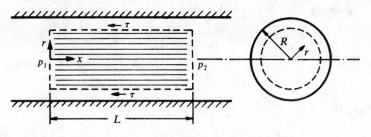

Fig. 6-12

Taking the element shown in Fig. 6-12, the forces are (1) shear on the cylindrical surface and (2) normal forces due to pressure on the ends.   There is no change in momentum since the velocity is the same at stations 1 and 2.   Therefore, a force balance gives

$$(p_1 - p_2)\pi r^2 = \tau(2\pi rL)$$

But

$$\tau = -\mu_f \frac{du}{dr}$$

so

$$\frac{du}{dr} = -\frac{p_1 - p_2}{2\mu_f L} r$$

Separating variables and integrating,

$$\int_u^0 du = -\frac{p_1 - p_2}{2\mu_f L} \int_r^R r\, dr$$

$$u = \frac{1}{4\mu_f} \frac{p_1 - p_2}{L} (R^2 - r^2)$$

which is equivalent to (6.32) with

$$\frac{dp}{dx} = -\frac{p_1 - p_2}{L}$$

6.18.   What is the pressure drop in a 60-ft length of smooth 3/8-in.-i.d. tubing when 85 °F benzene flows at an average velocity of 0.4 fps?

Assume fully developed flow.   The Reynolds number is

$$\mathbf{Re}_D = \frac{VD}{\nu} = \frac{(0.4 \text{ ft/sec})\left(\dfrac{0.375}{12}\text{ ft}\right)}{7 \times 10^{-6}\text{ ft}^2\text{/sec}} = 1786$$

where the kinematic viscosity is from Figure B-2.   Since $\mathbf{Re}_D < 2000$, (6.37) gives the friction factor as

$$f = \frac{64}{\mathbf{Re}_D} = \frac{64}{1786} = 0.03584$$

and the pressure drop is given by (6.35),

$$\Delta p = f \frac{L}{D} \frac{\rho V^2}{2g_c}$$

where $\rho = \mu_m/\nu = g_c\mu_f/\nu$.   From Figure B-1, $\mu_f = 1.2 \times 10^{-5}$ lbf-sec/ft^2; therefore,

$$\Delta p = (0.03584)\left(\frac{60 \times 12 \text{ in.}}{0.375 \text{ in.}}\right)\frac{1.2 \times 10^{-5}\dfrac{\text{lbf-sec}}{\text{ft}^2}}{2\left(7 \times 10^{-6}\dfrac{\text{ft}^2}{\text{sec}}\right)}\left(0.4\,\frac{\text{ft}}{\text{sec}}\right)^2 = 9.44 \text{ lbf/ft}^2$$

6.19. What volumetric flow rate of 50 °C water can be developed in 20 m of smooth 2-cm-i.d. tubing by a pump having a total head capacity of 0.60 N/m^2?

Assuming laminar flow, the Hagen-Poiseuille equation, (6.39), holds.   Using the data of Table B-3, the dynamic viscosity is

$$\mu_m = \rho\nu = [(61.80)(16.018)][(6.11 \times 10^{-6})(0.0929)] = 0.00056 \text{ kg/m-s}$$

or $\mu_f = 0.00056$ N-s/m^2.   Hence,

$$Q = \frac{\pi}{8(0.00056 \text{ N-s/m}^2)}\frac{(0.01 \text{ m})^4}{20 \text{ m}}(0.60 \text{ N/m}^2) = 4.2 \times 10^{-6} \text{ m}^3\text{/s}$$

or

$$Q = \left(4.2 \times 10^{-6}\,\frac{\text{m}^3}{\text{s}}\right)\left(\frac{1000 \text{ liter}}{\text{m}^3}\right)\left(\frac{3600 \text{ s}}{\text{hr}}\right) = 15.1 \text{ liter/hr}$$

To check the assumption of laminar flow, the mean velocity is required.

$$V = \frac{Q}{A} = \frac{4.2 \times 10^{-6} \text{ m}^3\text{/s}}{\dfrac{\pi}{4}(0.02 \text{ m})^2} = 0.0134 \text{ m/s}$$

and the kinematic viscosity is $\nu = 5.7 \times 10^{-7}$ m^2/s, from Table B-3, giving a Reynolds number of

$$\mathbf{Re}_D = \frac{VD}{\nu} = \frac{(0.0134 \text{ m/s})(0.02 \text{ m})}{5.7 \times 10^{-7} \text{ m}^2\text{/s}} = 470$$

Since $\mathbf{Re}_D < 2000$, the assumption of laminar flow was valid.

6.20. What is the pressure drop in 100 feet of 1/2-in. by 1-in. rectangular duct when 125 °F water flows at 0.2 fps?

Assuming the duct flows full, the hydraulic diameter is

$$D_h = \frac{4A}{P} = \frac{4(0.5 \text{ in.})(1 \text{ in.})}{3 \text{ in.}} = 0.667 \text{ in.}$$

The Reynolds number is

$$\mathbf{Re} = \frac{VD_h}{\nu} = \frac{\left(0.2\,\frac{\text{ft}}{\text{sec}}\right)\left(\frac{0.667}{12}\,\text{ft}\right)}{6 \times 10^{-6}\,\text{ft}^2/\text{sec}} = 1853$$

where the kinematic viscosity is taken from Figure B-2; the flow is laminar. Hence, the Darcy-Weisbach equation, (6.35), is valid when the friction factor is taken from Table 6-3 at the ratio 0.50, i.e.

$$f \cdot \mathbf{Re} = 62.19 \qquad \text{or} \qquad f = \frac{62.19}{1853} = 0.03356$$

Therefore,

$$\Delta p = f \frac{L}{D_h}\frac{\rho V^2}{2g_c} = (0.03356)\frac{100}{0.667/12}\frac{\left(62.4\,\frac{\text{lbm}}{\text{ft}^3}\right)\left(0.2\,\frac{\text{ft}}{\text{sec}}\right)^2}{2\left(32.2\,\frac{\text{lbm-ft}}{\text{lbf-sec}^2}\right)} = 2.340\,\text{psf}$$

6.21.   Derive an expression for Nusselt number for fully developed laminar flow in a tube with constant heat flux.

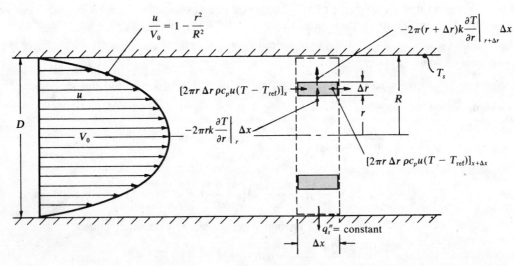

Fig. 6-13

Neglecting axial conduction, Fig. 6-13 shows the radial conduction and the axial enthalpy transport in an annular control element of length Δx and thickness Δr. An energy balance on the annular element gives

$$-2\pi rk\frac{\partial T}{\partial r}\bigg|_r \Delta x + (2\pi r\,\Delta r\,\rho c_p u\theta)_x = -2\pi(r+\Delta r)k\frac{\partial T}{\partial r}\bigg|_{r+\Delta r}\Delta x + (2\pi r\,\Delta r\,\rho c_p u\theta)_{x+\Delta x}$$

where $\theta = T - T_{\text{ref}}$. Rearranging, dividing by $2\pi\,\Delta x\,\Delta r$, and taking the limit as Δr and Δx approach zero, we obtain the governing energy equation

$$\frac{\partial}{\partial r}\left(r\frac{\partial T}{\partial r}\right) = \frac{ur}{\alpha}\frac{\partial T}{\partial x} \tag{1}$$

for constant fluid properties and $\alpha \equiv k/\rho c_p$.

In order to solve the energy equation, an expression for the local velocity u is needed. This is furnished by combining (6.32) with the expression for V_0, giving

$$\frac{u}{V_0} = 1 - \frac{r^2}{R^2} \tag{2}$$

where V_0 is the centerline velocity. We then have

$$\frac{\partial}{\partial r}\left(r\frac{\partial T}{\partial r}\right) = \frac{1}{\alpha}\frac{\partial T}{\partial x}V_0\left(1 - \frac{r^2}{R^2}\right)r \tag{3}$$

For a fully developed thermal field and constant heat flux, $\partial T/\partial x$ is constant. Thus (3) integrates to give

$$r\frac{\partial T}{\partial r} = \frac{1}{\alpha}\frac{\partial T}{\partial x}V_0\left(\frac{r^2}{2} - \frac{r^4}{4R^2}\right) + C_1 \tag{4}$$

$$T = \frac{1}{\alpha}\frac{\partial T}{\partial x}V_0\left(\frac{r^2}{4} - \frac{r^4}{16R^2}\right) + C_1\ln r + C_2 \tag{5}$$

The boundary conditions are

at $r = 0$: T is finite

at $r = R$: $-k\dfrac{\partial T}{\partial r} = q_s'' = $ constant

The first boundary condition requires $C_1 = 0$, whence $C_2 = T_c$, the centerline temperature. The second boundary condition could be used to determine T_c as a function of x (see Problem 6.31), but an explicit expression is not needed here. Therefore,

$$T - T_c = \frac{V_0}{\alpha}\frac{\partial T}{\partial x}\left(\frac{r^2}{4} - \frac{r^4}{16R^2}\right) \tag{6}$$

The bulk temperature T_b, which is needed in order to get the heat transfer coefficient, may be obtained using (6.46):

$$T_b = \frac{\displaystyle\int_0^R V_0\left(1 - \frac{r^2}{R^2}\right)\left[T_c + \frac{V_0}{\alpha}\frac{\partial T}{\partial x}\left(\frac{r^2}{4} - \frac{r^4}{16R^2}\right)\right]r\,dr}{\displaystyle\int_0^R V_0\left(1 - \frac{r^2}{R^2}\right)r\,dr} = T_c + \frac{7}{96}\frac{V_0R^2}{\alpha}\frac{\partial T}{\partial x} \tag{7}$$

The heat transfer coefficient may be determined from the fundamental convection and conduction equations,

$$q = hA(T_s - T_b) = kA\frac{\partial T}{\partial r}\bigg|_R \quad\text{or}\quad h = \frac{k}{T_s - T_b}\frac{\partial T}{\partial r}\bigg|_R \tag{8}$$

Making use of (6) and (7), the heat transfer coefficient is given by

$$h = \frac{\dfrac{k}{4}\dfrac{V_0R}{\alpha}\dfrac{\partial T}{\partial x}}{\left(T_c + \dfrac{3}{16}\dfrac{V_0R^2}{\alpha}\dfrac{\partial T}{\partial x}\right) - \left(T_c + \dfrac{7}{96}\dfrac{V_0R^2}{\alpha}\dfrac{\partial T}{\partial x}\right)} = \frac{24}{11}\frac{k}{R} = \frac{48}{11}\frac{k}{D}$$

or, in terms of Nusselt number,

$$\mathbf{Nu}_D = \frac{hD}{k} = 4.364$$

which is the asymptotic value cited in (6.41).

6.22.   Compare the hydrodynamic and thermal entry lengths of mercury (liquid metal) and a light oil flowing at 3.0 mm/s in a 25.0-mm-diameter smooth tube at a bulk temperature of

75 °C. The pertinent parameters of the fluids at that temperature are: $\nu_{Hg} = 1.0 \times 10^{-7}$ m²/s, $\nu_{oil} = 6.5 \times 10^{-6}$ m²/s; $Pr_{Hg} = 0.019$, $Pr_{oil} = 85$.

The Reynolds numbers based on the tube diameter are:

$$\mathbf{Re}_{Hg} = \frac{VD}{\nu_{Hg}} = \frac{(3.0 \times 10^{-3} \text{ m/s})(25.0 \times 10^{-3} \text{ m})}{1.0 \times 10^{-7} \text{ m}^2/\text{s}} = 750$$

$$\mathbf{Re}_{oil} = \frac{VD}{\nu_{oil}} = \frac{(3.0 \times 10^{-3} \text{ m/s})(25.0 \times 10^{-3} \text{ m})}{6.5 \times 10^{-6} \text{ m}^2/\text{s}} = 11.53$$

The hydrodynamic entry lengths are given by (6.31) as

$$x_{e_{Hg}} \approx (0.05)(750)(25.0 \text{ mm}) = 937.5 \text{ mm}$$

$$x_{e_{oil}} \approx (0.05)(11.53)(25.0 \text{ mm}) = 14.4 \text{ mm}$$

For the thermal entry lengths, (6.43) gives

$$x_{e,t_{Hg}} \approx (0.05)(750)(0.019)(25.0 \text{ mm}) = 17.8 \text{ mm}$$

$$x_{e,t_{oil}} \approx (0.05)(11.53)(85)(25.0 \text{ mm}) = 1225 \text{ mm}$$

The short thermal entry of mercury compared to the hydrodynamic entry gives rise to the assumption often made in solving liquid metal problems—that the flow is uniform across the tube when solving the temperature inlet problem.

6.23. Air at 200 °F and 30 psia flows through a heated 1/2-in.-diameter tube at a velocity of 5 fps. For a constant heat flux at the wall, what is the heat transfer per square foot if the wall is 30 °F above the bulk temperature of the air?

At 200 °F the properties of the air are

$$\rho = \frac{p}{RT} = \frac{30(144) \text{ lbf/ft}^2}{(53.3 \text{ ft-lbf/lbm-°R})(660 \text{ °R})} = 0.123 \text{ lbm/ft}^3$$

$$\mu_m = 1.44 \times 10^{-5} \text{ lbm/ft-sec} \qquad k = 0.018 \text{ Btu/hr-ft-°F} \qquad \mathbf{Pr} = 0.694$$

To ascertain the flow regime, the Reynolds number is necessary:

$$\mathbf{Re}_D = \frac{VD\rho}{\mu_m} = \frac{(5 \text{ ft/sec})(0.5/12 \text{ ft})(0.123 \text{ lbm/ft}^3)}{1.44 \times 10^{-5} \text{ lbm/ft-sec}} = 1780$$

and the flow is laminar.

Using values from Table 6-4 in (6.44), the local heat-transfer coefficient may be obtained at, say, 1 ft from the tube inlet:

$$\mathbf{Nu}_D \equiv \frac{hD}{k} = 4.36 + \frac{(0.036)\{[0.5/(12)(1)](1780)(0.694)\}}{1 + (0.0011)[(0.5/12)(1780)(0.694)]}$$

$$= 4.36 + 1.754 = 6.114$$

or

$$h = \frac{k}{D}(6.114) = \frac{0.0180 \text{ Btu/hr-ft-°F}}{0.5/12 \text{ ft}}(6.114) = 2.641 \text{ Btu/hr-ft}^2\text{-°F}$$

A developing velocity profile was assumed in the calculation, since the first foot of the tube was considered. As a matter of interest, the Nusselt number and heat transfer coefficient 2 ft from the inlet are

$$\mathbf{Nu}_2 = 4.36 + 0.901 = 5.261 \qquad h_2 = 2.273 \text{ Btu/hr-ft}^2\text{-°F}$$

Similarly, at 3 ft,

$$\mathbf{Nu}_3 = 4.36 + 0.606 = 4.966 \qquad h_3 = 2.145 \text{ Btu/hr-ft}^2\text{-°F}$$

Note the decrease in the second term of the equation for Nusselt number, approaching the asymptote 4.364 given by (6.41).

The local heat transfer is given by (6.45) with $\bar{h}$ replaced by h. At 1 ft,

$$\frac{q}{A} = h(T_s - T_b) = \left(2.641\,\frac{\text{Btu}}{\text{hr-ft}^2\text{-°F}}\right)(30\,°\text{F}) = 79.23\,\text{Btu/hr-ft}^2$$

6.24. For a fully developed velocity profile, approximate the length of 0.10-in.-i.d. tube required to raise the bulk temperature of benzene from 60 °F to 100 °F. The tube wall temperature is constant at 150 °F, and the average velocity is 1.6 fps. At an average bulk temperature of

$$T_b = \frac{T_1 + T_2}{2} = \frac{60 + 100}{2} = 80\,°\text{F}$$

the fluid properties are

$$\rho = 54.6\,\text{lbm/ft}^3 \qquad\qquad k = 0.092\,\text{Btu/hr-ft-°F} \qquad c_p = 0.42\,\text{Btu/lbm-°F}$$

$$\mu_m = 3.96 \times 10^{-4}\,\text{lbm/ft-sec} \qquad \textbf{Pr} = 6.5$$

The Reynolds number is

$$\textbf{Re}_D = \frac{VD\rho}{\mu_m} = \frac{\left(1.6\,\dfrac{\text{ft}}{\text{sec}}\right)\left(\dfrac{0.10}{12}\,\text{ft}\right)\left(54.6\,\dfrac{\text{lbm}}{\text{ft}^3}\right)}{3.96 \times 10^{-4}\,\dfrac{\text{lbm}}{\text{ft-sec}}} = 1838$$

which is laminar.

The average Nusselt number is given by (6.44), with the pertinent parameters taken from Table 6-4.

$$\overline{\textbf{Nu}}_D = 3.66 + \frac{(0.0668)[(D/L)\textbf{Re}_D\textbf{Pr}]}{1 + (0.04)[(D/L)\textbf{Re}_D\textbf{Pr}]^{2/3}}$$

Equating $\overline{\textbf{Nu}}_D$ to $\bar{h}D/k$,

$$\frac{\bar{h}\left(\dfrac{0.10}{12}\,\text{ft}\right)}{0.092\,\dfrac{\text{Btu}}{\text{hr-ft-°F}}} = 3.66 + \frac{(0.0668)\left[\left(\dfrac{0.10}{12L}\right)(1838)(6.5)\right]}{1 + (0.04)\left[\left(\dfrac{0.10}{12L}\right)(1838)(6.5)\right]^{2/3}}$$

we have an equation in two unknowns, $\bar{h}$ and L, which can be simplified to give

$$\bar{h} = 40.406 + \frac{\left(\dfrac{73.42}{L}\right)}{1 + (0.04)\left[\dfrac{99.56}{L}\right]^{2/3}} \tag{1}$$

where the units for $\bar{h}$ are Btu/hr-ft² when L has units of ft. Another equation must now be found. Making an energy balance on the fluid, we get

$$q = \dot{m}c_p\,\Delta T_b = \bar{h}\pi DL(T_s - T_b)$$

or

$$\bar{h} = \frac{\rho V(\pi D^2/4)c_p\,\Delta T_b}{\pi DL(T_s - T_b)} = \frac{\rho VDc_p\,\Delta T_b}{4L(T_s - T_b)}$$

$$= \frac{\left(54.6\,\dfrac{\text{lbm}}{\text{ft}^3}\right)\left(1.6\,\dfrac{\text{ft}}{\text{sec}}\right)\left(\dfrac{0.10}{12}\,\text{ft}\right)\left(0.42\,\dfrac{\text{Btu}}{\text{lbm-°F}}\right)(40\,°\text{F})}{4L[(150 - 80)\,°\text{F}]}\left(\frac{3600\,\text{sec}}{\text{hr}}\right)$$

$$= \frac{157.25}{L}\,\text{Btu/hr-ft-°F} \tag{2}$$

Eliminating $\bar{h}$ between (1) and (2), we get

$$\frac{157.25}{L} = 40.406 + \frac{\left(\dfrac{73.42}{L}\right)}{1 + 0.04\left[\dfrac{99.56}{L}\right]^{2/3}}$$

which can be readily solved by trial and error to give $L = 2.64$ ft.

Before leaving this problem, it is instructive to note the influence of the viscosity variation. At the surface temperature $T_s = 150\,°\text{F}$, the viscosity is $\mu_s = 2.60 \times 10^{-4}$ lbm/ft-sec; therefore,

$$\left(\frac{\mu}{\mu_s}\right)^{0.14} = \left(\frac{3.96 \times 10^{-4}}{2.60 \times 10^{-4}}\right)^{0.14} = 1.061$$

Multiplying the second term of (6.44) by this factor and solving simultaneously with the energy balance equation, the resulting length is 2.57 ft, 2.7 percent less than that given by ignoring the viscosity correction.

It is also interesting to note that the asymptotic solution (6.42),

$$\frac{\bar{h}D}{k} = 3.656 \qquad \text{or} \qquad \bar{h} = 40.406$$

coupled with the energy balance equation (2) gives $L = 3.89$ ft. It should be pointed out, however, that this result neglects axial conduction, buoyancy effects and the influence of viscosity variations.

The best answer, therefore, is 2.57 ft.

Supplementary Problems

6.25.   What is the drag on a thin model airfoil, of 1-ft chord length, in a 100 °F airstream moving at 20.0 fps?   Treat the airfoil as a flat plate.   *Ans.* 3.5×10^{-3} lbf/ft

6.26.   Glycerin at 104 °F flows over a flat plate at a free-stream velocity of 25 fps.   What is the thickness of the hydrodynamic boundary layer 2 ft from the leading edge?   *Ans.*   0.069 ft = 21.0 mm

6.27.   What is the thickness of the thermal boundary layer in Problem 6.26, at the same location?
Ans.   1.55 mm

6.28.   Using the linear velocity profile $u/V_\infty = y/\delta$ in the integral momentum equation, verify the expression for boundary layer thickness given in Table 6-1.

6.29.   For the linear velocity profile of Problem 6.28, verify the average skin-friction coefficient shown in Table 6-1.

6.30.   What is the heat loss from both sides of a 3-ft-square plate maintained at 200 °F when air at 150 °F flows parallel to it with a free-stream velocity of 20 fps?   *Ans.*   1585 Btu/hr

6.31.   For the flow of Problem 6.21 show that

$$T_c = T_0 - \frac{4\alpha q_s''}{kV_0R} x$$

where T_0 is the centerline temperature at $x = 0$.

6.32.   Glycerin at 32 °F flows at the rate of 50 ft/sec parallel to both sides of a 1-ft-square, thin flat plate at 68 °F, which is suspended from a balance. (*a*) What drag should be indicated by the balance? (*b*) Determine the heat transfer to the glycerin.   *Ans.*   (*a*) 207.8 lbf; (*b*) 9691 Btu/hr

Chapter 7

Forced Convection: Turbulent Flow

Turbulent flow is characterized by random motion of fluid particles, disrupting the fluid's movement in lamina as discussed in the preceding chapters. It is the most common type of motion, however, because of the minimal disturbances which might cause it to occur.

The time-average equation (5.13), written for velocity, is valid for any quantity ϕ which has a time-average value $\bar{\phi}$ and a fluctuating component ϕ', i.e. $\phi = \bar{\phi} + \phi'$. Thus

$$\bar{\phi} = \lim_{\Delta t \to \infty} \frac{1}{\Delta t} \int_{t_0}^{t} \phi \, dt \tag{7.1}$$

where $\Delta t = t - t_0$. Properties of the time average are:

$$\overline{\phi_1 + \phi_2} = \bar{\phi}_1 + \bar{\phi}_2$$

$$\overline{C\phi_1} = C\bar{\phi}_1$$

$$\overline{\bar{\phi}_1} = \bar{\phi}_1 \tag{7.2}$$

$$\overline{\frac{\partial \phi_1}{\partial s}} = \frac{\partial \bar{\phi}_1}{\partial s}$$

$$\overline{\phi'_1} = 0$$

where C is independent of t and s is any spatial coordinate. In addition, it is almost always the case that $\overline{\phi'_1 \phi'_2} \neq 0$ if ϕ_1 and ϕ_2 are turbulent flow properties.

7.1 EQUATIONS OF MOTION

By use of the boundary layer concept the general equations of motion, called the *Navier-Stokes equations* after their formulators, can be simplified to the point of being solved. The x-direction momentum equation for incompressible, laminar, boundary layer flow over a flat plate was derived in Problem 6.1. Since it is a simplified form of the more general x-direction Navier-Stokes equation, we shall extend it to the case of turbulent flow, i.e.

laminar: $\quad u\dfrac{\partial u}{\partial x} + v\dfrac{\partial u}{\partial y} = -\dfrac{g_c}{\rho}\dfrac{dp}{dx} + \nu\dfrac{\partial^2 u}{\partial y^2}$

$$\tag{7.3}$$

turbulent: $\quad (\bar{u} + u')\dfrac{\partial}{\partial x}(\bar{u} + u') + (\bar{v} + v')\dfrac{\partial}{\partial y}(\bar{u} + u')$

$$= -\frac{g_c}{\rho}\frac{d}{dx}(\bar{p} + p') + \nu\frac{\partial^2}{\partial y^2}(\bar{u} + u')$$

The instantaneous quantities in the laminar equation have been replaced by the sum of their average and fluctuating components in the turbulent equation. Inherent in this move is the assumption that the Navier-Stokes equations are valid for turbulent flow.

Along with the momentum equations, we may consider the two-dimensional incompressible continuity equation:

laminar:
$$\frac{\partial u}{\partial x} + \frac{\partial v}{\partial y} = 0$$

(7.4)

turbulent:
$$\frac{\partial}{\partial x}(\bar{u} + u') + \frac{\partial}{\partial y}(\bar{v} + v') = 0$$

Combining the turbulent equations (7.3) and (7.4), taking the time average of the resultant equation, and applying the rules (7.2), we obtain

$$\bar{u}\frac{\partial \bar{u}}{\partial x} + \bar{v}\frac{\partial \bar{u}}{\partial y} = -\frac{g_c}{\rho}\frac{d\bar{p}}{dx} + \nu\frac{\partial^2 \bar{u}}{\partial y^2} - \left(\frac{\partial \overline{u'u'}}{\partial x} + \frac{\partial \overline{v'u'}}{\partial y}\right)$$

(7.5)

which is the x-direction equation of motion for a viscous incompressible fluid with negligible body forces. This equation can be expected to hold in the turbulent portion of the boundary layer shown in Fig. 7-1(a). The laminar portion at the leading edge and the laminar sublayer are governed by the relations developed in Chapter 6. This chapter deals with the turbulent regime, which is often idealized as shown in Fig. 7-1(b), in which the critical length x_c is taken as that distance required to produce a Reynolds number of 500,000 (although this may vary from 300,000 to 2,800,000 depending upon such factors as surface roughness and free-stream turbulence).

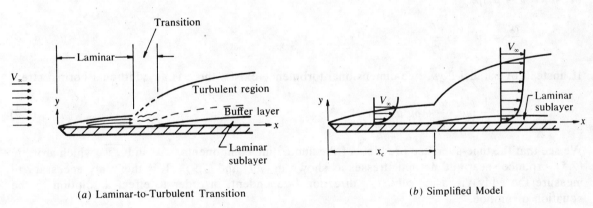

(a) Laminar-to-Turbulent Transition   (b) Simplified Model

Fig. 7-1

Upon comparing (7.5) with the laminar equation (7.3), we note that an additional term occurs in the equation for turbulent flow. The three terms on the right-hand side of (7.5) express the effects of pressure (normal stress), viscosity (shear stress), and turbulent fluctuations (*Reynolds*, or *apparent*, *stress*). The last term produces an apparent volume force

$$f_x\big|_{\text{apparent}} = -\frac{\rho}{g_c}\left(\frac{\partial \overline{u'u'}}{\partial x} + \frac{\partial \overline{v'u'}}{\partial y}\right) \qquad (7.6)$$

which will be interpreted in the next subsection.

Eddy Viscosity

Assume turbulent motion along the surface of Fig. 7-2. At a typical plane parallel to the surface, *A-A*, a lump of fluid designated 1 is

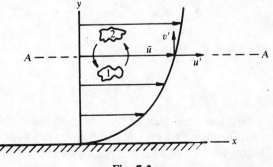

Fig. 7-2

moved by turbulent fluctuations to a region of increased velocity. It is replaced by lump 2 moving downward to a region of lower velocity. Since momentum is the product of mass and velocity, a change in momentum results. This change can be evaluated quantitatively by considering the velocity components

$$u = \bar{u} + u' \qquad v = 0 + v'$$

where $\bar{v} = 0$ results from zero mean flow in the y-direction. The instantaneous mass flow per unit area across plane A-A is $\rho v'$. Multiplying by the x-velocity deviation u' gives the rate of momentum change per unit area $\rho v' u'$, whose average value, $\overline{\rho v' u'}$, is negative. Thus there is a shear stress $-(\rho/g_c)\overline{v'u'}$ over and above the laminar shear $(\mu_m/g_c)(\partial\bar{u}/\partial y)$, i.e.

$$\tau = \tau_{\text{lam}} + \tau_{\text{turb}} = \frac{1}{g_c}\left(\mu_m \frac{\partial\bar{u}}{\partial y} - \rho\overline{v'u'}\right) \tag{7.7}$$

This result is not readily usable, except through measurement of the fluctuating components, since the fluctuating velocities are not related to fluid properties. It is convenient to assume that the viscosity effect is increased because of the turbulence and define an *eddy viscosity* ϵ such that

$$\tau = \frac{1}{g_c}(\mu_m + \rho\epsilon)\frac{\partial\bar{u}}{\partial y}$$

or, since $\nu = \mu_m/\rho$,

$$\frac{\tau g_c}{\rho} = (\nu + \epsilon)\frac{\partial\bar{u}}{\partial y} \tag{7.8}$$

If, instead of parallel flow, two-dimensional turbulence exists, there is an additional normal stress

$$\tau_x = \frac{1}{g_c}\left(\mu_m \frac{\partial\bar{u}}{\partial x} + \rho\overline{u'u'}\right) \tag{7.9}$$

We see that the time-averaged products of the fluctuating components, $\overline{v'u'}$ and $\overline{u'u'}$, which arose in (7.5), produce shear and normal stresses as shown in (7.7) and (7.9). It is then only necessary to measure the eddy viscosity, which is directional-dependent, in order to effect a solution to the equation of motion.

Eddy Diffusivity

An argument analogous to that of the preceding subsection can be made for the influence of turbulence on heat transfer. If a temperature fluctuation exists in the flow field of Fig. 7-2, so that $T = \bar{T} + T'$, the heat transfer per unit area would be given by

$$q'' = q''_{\text{lam}} + q''_{\text{turb}} = -k\frac{\partial\bar{T}}{\partial y} + \rho c_p \overline{v'T'} = -(k + \rho c_p \epsilon_H)\frac{\partial\bar{T}}{\partial y}$$

or, since $\alpha = k/\rho c_p$,

$$\frac{q''/c_p}{\rho} = -(\alpha + \epsilon_H)\frac{\partial\bar{T}}{\partial y} \tag{7.10}$$

where the time-averaged product of the fluctuating velocity v' and fluctuating temperature T' has been replaced by an *eddy diffusivity for heat*, ϵ_H. Equations (7.8) and (7.10) are now of identical form, which suggests the analogy between momentum transfer and heat transfer of the following section.

7.2 HEAT TRANSFER AND SKIN FRICTION: REYNOLDS' ANALOGY

In the vicinity of a surface the fluid is essentially stationary, requiring that the transfer of heat take place by conduction. Such is the case in the laminar sublayer, where the heat transfer and shear stress are in the ratio

$$\left(\frac{q''}{\tau}\right)_{\text{lam}} = \frac{-k\dfrac{\partial T}{\partial y}}{\dfrac{\mu_m}{g_c}\dfrac{\partial u}{\partial y}} = -\frac{kg_c}{\mu_m}\frac{dT}{du}\bigg|_{x=\text{const}} \tag{7.11}$$

At some distance from the surface, where the random fluctuations transport momentum and heat, the turbulent shear stress and heat transfer (at fixed x) have the instantaneous values

$$\tau_{\text{turb}} = -\frac{\rho}{g_c}\,v'\,\Delta u \qquad q''_{\text{turb}} = \rho c_p v'\,\Delta T$$

giving, in the limit,

$$\left(\frac{q''}{\tau}\right)_{\text{turb}} = -c_p g_c \frac{dT}{du}\bigg|_{x=\text{const}} \tag{7.12}$$

A comparison of (7.11) and (7.12) shows that when $k/\mu_m = c_p$, i.e. when

$$\mathbf{Pr} \equiv \frac{\mu_m c_p}{k} = 1$$

then the single equation

$$\frac{q''}{\tau} = -c_p g_c \frac{dT}{du}\bigg|_{x=\text{const}} \tag{7.13}$$

is valid through the entire boundary layer. *Reynolds' analogy* results from making the approximation that T and u change at proportional rates through the boundary layer. Then each side of (7.13) is a constant, i.e.

$$\frac{q''}{\tau} = \text{constant} = \frac{q''_s}{\tau_s} = -c_p g_c \frac{dT}{du}\bigg|_{x=\text{const}}$$

where q''_s and τ_s are the surface values. Integrating through the boundary layer and applying the known surface boundary condition ($u = 0$, $T = T_s$), we have:

$$-\frac{1}{c_p g_c}\frac{q''_s}{\tau_s}\int_0^{} du = \int_{T_s}^{} dT \tag{7.14}$$

The condition $\mathbf{Pr} = 1$ necessary for (7.14) is approximated by many real fluids.

The upper limits of integration in Reynolds' analogy will depend upon the flow configuration. The following sections will complete this relation between heat transfer and fluid friction for external (flat plate) and internal (pipe) flow configurations.

7.3 FLOW OVER A FLAT PLATE

Two unknowns, q''_s and τ_s, are present in (7.14). We shall at first concentrate on determining τ_s for the case of no temperature variation (isothermal). Then we shall use the isothermal result and (7.14) to determine q''_s for the nonisothermal case. The flow configuration is shown in Fig. 7-3.

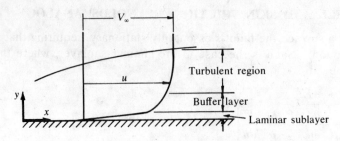

Fig. 7-3

Isothermal Flow

The integral momentum equation, (6.14), is equally valid for turbulent flow since no flow assumption was required when choosing the control volume. Therefore, in the incompressible case,

$$\tau_s = \frac{\rho}{g_c}\frac{d}{dx}\left[\int_0^\delta (V_\infty - \bar{u})\bar{u}\, dy\right] \tag{7.15}$$

Observe that the mean turbulent velocity $\bar{u}$ is used over the whole range of integration; this is tantamount to neglecting the laminar sublayer and the buffer zone.

In the fully turbulent portion of the boundary layer the velocity increases approximately as the one-seventh power of distance from the wall, giving

$$\frac{\bar{u}}{V_\infty} = \left(\frac{y}{\delta}\right)^{1/7} \tag{7.16}$$

which is known as the *one-seventh-power law*. For the same regime, Blasius experimentally deduced that the shear stress is related to the boundary layer thickness by

$$\tau_s = (0.0225)\frac{\rho V_\infty^2}{g_c}\left(\frac{\nu}{V_\infty\delta}\right)^{1/4} \tag{7.17}$$

for Reynolds numbers ranging from 5×10^5 to 10^7. Using (7.16) and (7.17) in (7.15), the boundary layer thickness and the local skin-friction coefficient are found to be

$$\blacksquare \qquad \frac{\delta}{x} = \frac{0.376}{(V_\infty x/\nu)^{1/5}} = \frac{0.376}{\mathbf{Re}_x^{1/5}} \tag{7.18}$$

$$\blacksquare \qquad c_f \equiv \frac{\tau_s}{\rho V_\infty^2/2g_c} = \frac{0.0576}{\mathbf{Re}_x^{1/5}} \tag{7.19}$$

and the average skin-friction coefficient is

$$C_f \equiv \frac{F_f}{(\rho V_\infty^2/2g_c)L} = \frac{0.072}{\mathbf{Re}_L^{1/5}} \tag{7.20}$$

where F_f is the friction drag per unit width.

These equations make no allowance for the laminar boundary layer, $0 < x < x_c$, which precedes the turbulent portion. They are quite accurate, however, beyond the critical length x_c when the length is taken as if the turbulent boundary layer begins at the leading edge of the plate. Both laminar and turbulent drag can be accounted for by subtracting the turbulent drag for the critical length and adding the laminar drag for that portion, i.e.

$$C_f = \frac{0.072}{\mathbf{Re}_L^{1/5}} - \frac{0.072}{\mathbf{Re}_{x_c}^{1/5}} \frac{x_c}{L} + \frac{1.328}{\sqrt{\mathbf{Re}_{x_c}}} \frac{x_c}{L} \tag{7.21}$$

where the last term is from the Blasius solution of Table 6-1. For a critical Reynolds number of 5×10^5, (7.21) simplifies to

$$\blacksquare \qquad C_f = \frac{0.072}{\mathbf{Re}_L^{1/5}} - (0.00334) \frac{x_c}{L} \tag{7.22}$$

Flow With Heat Transfer

Reynolds' analogy—flat plate. At the edge of the boundary layer $u = V_\infty$ and $T = T_\infty$, which permits the upper limits of integration to be added to (7.14). The integration yields

$$-\frac{1}{c_p g_c} \frac{q_s''}{\tau_s} V_\infty = T_\infty - T_s \qquad \text{or} \qquad \frac{-q_s''}{T_\infty - T_s} \frac{1}{\rho c_p V_\infty} = \frac{1}{2} \frac{\tau_s}{\rho V_\infty^2 / 2 g_c}$$

which, by use of (6.11) and (6.17), can be put in the form

$$\frac{h_x}{\rho c_p V_\infty} = \frac{c_f}{2} \tag{7.23}$$

The dimensionless group of terms on the left side of (7.23), called the *Stanton number*, **St**, is the Nusselt number divided by the product of the Reynolds and Prandtl numbers, i.e.

$$\frac{\mathbf{Nu}_x}{\mathbf{Re}_x \mathbf{Pr}} \equiv \mathbf{St}_x = \frac{c_f}{2} \tag{7.24}$$

which is Reynolds' analogy for the flat plate, relating the skin friction to the heat transfer. A. P. Colburn showed that the analogy may be modified to pertain to Prandtl numbers ranging from 0.6 to 50 by

$$\mathbf{j}_H \equiv \mathbf{St}_x \mathbf{Pr}^{2/3} = \frac{c_f}{2} \tag{7.25}$$

where $\mathbf{j}_H$ is known as the *Colburn factor*, or simply the *j-factor*, for heat transfer.

Taking average values over $0 < x < L$, we obtain

$$\frac{\bar{h}}{\rho c_p V_\infty} = \frac{C_f}{2} \tag{7.26}$$

$$\blacksquare \qquad \bar{\mathbf{j}}_H = \overline{\mathbf{St}} \, \mathbf{Pr}^{2/3} = \frac{C_f}{2} \tag{7.27}$$

The appropriate expression for the average skin-friction coefficient, either (7.20) or (7.21) or (7.22), may be substituted into the Colburn equation (7.27) to get the average convective heat-transfer coefficient. The third choice gives (for a critical Reynolds number of 500,000)

$$\blacksquare \qquad \overline{\mathbf{Nu}} \equiv \frac{\bar{h}L}{k} = \mathbf{Pr}^{1/3}(0.036 \mathbf{Re}_L^{0.8} - 836) \tag{7.28}$$

Thus far, we have ignored the laminar sublayer and the buffer zone. Von Kármán, including both of these, found the local Stanton number for flow over a plane surface to be

$$\mathbf{St}_x \equiv \frac{\mathbf{Nu}_x}{\mathbf{Re}_x \mathbf{Pr}} = \frac{c_f/2}{1 + 5\sqrt{c_f/2} \left[(\mathbf{Pr} - 1) + \ln\left(\frac{5\mathbf{Pr} + 1}{6}\right) \right]} \tag{7.29}$$

which reduces to Reynolds' analogy for $\mathbf{Pr} = 1$. If the one-seventh-power friction law, (7.19), is introduced into (7.29), the result is

$$\blacksquare \qquad \mathbf{Nu}_x \equiv \frac{h_x x}{k} = \frac{(0.0288)\,\mathbf{Re}_x^{0.8}\,\mathbf{Pr}}{1 + (0.849)\,\mathbf{Re}_x^{-0.10}\left[(\mathbf{Pr} - 1) + \ln\left(\dfrac{5\mathbf{Pr} + 1}{6}\right)\right]} \qquad (7.30)$$

In all the equations of this section the pertinent parameters should be evaluated at the film temperature, $T_f = (T_s + T_\infty)/2$.

7.4 FLOW IN PIPES

Isothermal Flow

Smooth pipes. From a series of experiments, Nikuradse concluded that the velocity distribution for turbulent flow in smooth pipes is of the power-law form, i.e.

$$\frac{u}{V_{\max}} = \left(\frac{y}{R}\right)^{1/n} \qquad (7.31)$$

where u is the local time-average velocity, $V_{\max}$ is the time-average velocity at the centerline, R is the pipe radius and $y = R - r$ is the distance from the pipe wall. (For simplicity we shall omit the overbar denoting the time average.) Table 7-1 gives the values of n for several Reynolds numbers. Note that for $\mathbf{Re} = 1.1 \times 10^5$ the exponent is 1/7, making the profile of identical form to that of (7.16) for flow over a flat plate.

Table 7-1

Re	n
4.0×10^3	6.0
2.3×10^4	6.6
1.1×10^5	7.0
1.1×10^6	8.8
2.0×10^6	10.0
3.2×10^6	10.0

The Darcy-Weisbach equation (6.36),

$$h_L \equiv \frac{\Delta p}{\rho} = f\,\frac{L}{D}\,\frac{V^2}{2g_c}$$

is equally valid for turbulent flow, but the friction factor f must be determined experimentally, rather than analytically as for laminar flow. In this equation V is the average velocity over the cross section, which for the power-law profile is expressed by

$$\frac{V}{V_{\max}} = \frac{2n^2}{(2n + 1)(n + 1)} \qquad (7.32)$$

(see Problem 7.6).

For $10,000 < \mathbf{Re}_D < 100,000$, the friction factor f is well represented by

$$f = (0.184)\,\mathbf{Re}_D^{-0.2} \qquad (7.33)$$

beyond the dimensionless distance given by H. Latzko:

$$\left.\frac{L}{D}\right|_c = (0.623)\,\mathbf{Re}_D^{1/4} \tag{7.34}$$

This is the distance required in turbulent flow for the friction factor to become constant, which is much less than the 40 to 50 diameters required for the turbulent velocity profile to develop.

Commercial (rough) pipes.   In rough pipes, where surface imperfections extend beyond the laminar sublayer, the friction factor f depends upon both $\mathbf{Re}_D$ and the *roughness height e.*   Figure 7-4, commonly referred to as *Moody's diagram*, is a plot of friction factor versus Reynolds number, with the relative roughness e/D as parameter.   Included are the results for hydraulically smooth pipes, discussed earlier in this section, as well as the straight-line, laminar flow relation (6.37).   In the region of complete turbulence (high Reynolds number and/or large e/D) the friction factor depends predominantly upon the relative roughness, as shown by the flatness of the curves. Typical values of e for various kinds of new (i.e. rougher for being unworn) commercial piping are given in Table 7-2.

Pipe fittings and minor losses.   In piping systems having short lengths of pipes the pressure drop is sometimes more significant in the fittings than in the straight piping itself.   Such *minor losses*, although they are not always small as the name might suggest, are due primarily to flow separation and form drag (Section 7.5).

For convenience, head-loss experimental data are often expressed in the form of dimensionless *loss coefficients* k_L, where

$$h_L = k_L \frac{V^2}{2g_c} \tag{7.35}$$

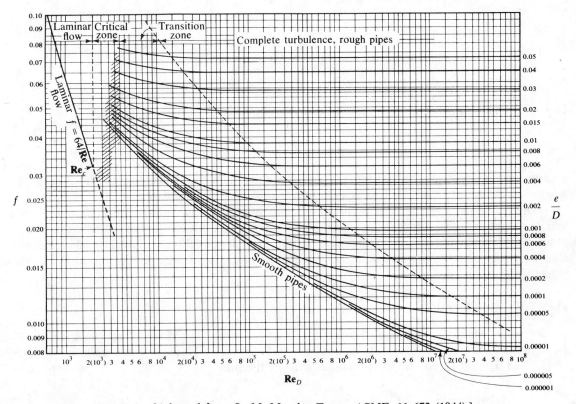

Fig. 7-4.   [Adapted from L. M. Moody, *Trans. ASME*, **66**:672 (1944).]

Table 7-2

Type	e, in.
Drawn tubing	0.00006
Brass, lead, glass, spun cement	≈ 0.0003
Commercial steel or wrought iron	0.0018
Cast iron (asphalt dipped)	0.0048
Galvanized iron	0.0060
Wool stave	0.0072 to 0.036
Cast iron (uncoated)	0.0102
Concrete	0.012 to 0.12
Riveted steel	0.036 to 0.36

Table 7-3

Item	k_L
Angle valve, fully open	3.1 to 5.0
Ball check valve, fully open	4.5 to 7.0
Gate valve, fully open	0.19
Globe valve, fully open	10
Swing check valve, fully open	2.3 to 3.5
Regular-radius elbow, screwed	0.9
Flanged	0.3
Long-radius elbow, screwed	0.6
Flanged	0.23
Close return bend, screwed	2.2
Flanged return bend, two elbows, regular radius	0.38
Long radius	0.25
Standard tee, screwed, flow through run	0.6
Flow through side	1.8

Use of an *equivalent length* of pipe,

$$L_{\text{eq}} \equiv \frac{k_L D}{f} \tag{7.36}$$

permits minor losses to be accounted for by adding equivalent lengths to the straight pipe length. In any case, the head loss is incorporated in the energy equation, (5.31).

Table 7-3 gives loss coefficients for some common valves and fittings. The values given are valid only when fully turbulent flow exists upstream. Since 40 to 50 pipe diameters is required downstream of any obstruction for the flow to become fully developed, when fittings or valves are close together actual loss coefficients are less than those tabulated, resulting in conservative calculated values for pressure drop.

Flow With Heat Transfer

Reynolds' analogy—tube flow. Returning to (7.14) for the case of turbulent flow in a tube (Fig. 7-5), we have as the upper limits of integration $u = V$, $T = T_b$. Here, V is the mean velocity [as in (7.32)] and T_b is the bulk, or mean, temperature of the fluid:

$$T_b = \frac{T_i + T_o}{2} \tag{7.37}$$

where T_i and T_o are the average fluid temperatures at the inlet and outlet.   Integrating (7.14), we get, analogous to (7.23),

$$\frac{h_x}{\rho c_p V} = \frac{\tau_s}{\rho V^2/g_c} \tag{7.38}$$

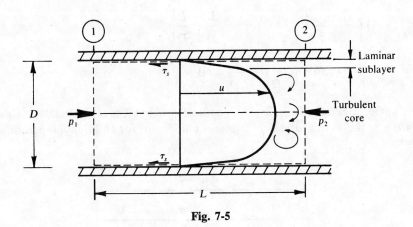

Fig. 7-5

A simple force balance on a cylindrical control volume of length L and diameter D gives

$$\tau_s = \frac{(p_1 - p_2)D}{4L}$$

which can be combined with (6.35) to give

$$\tau_s = f\frac{\rho V^2}{8g_c} \tag{7.39}$$

Substituting (7.39) into (7.38), we get

■ $$\mathbf{St}_x \equiv \frac{h_x}{\rho c_p V} = \frac{f}{8} \tag{7.40}$$

The same form is also valid for average values, i.e.

$$\overline{\mathbf{St}} = \frac{f}{8} \tag{7.41}$$

Equations (7.40) and (7.41) are, of course, restricted to **Pr** = 1. The Colburn modification is applicable for the case of internal tube flow, giving

■ $$\bar{j}_H = \overline{\mathbf{St}}\,\mathbf{Pr}^{2/3} = \frac{f}{8} \tag{7.42}$$

for fluids with Prandtl numbers from 0.5 to 100.
Substituting the expression (7.33) for f into (7.42) yields a working relation for Nusselt number:

■ $$\overline{\mathbf{Nu}_D} \equiv \frac{\bar{h}D}{k} = (0.023)\mathbf{Re}_D^{0.8}\mathbf{Pr}^{1/3} \tag{7.43}$$

valid for 10,000 < **Re**$_D$ < 100,000, 0.5 < **Pr** < 100 and L/D > 60.   Fluid properties, except specific heat, are evaluated at the average film temperature,

$$T_f = \frac{T_s + T_b}{2} \tag{7.44}$$

while specific heat is evaluated at the bulk temperature T_b.

Entrance-region modifications. For short tubes Latzko recommended the following relations for the average heat-transfer coefficient in the entry region, $\bar{h}_e$:

$$\frac{\bar{h}_e}{\bar{h}} = (1.11) \left[\frac{\mathbf{Re}_D^{1/5}}{(L/D)^{4/5}} \right]^{0.275} \quad \text{for} \quad \frac{L}{D} < \frac{L}{D} \bigg|_c \tag{7.45}$$

$$\frac{\bar{h}_e}{\bar{h}} = 1 + \frac{C}{L/D} \quad \text{for} \quad \frac{L}{D} \bigg|_c < \frac{L}{D} < 60 \tag{7.46}$$

Table 7-4 gives some selected values for the coefficient C in (7.46), for $26{,}000 < \mathbf{Re}_D < 56{,}000$. In (7.45) and (7.46) $\bar{h}$ is the asymptotic heat-transfer coefficient for fully developed flow and $(L/D)_c$ is given by (7.34).

Table 7-4

Inlet Configuration	C
Bell-mouthed with screen	1.4
Calming section, $L/D = 11.2$	1.4
$L/D = 2.8$	3.0
45° bend	5.0
90° bend	7.0

Design equations. A relation in which the fluid properties are evaluated at the bulk temperature T_b, making it much easier to use than (7.43), is the widely used *Dittus-Boelter equation*:

■ $$\overline{\mathbf{Nu}}_D \equiv \frac{\bar{h}D}{k} = (0.023)\,\mathbf{Re}_D^{0.8}\,\mathbf{Pr}^n \tag{7.47}$$

where

$$n = \begin{cases} 0.4 & \text{for heating the fluid} \\ 0.3 & \text{for cooling the fluid} \end{cases}$$

This equation is valid for $10{,}000 < \mathbf{Re}_D < 120{,}000$, $0.7 < \mathbf{Pr} < 120$, and $L/D > 60$. Use of this equation should be limited to cases where the pipe surface temperature and the bulk fluid temperature differ by no more than 10 °F for liquids and 100 °F for gases.

For higher Prandtl numbers, $0.7 < \mathbf{Pr} < 16{,}700$, and bigger temperature differences the *Sieder-Tate equation*, which accounts for large changes in viscosity, is recommended.

■ $$\overline{\mathbf{Nu}}_D \equiv \frac{\bar{h}D}{k} = (0.023)\,\mathbf{Re}_D^{0.8}\,\mathbf{Pr}^{1/3} \left(\frac{\mu_b}{\mu_s} \right)^{0.14} \tag{7.48}$$

Valid for $\mathbf{Re}_D > 10{,}000$ and $L/D > 60$, this equation may be used for both heating and cooling. Except for μ_s, which is evaluated at the surface temperature, all properties are evaluated at the bulk temperature.

The equations of this section are valid for noncircular ducts when the duct diameter D is replaced by the hydraulic diameter D_h given by (6.40).

7.5 EXTERNAL FLOW OVER SUBMERGED BODIES

Section 7.3 covered a special case of external flow in which the boundary layer remains attached to the surface and grows throughout its length.   The results were obtained for a uniform velocity field, V_∞ = constant, producing fully developed flow, $dp/dx = 0$.   Such is not the case when the fluid path is oblique to the surface.

For the blunt-nosed body of Fig. 7-6(a) the boundary layer detaches, separating from the surface at the upstream end and producing a *wake*.   The boundary layer remains attached to the round-nosed body, Fig. 7-6(b).   The airfoil of Fig. 7-7 experiences accelerating flow from A to B and deceleration from B to the trailing edge.   At point C, known as the *separation point*, the velocity gradient is zero, i.e.

$$\left.\frac{\partial u}{\partial y}\right|_{y=0} = 0$$

while the flow is actually reversed between C and D.   The corresponding pressure gradients shown in the figure come from Bernoulli's equation of potential flow theory.   Separation can only occur in decelerating flow.   Beyond the point of separation the pressure gradient is said to be *adverse*.

Isothermal Flow

The drag on submerged bodies, such as those shown in Figs. 7-6 and 7-7, is made up of two components: *skin-friction drag*, F_f, and *form*, or *profile*, *drag*, F_p.

$$F_D = F_f + F_p \tag{7.49}$$

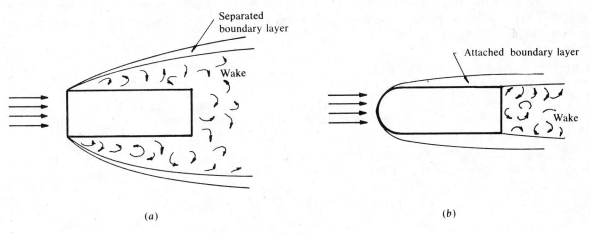

(a)   (b)

Fig. 7-6

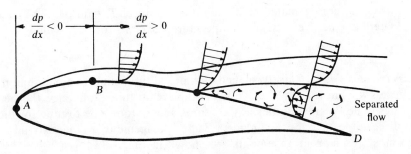

Fig. 7-7

Profile drag is not as amenable to analysis as skin-friction drag, particularly in the case of turbulent flow. Because of this, it is common practice in engineering to evaluate the total drag in terms of an empirical drag coefficient C_D:

■
$$F_D = C_D \left(\frac{\rho V_\infty^2}{2 g_c} \right) A \tag{7.50}$$

where A is the projected frontal area. Drag coefficients for flow over some common submerged bodies are given in Table 7-5. Flow over cylinders is laminar for Reynolds numbers up to approximately 5×10^5, compared to about 3×10^5 for spheres. The drag for a streamlined body is strongly dependent upon Reynolds number, but for a bluff body is essentially constant over a wide range of Reynolds numbers.

Flow with Heat Transfer

Many heat exchangers are designed to transfer heat from cylinders subjected to cross flow. Pebble-bed heaters involve heat exchange in flow over spherical, or near-spherical, particles. Convective heat exchange in such cases is complicated by the flow separation discussed in the preceding subsection. Reynolds' analogy, which permits the calculation of heat transfer from the skin-friction factor, does not apply, since the profile drag may be substantially larger than the shear drag. Hence, most heat transfer calculations for problems of this type are based on empirical correlation equations. The most common cases are considered in the following paragraphs.

Single cylinder in cross flow. The local heat-transfer coefficient (Nusselt number) for a single cylinder subjected to the cross flow of air at a uniform approach temperature and velocity varies widely from point to point. The average Nusselt number is well represented by the empirical correlation

$$\overline{\mathbf{Nu}_{Df}} \equiv \frac{\bar{h} D}{k_f} = C \, \mathbf{Pr}_f^{1/3} \, \mathbf{Re}_{Df}^n \tag{7.51}$$

where the constants for the various Reynolds number ranges are given in Table 7-6. As indicated by the subscript f, the parameters are evaluated at the film temperature, $T_f = (T_\infty + T_s)/2$.

Single sphere. For Reynolds numbers less than unity and for $\mathbf{Pr} = 1$, $\overline{\mathbf{Nu}_D} = 2$. For more common flow conditions, the following relations are recommended.

gases:
$$\frac{\bar{h} D}{k_f} = (0.37)(\mathbf{Re}_{Df})^{0.6} \qquad (17 < \mathbf{Re}_{Df} < 70{,}000) \tag{7.52}$$

liquids:
$$\frac{\bar{h} D}{k_\infty} = [1.2 + (0.53)(\mathbf{Re}_{D\infty})^{0.54}] \, \mathbf{Pr}^{0.3} \left(\frac{\mu_\infty}{\mu_s} \right)^{0.25} \qquad (1 < \mathbf{Re}_{D\infty} < 200{,}000) \tag{7.53}$$

The parameters of (7.52) are evaluated at the film temperature, $T_f = (T_\infty + T_s)/2$, whereas those of (7.53) are evaluated at the free-stream temperature, except for μ_s, which is at the surface temperature.

Tube bundles in cross flow. Closely spaced cylindrical tube bundles are commonly used in heat exchangers. In this situation the wakes from the upstream tubes influence the heat transfer and flow characteristics at downstream tubes. Variations occur from tube to tube for the first 10 tubes, after which no discernible changes take place. Tube arrangement is obviously another influencing factor; the two most common arrangements are shown in Fig. 7-8.

The results of several investigators were evaluated by E. D. Grimson, who found that the

Table 7-5

Configuration	L/D	$\mathbf{Re}_D = V_\infty D/\nu$	C_D
Circular cylinder, axis perpendicular to the flow $V_\infty \longrightarrow \bigcirc \updownarrow D$	1 5 20 ∞ 5 ∞	10^5 $>5 \times 10^5$	0.63 0.74 0.90 1.20 0.35 0.33
Circular cylinder, axis parallel to the flow $V_\infty \longrightarrow \overleftrightarrow{L} \updownarrow D$	0 1 2 4 7	$>10^3$	1.12 0.91 0.85 0.87 0.99
Elliptical cylinder $\quad\bigcirc\quad (2:1)^*$ $V_\infty \longrightarrow \bigcirc\quad (4:1)^*$ $\longrightarrow \bigcirc\quad (8:1)^*$		4×10^4 10^5 2.5×10^4 to 10^5 2.5×10^4 2×10^5	0.6 0.46 0.32 0.29 0.20
Airfoil $\qquad\qquad (1:3)^\dagger$	∞	4×10^4	0.07
Rectangular plate, normal to the flow $\quad L$ = length $\quad D$ = width	1 5 20 ∞	$>10^3$	1.16 1.20 1.50 1.90
Square cylinder $\quad V_\infty \longrightarrow \square \updownarrow D$ $\longrightarrow \diamondsuit D$		3.5×10^4 10^4 to 10^5	2.0 1.6
Triangular cylinder $120°$ $V_\infty \quad 60°$ $30°$		$>10^4$ $>10^5$	2.0 1.72 2.20 1.39 1.80 1.0
Hemispherical shell $\quad V_\infty \longrightarrow D$		$>10^3$ 10^3 to 10^5	1.33 0.4
Circular disk, normal to the flow $\quad V_\infty \longrightarrow D$		$>10^3$	1.12
Tandem disks $\quad V_\infty \longrightarrow \|\| D$	0 1 2 3	$>10^3$	1.12 0.93 1.04 1.54

*Ratio of major axis to minor axis
†Ratio of chord to span at zero angle of attack

Table 7-6. (From M. Jakob, *Heat Transfer*, Vol. I, John Wiley & Sons, Inc., New York, 1959, and J. G. Knudsen and D. L. Katz, *Fluid Dynamics and Heat Transfer*, McGraw-Hill Book Co., New York, 1958, by permission.)

Configuration	$\mathbf{Re}_{Df}$	C	n
$V_\infty \rightarrow$ ◯ D	0.4 to 4	0.989	0.330
	4 to 40	0.911	0.385
	40 to 4,000	0.683	0.466
	4,000 to 40,000	0.193	0.618
	40,000 to 400,000	0.0266	0.805
$\rightarrow$ ◇ D	2,500 to 7,500	0.261	0.624
	5,000 to 100,000	0.222	0.588
$\rightarrow$ ▢ D	2,500 to 8,000	0.160	0.699
	5,000 to 100,000	0.092	0.675
$\rightarrow$ ⬡ D	5,000 to 19,500	0.144	0.638
	19,500 to 100,000	0.035	0.782
$\rightarrow$ ⬡ D	5,000 to 100,000	0.138	0.638
$\rightarrow$ | D	4,000 to 15,000	0.205	0.731
$\rightarrow$ ⬭ D	2,500 to 15,000	0.224	0.612
$\rightarrow$ ⬯ D	3,000 to 15,000	0.085	0.804

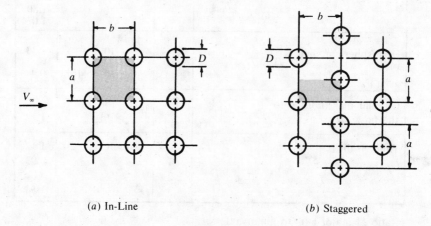

(*a*) In-Line   (*b*) Staggered

Fig. 7-8

average heat-transfer coefficient for bundles at least 10 tubes deep in the flow direction is given by

$$\frac{\bar{h}D}{k_f} = C_1 \, (\mathbf{Re}_{max})^n \qquad\qquad (7.54)$$

where $\mathbf{Re}_{max} = V_{max}D/\nu_f$ and the constants, C_1 and n, are given in Table 7-7.   The maximum velocity, V_{max}, occurs at the minimum flow passage.   Referring to the shaded unit cells of Fig. 7-8, we see that the minimum passage for in-line bundles is $a - D$, so that, by continuity,

$$V_{max} = \frac{V_{\infty}a}{a - D}$$

For staggered bundles, the minimum passage is the smaller of $(a - D)/2$ and $\sqrt{(a/2)^2 + b^2} - D$ (the diagonal), and V_{max} is $(V_{\infty}a/2)$ divided by this smaller value.

For tube bundles having less than 10 tubes in the flow direction, Kays and Lo obtained the correction coefficients given in Table 7-8, where $\bar{h}_{10}$ is given by (7.54).

Table 7-7

$\dfrac{b}{D}$	a/D							
	1.25		1.5		2		3	
	C_1	n	C_1	n	C_1	n	C_1	n
In-line tubes:								
1.25	0.348	0.592	0.275	0.608	0.100	0.704	0.0633	0.752
1.5	0.367	0.586	0.250	0.620	0.101	0.702	0.0678	0.744
2	0.418	0.570	0.299	0.602	0.229	0.632	0.198	0.648
3	0.290	0.601	0.357	0.584	0.374	0.581	0.286	0.608
Staggered tubes:								
0.6							0.213	0.636
0.9					0.446	0.571	0.401	0.581
1			0.497	0.558				
1.125					0.478	0.565	0.518	0.560
1.25	0.518	0.556	0.505	0.554	0.519	0.556	0.522	0.562
1.5	0.451	0.568	0.460	0.562	0.452	0.568	0.488	0.568
2	0.404	0.572	0.416	0.568	0.482	0.556	0.449	0.570
3	0.310	0.592	0.356	0.580	0.440	0.562	0.421	0.574

Table 7-8.   Ratio of $\bar{h}/\bar{h}_{10}$

	Number of Tubes								
	1	2	3	4	5	6	7	8	9
Staggered	0.68	0.75	0.83	0.89	0.92	0.95	0.97	0.98	0.99
In-line	0.64	0.80	0.87	0.90	0.92	0.94	0.96	0.98	0.99

7.6   HEAT TRANSFER TO LIQUID METALS

Because of their high thermal conductivity and low viscosity, liquid metals, such as mercury, sodium and lead-bismuth alloys, are ideally suited to transferring large amounts of heat in small spaces.

Flat Plate Analysis

Since the thermal conductivity is so high, the principal mode of energy transfer to liquid metals is conduction, both in laminar and in turbulent flow.   Since convection is secondary to conduction, turbulent fluctuations add little to the transport mechanism.   This being the case, we may recall from laminar flow over a flat plate that the hydrodynamic and thermal boundary layer thicknesses are related by

$$\delta_t = \frac{\delta}{\mathbf{Pr}^{1/3}}$$

Since the Prandtl numbers of common liquid metals range from 0.004 to 0.029, the boundary layers may be represented by the relative magnitudes shown in Fig. 7-9.   The velocity profile is very blunt over most of the thermal profile.   We may, therefore, assume slug flow as a first approximation, i.e. $u = V_\infty$.   Using this value of uniform velocity together with the cubic temperature profile

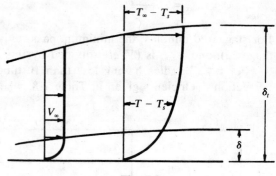

Fig. 7-9

$$\theta = \frac{T - T_s}{T_\infty - T_s} = \frac{3}{2}\frac{y}{\delta_t} - \frac{1}{2}\left(\frac{y}{\delta_t}\right)^3$$

in the integral energy equation (*6.29*), we solve to find:

$$\delta_t = \sqrt{\frac{8\alpha x}{V_\infty}} \tag{7.55}$$

whence

$$h_x = \frac{-k(\partial T/\partial y)_{y=0}}{T_s - T_\infty} = \frac{3k}{2\delta_t} = \frac{3\sqrt{2}}{8}k\sqrt{\frac{V_\infty}{\alpha x}} \tag{7.56}$$

or, in dimensionless form,

$$\mathbf{Nu}_x \equiv \frac{h_x x}{k} = (0.530)\sqrt{\mathbf{Re}_x\,\mathbf{Pr}} = (0.530)\sqrt{\mathbf{Pe}_x} \tag{7.57}$$

Here, the *Peclet number*, $\mathbf{Pe} \equiv \mathbf{Re} \cdot \mathbf{Pr}$, is a measure of the ratio of energy transport by convection to that by conduction.

Flow Inside Tubes

A number of correlation equations are available for various flow conditions.   Among them, the following are representative.

For constant heat flux at the wall, with fluid properties evaluated at the average bulk temperature, $T_b = (T_i + T_o)/2$,

$$\overline{\mathbf{Nu}_D} = \frac{\bar{h}D}{k_b} = 4.82 + (0.0185)\mathbf{Pe}_D^{0.827} \tag{7.58}$$

which is valid for $3600 < \mathbf{Re}_D < 9.05 \times 10^5$ and $100 < \mathbf{Pe}_D < 10^4$.

For constant wall temperature,

$$\overline{\mathbf{Nu}_D} = \frac{\bar{h}D}{k_b} = 5 + (0.025)\mathbf{Pe}_D^{0.8} \tag{7.59}$$

for $\mathbf{Pe}_D > 100$ and $L/D > 60$.   Fluid properties are evaluated at the average bulk temperature.

Cross Flow Over Tube Banks

For the flow of mercury ($\textbf{Pr} = 0.022$) over a staggered tube bank 10 tubes deep in the flow direction, with 1/2-in.-o.d. tubes arranged in an equilateral triangular array with a 1.375 pitch-to-diameter ratio, the following equation has been recommended:

$$\overline{\textbf{Nu}_D} = \frac{\bar{h}D}{k_f} = 4.03 + (0.228)\textbf{Pe}_{D\text{max}}^{0.67} \tag{7.60}$$

where $20{,}000 < \textbf{Re}_{D\text{max}} < 80{,}000$ and fluid properties are evaluated at the film temperature, $T_f = (T_\infty + T_s)/2$. The maximum Reynolds number is based on the maximum velocity in the passage.

Solved Problems

7.1. Verify that $\overline{\phi_1'\phi_2'} \neq 0$ for the functions of Fig. 7-10.

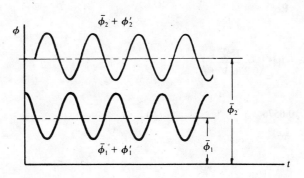

Fig. 7-10

Here, $\phi_2' \equiv -\phi_1'$, so that

$$\overline{\phi_1'\phi_2'} = -\overline{(\phi_1')^2} < 0$$

since ϕ_1' vanishes at only a finite number of points in any finite time interval.

7.2. By using the one-seventh-power law, (7.16), and the experimental expression (7.17) in the integral momentum equation, (7.15), derive expressions for the boundary layer thickness and the skin-friction coefficient for turbulent flow over a flat plate.

The integral momentum equation becomes

$$(0.0225)\frac{\rho V_\infty^2}{g_c}\left(\frac{\nu}{V_\infty\delta}\right)^{1/4} = \frac{\rho}{g_c}\frac{d}{dx}\left[\int_0^\delta V_\infty^2\left[\left(\frac{y}{\delta}\right)^{1/7} - \left(\frac{y}{\delta}\right)^{2/7}\right]dy\right]$$

Integrating and simplifying, we get

$$(0.0225)\left(\frac{\nu}{V_\infty\delta}\right)^{1/4} = \frac{7}{72}\frac{d\delta}{dx}$$

Separating variables and integrating:

$$\left(\frac{\nu}{V_\infty}\right)^{1/4} \int_0^x dx = (4.321) \int_0^\delta \delta^{1/4}\, d\delta$$

or

$$\frac{\delta}{x} = \frac{0.376}{(V_\infty x/\nu)^{1/5}} = \frac{0.376}{\mathbf{Re}_x^{1/5}} \qquad (1)$$

which is (7.18). It should be noted that in the limits of integration we have tacitly assumed that the turbulent boundary layer begins at the leading edge.

By definition, the skin-friction coefficient is

$$c_f \equiv \frac{\tau_s}{\rho V_\infty^2/2g_c}$$

Using (7.17) for τ_s,

$$c_f = \frac{(0.0225)(\rho V_\infty^2/g_c)(\nu/V_\infty \delta)^{1/4}}{\rho V_\infty^2/2g_c} = (0.045)\left(\frac{\nu}{V_\infty \delta}\right)^{1/4}$$

But from (1),

$$\delta = \frac{0.376\, x}{\mathbf{Re}_x^{1/5}}$$

Hence

$$c_f = 0.045\left(\frac{\nu\, \mathbf{Re}_x^{1/5}}{0.376\, V_\infty x}\right)$$

and

$$c_f = \frac{0.0576}{\mathbf{Re}_x^{1/5}}$$

which is (7.19).

7.3. Hydrogen at 140 °F and a pressure of 1 atm flows along a flat plate at 400 fps. If the plate is at 200 °F and 4 ft long, determine the following quantities, assuming a critical Reynolds number of 500,000: (a) the thickness of the hydrodynamic boundary layer at the end of the plate, (b) the local skin-friction coefficient at the end of the plate, (c) the average skin-friction coefficient, (d) the drag force per foot of plate width, (e) the local convective heat-transfer coefficient at the end of the plate, (f) the average heat-transfer coefficient, (g) the heat transfer from the plate to the hydrogen per foot of width.

At the film temperature,

$$T_f = \frac{T_s + T_\infty}{2} = \frac{200 + 140}{2} = 170 \text{ °F}$$

the appropriate parameters from Table B-4 are:

$$\rho = 0.00438 \text{ lbm/ft}^3 \qquad \nu = 152.7 \times 10^{-5} \text{ ft}^2/\text{sec} \qquad \mathbf{Pr} = 0.697$$
$$c_p = 3.448 \text{ Btu/lbm-°F} \qquad k = 0.119 \text{ Btu/hr-ft-°F}$$
$$\mu_m = 6.689 \times 10^{-6} \text{ lbm/sec-ft} \qquad \alpha = 7.87 \text{ ft}^2/\text{hr}$$

At the end of the plate the Reynolds number is

$$\mathbf{Re}_L = \frac{V_\infty L}{\nu} = \frac{(400 \text{ ft/sec})(4 \text{ ft})}{1.527 \times 10^{-3} \text{ ft}^2/\text{sec}} = 1.048 \times 10^6$$

and the transition from laminar to turbulent flow occurs at

$$x_c = \frac{\mathbf{Re}_c \nu}{V_\infty} = \frac{(500{,}000)(1.527 \times 10^{-3})}{400} = 1.909 \text{ ft}$$

(a) Assuming the boundary layer grows turbulently throughout the length of the plate (which gives good results in the turbulent regime), the hydrodynamic boundary layer thickness is given by (7.18).

$$\delta = \frac{(0.376)L}{\mathbf{Re}_L^{1/5}} = \frac{(0.376)(4 \text{ ft})}{(1.048 \times 10^6)^{1/5}} = 0.094 \text{ ft}$$

(b) Equation (7.19) gives

$$c_f = \frac{0.0576}{\mathbf{Re}_L^{1/5}} = \frac{0.0576}{(1.048 \times 10^6)^{1/5}} = 0.0036$$

(c) Accounting for both the laminar and the turbulent portions of the boundary layer, the average skin-friction coefficient is given by (7.22).

$$C_f = \frac{0.072}{(1.048 \times 10^6)^{1/5}} - (0.00334)\left(\frac{1.909}{4}\right) = 0.002906$$

(d) Using C_f from (c), the total drag per unit width on one side of the plate is, by (7.20),

$$F_f = C_f \frac{\rho V_\infty^2 L}{2g_c} = (0.002906) \frac{\left(0.00438 \ \dfrac{\text{lbm}}{\text{ft}^3}\right)\left(400 \ \dfrac{\text{ft}}{\text{sec}}\right)^2 (4 \text{ ft})}{2\left(32.2 \ \dfrac{\text{lbm-ft}}{\text{lbf-sec}^2}\right)} = 0.1262 \text{ lbf/ft}$$

(e) The Colburn equation, (7.25), gives the local convective heat-transfer coefficient:

$$\mathbf{St}_x \ \mathbf{Pr}^{2/3} = \frac{c_f}{2} \quad \text{or} \quad \frac{h_x x}{k} = \mathbf{Re}_x \ \mathbf{Pr}^{1/3} \frac{c_f}{2}$$

Thus

$$h_4 = \frac{0.119 \ \dfrac{\text{Btu}}{\text{hr-ft-}°\text{F}}}{4 \text{ ft}} (1.048 \times 10^6)(0.697)^{1/3} \frac{(0.0036)}{2} = 49.76 \text{ Btu/hr-ft}^2\text{-}°\text{F}$$

Because this result ignores the effect of the laminar sublayer and the buffer zone, it is appropriate to check their influence by use of (7.30).

$$h_4\bigg|_K = \frac{0.119 \ \dfrac{\text{Btu}}{\text{hr-ft-}°\text{F}}}{4 \text{ ft}} \frac{(0.0288)(1.048 \times 10^6)^{0.8}(0.697)}{1 + (0.849)(1.048 \times 10^6)^{-0.10}\left[(0.697 - 1) + \ln\left(\dfrac{5(0.697 + 1)}{6}\right)\right]}$$

$$= 44.76 \text{ Btu/hr-ft}^2\text{-}°\text{F}$$

which is approximately 10 percent smaller than the value given by the Colburn equation.

(f) Since the von Kármán equation is too formidable to integrate, except numerically, we shall be content with the average heat-transfer coefficient given by (7.28).

$$\frac{\bar{h}L}{k} = \mathbf{Pr}^{1/3}(0.036 \ \mathbf{Re}_L^{0.8} - 836)$$

$$\bar{h} = \frac{0.119 \ \dfrac{\text{Btu}}{\text{hr-ft-}°\text{F}}}{4 \text{ ft}} (0.697)^{1/3}[(0.036)(1.048 \times 10^6)^{0.8} - 836] = 40.153 \text{ Btu/hr-ft}^2\text{-}°\text{F}$$

(g) The heat transfer per foot of width from one side of the plate is given by

$$\frac{q}{W} = \bar{h}L(T_s - T_\infty) = \left(40.153 \ \frac{\text{Btu}}{\text{hr-ft}^2\text{-}°\text{F}}\right)(4 \text{ ft})[(200 - 140) \ °\text{F}] = 9637 \text{ Btu/hr-ft}$$

7.4. A smooth, thin, model airfoil is to be tested for drag in a wind tunnel. It is symmetrical and can be approximated by a flat plate. Its chord length is 150 mm. At an airstream velocity of 200 m/s and a temperature of 27 °C, what is the drag per unit width?

Assuming the pressure to be approximately atmospheric, the pertinent parameters, in SI units from Table B-4, are:

$$\rho = (16.01846)(0.0735) = 1.177 \text{ kg/m}^3$$
$$\nu = (0.0929)(16.88 \times 10^{-5}) = 1.57 \times 10^{-5} \text{ m}^2/\text{s}$$

At the trailing edge the Reynolds number is

$$\mathbf{Re}_L = \frac{V_\infty L}{\nu} = \frac{(200 \text{ m/s})(0.15 \text{ m})}{1.57 \times 10^{-5} \text{ m}^2/\text{s}} = 1.911 \times 10^6$$

which is turbulent. Assuming the transition from laminar to turbulent occurs at a Reynolds number of 500,000, the critical length is

$$x_c = \frac{\nu}{V_\infty} \mathbf{Re}_c = \frac{1.57 \times 10^{-5} \text{ m}^2/\text{s}}{200 \text{ m/s}} (500,000) = 39.25 \text{ mm}$$

The average skin-friction coefficient is given by (*7.22*).

$$C_f = \frac{0.072}{\mathbf{Re}_L^{1/5}} - (0.00334) \frac{x_c}{L}$$

$$= \frac{0.072}{(1.911 \times 10^6)^{1/5}} - (0.00334) \left(\frac{3.925}{15}\right) = 0.00399 - 0.00087 = 0.00312$$

If the boundary layer had been assumed turbulent from the leading edge, the average skin-friction coefficient,

$$C_f \bigg|_{\text{turb}} \approx 0.004$$

would be 27 percent higher than the more accurate value obtained by accounting for the influence of the laminar portion. The drag force per unit width is given by

$$\frac{F_f}{W} = 2C_f \frac{\rho V_\infty^2}{2g_c} L = (0.00312) \left(1.177 \frac{\text{kg}}{\text{m}^3}\right) \left(200 \frac{\text{m}}{\text{s}}\right)^2 (0.15 \text{ m}) \left(\frac{\text{N·s}^2}{1 \text{ kg-m}}\right) = 22.03 \text{ N/m}$$

where the factor of 2 accounts for both sides of the airfoil.

7.5. (*a*) In a chemical processing plant glycerin flows over a 1-m-long heated flat plate at free-stream conditions $V_\infty = 3$ m/s and $T_\infty = 10$ °C. If the plate is held at 30 °C, determine the heat transfer per unit width, assuming $\mathbf{Re}_c = 500,000$. (*b*) Repeat for ammonia.

(*a*) At a film temperature of

$$T_f = \frac{T_\infty + T_s}{2} = \frac{10 + 30}{2} = 20 \text{ °C}$$

the fluid properties for glycerin, given in Table B-3, are:

$$\rho = (16.01846)(78.91) = 1264 \text{ kg/m}^3 \qquad k = (1.729577)(0.165) = 0.2854 \text{ W/m-K}$$
$$\nu = (0.0929)(0.0127) = 0.00118 \text{ m}^2/\text{s} \qquad \mathbf{Pr} = 12,500$$

The critical length is

$$x_c = \frac{\nu}{V_\infty} \mathbf{Re}_c = \frac{0.00118 \text{ m}^2/\text{s}}{3 \text{ m/s}} (500,000) = 196.7 \text{ m}$$

Therefore, the flow is laminar throughout, and the average heat-transfer coefficient is given by (*6.28*) as

$$\frac{\bar{h}L}{k} = (0.664)\mathbf{Re}_L^{1/2}\,\mathbf{Pr}^{1/3}$$

The Reynolds number at the end of the plate is

$$\mathbf{Re}_L = \frac{V_\infty L}{\nu} = \frac{(3\ \text{m/s})(1\ \text{m})}{0.00118\ \text{m}^2/\text{s}} = 2542$$

Hence

$$\bar{h} = \frac{0.2854\ \text{W/m-K}}{1\ \text{m}}(2542)^{1/2}(12{,}500)^{1/3} = 333.9\ \text{W/m}^2\text{-K}$$

and the heat transfer is given by

$$\frac{q}{W} = \bar{h}L(T_s - T_\infty) = \left(333.9\ \frac{\text{W}}{\text{m}^2\text{-K}}\right)(1\ \text{m})[(30 - 10)\ \text{K}] = 6.679\ \text{kW/m}$$

(b)   For ammonia at 20 °C,

$$\rho = (16.01846)(38.19) = 611.75\ \text{kg/m}^3 \qquad\qquad k = (1.729577)(0.301) = 0.521\ \text{W/m-K}$$
$$\nu = (0.0929)(0.386 \times 10^{-5}) = 3.59 \times 10^{-7}\ \text{m}^2/\text{s} \quad \mathbf{Pr} = 2.02$$

and the critical length is

$$x_c = \frac{\nu}{V_\infty}\mathbf{Re}_c = \frac{3.59 \times 10^{-7}\ \text{m}^2/\text{s}}{3\ \text{m/s}}(500{,}000) = 0.0598\ \text{m}$$

The Reynolds number at the end of the plate is

$$\mathbf{Re}_L = \frac{V_\infty L}{\nu} = \frac{(3\ \text{m/s})(1\ \text{m})}{3.59 \times 10^{-7}\ \text{m}^2/\text{s}} = 8.36 \times 10^6$$

and the average heat-transfer coefficient is given by (7.28).

$$\frac{\bar{h}L}{k} = \mathbf{Pr}^{1/3}(0.036\ \mathbf{Re}_L^{0.8} - 836)$$

$$\bar{h} = \frac{0.521\ \text{W/m-K}}{1\ \text{m}}(2.02)^{1/3}[(0.036)(8.36 \times 10^6)^{0.8} - 836]$$

$$= (0.6586)[12{,}418 - 836] = 7.628\ \text{kW/m}^2\text{-K}$$

In this case the reduction due to the laminar portion at the leading edge is minor, being only 7.2 percent of the total.
    The heat transfer is

$$\frac{q}{W} = \bar{h}L(T_s - T_\infty) = \left(7.628\ \frac{\text{kW}}{\text{m}^2\text{-K}}\right)(1\ \text{m})[(30 - 10)\ \text{K}] = 153\ \text{kW/m}$$

This problem illustrates extremes. With glycerin, the flow was laminar throughout. With ammonia, very little of it was laminar—only 5.98 percent.

7.6.   (a) For the power-law velocity profile, (7.31), determine an expression for the ratio of the average velocity to the maximum velocity.   (b) Compare the ratio for turbulent flow, with $n = 7$, to that for laminar flow.

(a)   In terms of the pipe radius the power-law profile is

$$\frac{u}{V_{\max}} = \left(\frac{y}{R}\right)^{1/n} = \left(\frac{R - r}{R}\right)^{1/n}$$

where r is the radius and R is the pipe radius. The average velocity may be obtained by integrating this velocity over the flow area, i.e.

$$V = \frac{\int_0^R \left(\frac{R-r}{R}\right)^{1/n} V_{max} 2\pi r\, dr}{\int_0^R 2\pi r\, dr} = \frac{2V_{max}}{R^2}\int_0^R \left(1 - \frac{r}{R}\right)^{1/n} r\, dr$$

Integrating, we get

$$\frac{V}{V_{max}} = \frac{2}{R^2}\left[\frac{1}{\frac{1}{R^2}\left(\frac{1}{n}+2\right)}\left(1 - \frac{r}{R}\right)^{(1/n)+2} - \frac{1}{\frac{1}{R^2}\left(\frac{1}{n}+1\right)}\left(1 - \frac{r}{R}\right)^{(1/n)+1}\right]_0^R$$

$$= -2\left[\frac{n}{2n+1} - \frac{n}{n+1}\right] = \frac{2n^2}{(2n+1)(n+1)}$$

(b)   For turbulent flow with $n = 7$,

$$\frac{V}{V_{max}} = \frac{2(7)^2}{[2(7)+1](7+1)} = \frac{49}{60}$$

which is considerably greater than the ratio for laminar flow given by (6.33):

$$\frac{V}{V_{max}} = \frac{1}{2}$$

7.7.   Water at 68 °F flows at the rate of 1.0 ft³/min in a 1-in.-i.d. smooth, drawn copper tube 200 ft long. (a) Determine the friction factor and the length required for it to reach a constant value. (b) What is the pressure drop? (c) How would three globe valves, equally spaced in the line, influence the pressure drop?

The average velocity is

$$V = \frac{Q}{A} = \frac{\left(1.0\, \frac{ft^3}{min}\right)\left(\frac{1\, min}{60\, sec}\right)}{\frac{\pi}{4}\left(\frac{1}{12}\right)^2\, ft^2} = 3.056\, ft/sec$$

which gives a Reynolds number of

$$\mathbf{Re} = \frac{VD}{\nu} = \frac{\left(3.056\, \frac{ft}{sec}\right)\left(\frac{1}{12}\, ft\right)}{1.083 \times 10^{-5}\, ft^2/sec} = 23,515$$

(a)   Equation (7.33) gives a friction factor of

$$f = (0.184)\mathbf{Re}_D^{-0.2} = (0.184)(23,515)^{-0.2} = 0.0246$$

which compares very favorably with the value given by Moody's diagram, Fig. 7-4.

    The distance required for the friction factor to reach a constant value is given by the Latzko equation, (7.34), as

$$\left.\frac{L}{D}\right|_c = (0.623)(23,515)^{1/4} = 7.715$$

or

$$L_c = (7.715)D = 7.715\left(\frac{1}{12}\, ft\right) = 0.64\, ft$$

a negligible amount in a total of 200 ft; therefore, we shall assume that $f = 0.0246$ throughout.

(b) The pressure drop is given by the Darcy-Weisbach equation, (6.35).

$$\Delta p = f \frac{L}{D} \frac{\rho V^2}{2g_c} = (0.0246) \left(\frac{200 \text{ ft}}{\frac{1}{12} \text{ ft}} \right) \frac{\left(62.46 \frac{\text{lbm}}{\text{ft}^3} \right) \left(3.056 \frac{\text{ft}}{\text{sec}} \right)^2}{2 \left(32.2 \frac{\text{lbm-ft}}{\text{lbf-sec}^2} \right)}$$

$$= 534.8 \text{ lbf/ft}^2 = 3.71 \text{ psi}$$

(c) By (7.36), adding three globe valves in the line would be equivalent to adding tubing of length

$$(3) \frac{k_L D}{f} = (3) \frac{10 \left(\frac{1}{12} \text{ ft} \right)}{0.0246} = 101.6 \text{ ft} = L_{eq}$$

where k_L is from Table 7-3. This increases the pressure drop by 50.8 percent to

$$\Delta p = \frac{200 + 101.6}{200} (3.71 \text{ psi}) = 5.60 \text{ psi}$$

In this case, the globe valves cause losses which cannot be ignored.

7.8. What is the pressure drop in a 2000-ft length of 6-in.-i.d. galvanized iron pipe when 40 °C water flows through it at a velocity of 0.566 ft/sec?

From Table 7-2, the relative roughness is

$$\frac{e}{D} = \frac{0.0060}{6} = 0.001$$

The Reynolds number is

$$\mathbf{Re}_D = \frac{VD}{\nu} = \frac{(0.566 \text{ ft/sec})(0.5 \text{ ft})}{0.708 \times 10^{-5} \text{ ft}^2/\text{sec}} = 4 \times 10^4$$

With these values of relative roughness and Reynolds number, Moody's diagram, Fig. 7-4, gives $f = 0.025$, and (6.35) yields the pressure drop.

$$\Delta p = f \frac{L}{D} \frac{\rho V^2}{2g_c}$$

$$= (0.025) \left(\frac{2000}{0.5} \right) \frac{\left(62.09 \frac{\text{lbm}}{\text{ft}^3} \right) \left(0.566 \frac{\text{ft}}{\text{sec}} \right)^2}{2 \left(32.2 \frac{\text{lbm-ft}}{\text{lbf-sec}^2} \right)} = 30.89 \text{ lbf/ft}^2$$

7.9. Compare the heat transfer from a 2-in.-o.d. rod with atmospheric air flowing parallel to it (external flow) with that from a 2-in.-i.d. tube with air flowing through it (internal flow). In both cases the air velocity is 100 ft/sec, and the air temperature is 60 °F. The heated portions of the rod and the tube are each 2 ft long, and both are maintained at 100 °F. The heated section of the tube is located sufficiently downstream that fully developed flow occurs.

Rod: External Flow
    For the case of the rod the fluid properties are evaluated at the film temperature,

$$T_f = \frac{T_\infty + T_s}{2} = \frac{60 + 100}{2} = 80 \text{ °F}$$

and the required properties, from Table B-4, are:

$$\nu = 16.88 \times 10^{-5} \text{ ft}^2/\text{sec} \qquad k = 0.01516 \text{ Btu/hr-ft-}°\text{F} \qquad \textbf{Pr} = 0.708$$

If the thickness of the boundary layer at the end of the rod is of the same order of magnitude as (or preferably, an order of magnitude less than) the radius of the rod, we may neglect the effects of curvature and treat the external surface of the rod as a flat plate.   Assuming that it can be treated as a flat plate, the Reynolds number at the end of the rod is

$$\textbf{Re}_L = \frac{VL}{\nu} = \frac{\left(100 \dfrac{\text{ft}}{\text{sec}}\right)(2 \text{ ft})}{16.88 \times 10^{-5} \text{ ft}^2/\text{sec}} = 1.185 \times 10^6$$

which gives a boundary layer thickness, from (7.18), of

$$\delta = \frac{(0.376)L}{\textbf{Re}_L^{1/5}} = \frac{(0.376)(2 \text{ ft})}{(1.185 \times 10^6)^{1/5}} = 0.046 \text{ ft} = 0.55 \text{ in.}$$

Since $\delta < R$, we shall use the flat-plate equation (7.28) to get the average Nusselt number.

$$\overline{\textbf{Nu}} = \frac{\bar{h}L}{k} = \textbf{Pr}^{1/3}(0.036 \textbf{ Re}_L^{0.8} - 836)$$

or

$$\bar{h} = \frac{0.01516 \text{ Btu/hr-ft-}°\text{F}}{2 \text{ ft}} (0.708)^{1/3}[(0.036)(1.185 \times 10^6)^{0.8} - 836]$$

$$= 11.93 \text{ Btu/hr-ft}^2\text{-}°\text{F}$$

Since the surface area is

$$A_s = \pi \left(\frac{2}{12} \text{ ft}\right)(2 \text{ ft}) = 1.047 \text{ ft}^2$$

the heat transfer is

$$q = \bar{h}A_s(T_s - T_\infty) = \left(11.93 \frac{\text{Btu}}{\text{hr-ft}^2\text{-}°\text{F}}\right)(1.047 \text{ ft}^2)[(100 - 60) °\text{F}] = 499.7 \text{ Btu/hr}$$

Tube: Internal Flow

The case of flow through the tube is not as straightforward as that for the rod above since the tube equations require that properties be evaluated at the mean bulk temperature $T_b = (T_i + T_o)/2$ or at an average film temperature $T_f = (T_s + T_b)/2$. In either case the outlet temperature T_o is unknown, requiring a trial-and-error solution to the heat balance equation.

The Dittus-Boelter equation, (7.47), gives the heat transfer coefficient

$$\bar{h} = \frac{k}{D}(0.023)\textbf{Re}_D^{0.8} \textbf{ Pr}^{0.4} \qquad\qquad (1)$$

required in the heat balance

$$q = \bar{h}A_s(T_s - T_b) = \dot{m}c_p(T_o - T_i)$$

or

$$\bar{h}A_s\left[T_s - \left(\frac{T_i + T_o}{2}\right)\right] = \dot{m}c_p(T_o - T_i)$$

which may be rearranged to give the outlet temperature

$$T_o = \frac{\bar{h}A_sT_s + \left(\dot{m}c_p - \dfrac{\bar{h}A_s}{2}\right)T_i}{\left(\dot{m}c_p + \dfrac{\bar{h}A_s}{2}\right)} \qquad\qquad (2)$$

Assuming an outlet temperature $T_o = 70$ °F, the fluid properties, evaluated at $T_b = 65$ °F, are (from Table B-4):

$$\rho = 0.076 \text{ lbm/ft}^3 \qquad \nu = 15.77 \times 10^{-5} \text{ ft}^2/\text{sec} \qquad \mathbf{Pr} = 0.71$$
$$c_p = 0.24 \text{ Btu/lbm-°F} \qquad k = 0.0148 \text{ Btu/hr-ft-°F}$$

Using these properties, the pertinent parameters are:

$$\mathbf{Re}_D = \frac{VD}{\nu} = \frac{\left(100 \, \dfrac{\text{ft}}{\text{sec}}\right)\left(\dfrac{2}{12} \text{ ft}\right)}{15.77 \times 10^{-5} \text{ ft}^2/\text{sec}} = 1.057 \times 10^5$$

$$\dot{m} = \rho A V = \left(0.076 \, \frac{\text{lbm}}{\text{ft}^3}\right)\left[\frac{\pi}{4}\left(\frac{2}{12} \text{ ft}\right)^2\right]\left(100 \, \frac{\text{ft}}{\text{sec}}\right)\left(\frac{3600 \text{ sec}}{\text{hr}}\right) = 596.9 \text{ lbm/hr}$$

$$\bar{h} = \frac{\left(0.0148 \, \dfrac{\text{Btu}}{\text{hr-ft-°F}}\right)}{\left(\dfrac{2}{12} \text{ ft}\right)} (0.023)(1.057 \times 10^5)^{0.8}(0.71)^{0.4} = 18.62 \, \frac{\text{Btu}}{\text{hr-ft}^2\text{-°F}}$$

From (2),

$$T_o = \frac{(18.62)(1.047)(100) + [(596.9)(0.24) - \frac{1}{2}(18.62)(1.047)](60)}{(596.9)(0.24) + \frac{1}{2}(18.62)(1.047)} = 65.09 \text{ °F}$$

which is approximately 5 °F lower than assumed.

For a second trial, take $T_o = 65$ °F, giving $T_b = 62.5$ °F. Properties from Table B-4 are:

$$\rho = 0.0764 \text{ lbm/ft}^3 \qquad \nu = 15.585 \times 10^{-5} \text{ ft}^2/\text{sec} \qquad \mathbf{Pr} = 0.7107$$
$$c_p = 0.24 \text{ Btu/lbm-°F} \qquad k = 0.01471 \text{ Btu/hr-ft-°F}$$

These values give

$$\mathbf{Re}_D = 1.07 \times 10^5 \qquad \dot{m} = 600.0 \text{ lbm/hr} \qquad \bar{h} = 18.69 \text{ Btu/hr-ft}^2\text{-°F}$$

Calculating the outlet temperature, we get

$$T_o = \frac{(18.69)(1.047)(100) + [(600)(0.24) - \frac{1}{2}(18.69)(1.047)](60)}{(600)(0.24) + \frac{1}{2}(18.69)(1.047)} = 65.08 \text{ °F}$$

We note that the calculated outlet temperature is relatively insensitive to the assumed value because of the small difference in fluid properties.

Taking T_o to be 65.08 °F,

$$T_b = \frac{60 + 65.08}{2} = 62.54 \text{ °F}$$

and the heat transfer is

$$q = \bar{h} A_s (T_s - T_b) = \left(18.69 \, \frac{\text{Btu}}{\text{hr-ft}^2\text{-°F}}\right)(1.047 \text{ ft}^2)[(100 - 62.54) \text{ °F}] = 733.0 \text{ Btu/hr}$$

Therefore

$$\frac{q_{\text{tube}}}{q_{\text{rod}}} = \frac{733.0}{499.7} = 1.47$$

7.10. Ethylene glycol enters a 5-m length of 100-mm-diameter, hard-drawn copper tube in a cooling system at a velocity of 5 m/s. What is the heat transfer rate if the average bulk fluid temperature is 20 °C and the tube wall is maintained at 100 °C?

At a mean bulk temperature of 20 °C, the fluid properties, from Table B-2, are:

$$\nu = (20.64 \times 10^{-5})(9.29 \times 10^{-2}) = 1.92 \times 10^{-5} \ \text{m}^2/\text{s}$$

$$k = (0.144)(1.7296) = 0.249 \ \text{W/m-K}$$

$$\mathbf{Pr} = 204$$

The Reynolds number is

$$\mathbf{Re}_D = \frac{VD}{\nu} = \frac{(5 \ \text{m/s})(0.10 \ \text{m})}{1.92 \times 10^{-5} \ \text{m}^2/\text{s}} = 2.6 \times 10^4$$

Since the Prandtl number is large, the Sieder-Tate equation,

$$\overline{\mathbf{Nu}_D} = \frac{\bar{h}D}{k} = (0.023)\mathbf{Re}_D^{0.8} \ \mathbf{Pr}^{1/3} \left(\frac{\mu_b}{\mu_s}\right)^{0.14}$$

shall be used.   Table B-4 gives

$$\frac{\mu_b}{\mu_s} = \frac{\nu_b \rho_b}{\nu_s \rho_s} = \frac{(20.64 \times 10^{-5})(69.71)}{(2.18 \times 10^{-5})(66.08)} = 9.988$$

and the Sieder-Tate equation yields

$$\bar{h} = \frac{0.249 \ \text{W/m-K}}{0.10 \ \text{m}}(0.023)(2.6 \times 10^4)^{0.8}(204)^{1/3}(9.988)^{0.14} = 1583.8 \ \text{W/m}^2\text{-K}$$

The heat transfer is given by

$$q = \bar{h}A_s(T_s - T_b) = \left(1583.8 \ \frac{\text{W}}{\text{m}^2\text{-K}}\right)[\pi(0.100 \ \text{m})(5 \ \text{m})][(100 - 20) \ \text{K}] = 1.99 \times 10^5 \ \text{W}$$

or 0.199 MW.

7.11.   Estimate the heat transfer rate from water at a mean bulk temperature of 68 °F to a 3-ft-long by 6-in.-dia. pipe at 32 °F.   The velocity is 0.5 ft/sec.

At the bulk temperature the fluid properties, from Table B-3, are:

$$\nu = 1.083 \times 10^{-5} \ \text{ft}^2/\text{sec} \qquad k = 0.345 \ \text{Btu/hr-ft-}°\text{F} \qquad \mathbf{Pr} = 7.02$$

The Reynolds number is

$$\mathbf{Re}_D = \frac{VD}{\nu} = \frac{(0.5 \ \text{ft/sec})(0.5 \ \text{ft})}{1.083 \times 10^{-5} \ \text{ft}^2/\text{sec}} = 2.3 \times 10^4$$

This is a short-tube case since $L/D = 6$.   Equation (7.34) gives the length-to-diameter ratio required to ascertain whether to use (7.45) or (7.46).

$$\left.\frac{L}{D}\right|_c = (0.623)\mathbf{Re}_D^{1/4} = (0.623)(2.3 \times 10^4)^{1/4} = 7.67 > \frac{L}{D}$$

making (7.45) applicable.

$$\frac{\bar{h}_e}{\bar{h}} = (1.11)\left[\frac{\mathbf{Re}_D^{1/5}}{(L/D)^{4/5}}\right]^{0.275} = (1.11)\left[\frac{(2.3 \times 10^4)^{1/5}}{6^{4/5}}\right]^{0.275} = 1.30$$

For $\bar{h}$ we shall use (7.47):

$$\bar{h} = \frac{k}{D}(0.023)\mathbf{Re}_D^{0.8} \ \mathbf{Pr}^{0.3}$$

$$= \frac{0.345 \ \dfrac{\text{Btu}}{\text{hr-ft-}°\text{F}}}{0.5 \ \text{ft}}(0.023)(2.3 \times 10^4)^{0.8}(7.02)^{0.3} = 87.87 \ \text{Btu/hr-ft}^2\text{-}°\text{F}$$

Therefore, the heat transfer coefficient in this *entrance region* pipe is

$$\bar{h}_e = (1.30)(87.87) = 114.24 \text{ Btu/hr-ft}^2\text{-}°F$$

The heat transfer from the water is given by

$$q = \bar{h}_e A_s (T_b - T_s) = \left(114.24 \frac{\text{Btu}}{\text{hr-ft}^2\text{-}°F}\right) [\pi(0.5 \text{ ft})(3 \text{ ft})][(68 - 32) °F] = 19,380 \text{ Btu/hr}$$

7.12. What force is exerted by a 30 mph wind blowing normal to a 5-ft-dia. disk-shaped sign mounted on a 6-in.-dia. by 10-ft-long post? Air temperature is 80 °F.

At 80 °F, $\nu = 16.88 \times 10^{-5} \text{ ft}^2/\text{sec}$. The total drag force is found by summing that on the cylindrical post and that on the circular disk, using (7.50).

$$F_D = \frac{\rho V_\infty^2}{2g_c} [(C_D A)_{\text{post}} + (C_D A)_{\text{disk}}]$$

For the post (60 mph = 88 fps),

$$\mathbf{Re} = \frac{V_\infty D}{\nu} = \frac{\left(44 \dfrac{\text{ft}}{\text{sec}}\right)\left(\dfrac{1}{2} \text{ ft}\right)}{16.88 \times 10^{-5} \text{ ft}^2/\text{sec}} = 1.3 \times 10^5$$

and the drag coefficient for the post is 0.90 (from Table 7-5 with $L/D = 20$). For the disk,

$$\mathbf{Re} = \frac{V_\infty D}{\nu} = \frac{\left(44 \dfrac{\text{ft}}{\text{sec}}\right)(5 \text{ ft})}{16.88 \times 10^{-5} \text{ ft}^2/\text{sec}} = 1.303 \times 10^6$$

Table 7-5 gives $C_D = 1.12$.

The total drag is, therefore,

$$F_D = \frac{\left(0.0735 \dfrac{\text{lbm}}{\text{ft}^3}\right)\left(44 \dfrac{\text{ft}}{\text{sec}}\right)^2}{2\left(32.2 \dfrac{\text{lbm-ft}}{\text{lbf-sec}^2}\right)} \left[(0.90)\left(\frac{1}{2} \text{ ft}\right)(10 \text{ ft}) + (1.12)\frac{\pi}{4}(5 \text{ ft})^2\right] = 58.5 \text{ lbf}$$

7.13. A 6-in.-dia. by 2-ft-long cylinder is located axially with a stream of 68 °F water flowing at 10 fps. Compare its profile drag to its skin-friction drag.

From Table B-3, $\rho = 62.46 \text{ lbm/ft}^3$, $\nu = 1.083 \times 10^{-5} \text{ ft}^2/\text{sec}$, and

$$\mathbf{Re}_D = \frac{V_\infty D}{\nu} = \frac{\left(10 \dfrac{\text{ft}}{\text{sec}}\right)\left(\dfrac{1}{2} \text{ ft}\right)}{1.083 \times 10^{-5} \text{ ft}^2/\text{sec}} = 4.62 \times 10^5$$

giving a drag coefficient of 0.87, from Table 7-5, since $L/D = 4$. The total drag is given by (7.50).

$$F_D = C_D \left(\frac{\rho V_\infty^2}{2g_c}\right) A$$

$$= (0.87) \left[\frac{\left(62.46 \dfrac{\text{lbm}}{\text{ft}^3}\right)\left(100 \dfrac{\text{ft}^2}{\text{sec}^2}\right)}{2\left(32.2 \dfrac{\text{lbm-ft}}{\text{lbf-sec}^2}\right)}\right] \frac{\pi}{4}\left(\frac{1}{2} \text{ ft}\right)^2 = 16.57 \text{ lbf}$$

Considering the shear, the appropriate Reynolds number is

$$\mathbf{Re}_L = \frac{V_\infty L}{\nu} = \frac{\left(10\,\dfrac{\text{ft}}{\text{sec}}\right)(2\text{ ft})}{1.083 \times 10^{-5}\text{ ft}^2/\text{sec}} = 1.85 \times 10^6$$

and the flow is turbulent. Treating the cylindrical surface as a flat plate, the boundary layer thickness at the end of the cylinder is given by (7.18) as

$$\delta = \frac{(0.376)L}{\mathbf{Re}_L^{1/5}} = \frac{(0.376)(2\text{ ft})}{(1.85 \times 10^6)^{0.2}} = 0.042\text{ ft}$$

and we may neglect the effects of curvature since $\delta \ll R$. The average skin-friction drag coefficient is given by (7.20):

$$C_f = \frac{0.072}{\mathbf{Re}_L^{1/5}} = \frac{0.072}{(1.85 \times 10^6)^{0.2}} = 0.00402$$

where for simplicity we have assumed a negligible laminar leading edge. The skin-friction drag is

$$F_f = C_f \left(\frac{\rho V_\infty^2}{2g_c}\right) A$$

$$= (0.00402)\left[\frac{\left(62.46\,\dfrac{\text{lbm}}{\text{ft}^3}\right)\left(100\,\dfrac{\text{ft}^2}{\text{sec}^2}\right)}{2\left(32.2\,\dfrac{\text{lbm-ft}}{\text{lbf-sec}^2}\right)}\right]\pi\left(\frac{1}{2}\text{ ft}\right)(2\text{ ft}) = 1.22\text{ lbf}$$

The profile drag may now be determined as

$$F_p = F_D - F_f = 16.57 - 1.22 = 15.35\text{ lbf}$$

and

$$\frac{F_p}{F_f} = \frac{15.35}{1.22} = 12.58$$

which gives us some feel for the influence of the wake in Fig. 7-6(a).

7.14. Air at 27 °C flows normal to a 77 °C, 30-mm-o.d. water pipe. The air moves at 1.0 m/sec. Estimate the heat transfer per unit length.

Assuming atmospheric pressure, the required properties (from Table B-4), evaluated at

$$T_f = \frac{T_\infty + T_s}{2} = 52\ °\text{C}$$

are:

$$\nu_f = (19.63 \times 10^{-5})(0.0929) = 1.824 \times 10^{-5}\text{ m}^2/\text{s}$$
$$k_f = (0.016255)(1.7296) = 0.0281\text{ W/m-K} \qquad \mathbf{Pr}_f = 0.702$$

The Reynolds number is

$$\mathbf{Re}_{Df} = \frac{V_\infty D}{\nu_f} = \frac{(1\text{ m/s})(0.030\text{ m})}{1.824 \times 10^{-5}\text{ m}^2/\text{s}} = 1645$$

which defines the appropriate constants from Table 7-6 to be used in (7.51); therefore,

$$\bar{h} = \frac{0.0281\text{ W/m-K}}{0.030\text{ m}}(0.683)(0.702)^{1/3}(1645)^{0.466} = 17.93\text{ W/m}^2\text{-K}$$

Hence, the heat transfer is

$$\frac{q}{L} = \bar{h}\pi D(T_s - T_\infty) = \left(17.93\,\frac{\text{W}}{\text{m}^2\text{-K}}\right)\pi(0.030\text{ m})(50\text{ K}) = 84.49\text{ W/m}$$

7.15.   Estimate the heat transfer from a 40-watt incandescent bulb at 127 °C to a 27 °C airstream
moving at 0.3 m/s.   Approximate the bulb as a 50-mm-diameter sphere.   What percent of
the power is lost by convection?

From Table B-4 the required parameters, evaluated at $T_f = (T_s + T_\infty)/2 = 77$ °C, are:

$$\nu_f = (22.38 \times 10^{-5})(0.0929) = 2.079 \times 10^{-5} \text{ m}^2/\text{s}$$

$$k_f = (0.01735)(1.729) = 0.0300 \text{ W/m-K} \qquad \mathbf{Pr} = 0.697$$

The Reynolds number is

$$\mathbf{Re}_{Df} = \frac{V_\infty D}{\nu_f} = \frac{(0.3 \text{ m/s})(0.050 \text{ m})}{2.079 \times 10^{-5} \text{ m}^2/\text{s}} = 721.5$$

and (7.52) gives the average heat-transfer coefficient as

$$\bar{h} = \frac{k_f}{D}(0.37)(\mathbf{Re}_{Df})^{0.6}$$

$$\bar{h} = \frac{0.0300 \text{ W/m-K}}{0.050 \text{ m}}(0.37)(721.5)^{0.6} = 11.52 \text{ W/m}^2\text{-K}$$

The heat transfer is given by

$$q = \bar{h}A(T_s - T_\infty) = \left(11.52 \frac{\text{W}}{\text{m}^2\text{-K}}\right)\left(\frac{\pi}{4}\right)(0.050 \text{ m})^2[(127 - 27) \text{ K}] = 2.26 \text{ W}$$

The percentage lost by forced convection is therefore

$$\frac{2.26}{40}(100\%) = 5.65\%$$

In Problem 8.8 it will be shown that the loss by free convection is more than this forced convection
loss; hence, both mechanisms should be considered.

7.16.   A 2-in.-square shaft rotates slowly at a constant velocity.   The shaft temperature is 440 °F,
and 80 °F atmospheric air flows normal to it at 15 mph.   Estimate the heat transfer.

From Table B-4, at the film temperature $T_f = (T_\infty + T_s)/2 = 260$ °F, the required properties are:

$$\nu_f = 27.88 \times 10^{-5} \text{ ft}^2/\text{sec} \qquad k_f = 0.01944 \text{ Btu/hr-ft-°F} \qquad \mathbf{Pr}_f = 0.689$$

Since the shaft is rotating, we shall take the average of the two $\bar{h}$'s given by (7.51) for the
second and third configuration rows of Table 7-6.   With the flat face normal, the Reynolds
number is

$$\mathbf{Re}_{Df} = \frac{V_\infty D}{\nu_f} = \frac{\left(22 \frac{\text{ft}}{\text{sec}}\right)\left(\frac{2}{12} \text{ ft}\right)}{27.88 \times 10^{-5} \text{ ft}^2/\text{sec}} = 13,152$$

With the shaft in the diamond configuration,

$$\mathbf{Re}_{Df} = \frac{V_\infty D}{\nu_f} = \frac{\left(22 \frac{\text{ft}}{\text{sec}}\right)\left(\frac{2}{12}\sqrt{2} \text{ ft}\right)}{27.88 \times 10^{-5} \text{ ft}^2/\text{sec}} = 18,599$$

The respective convective coefficients are:

$$\bar{h}_{\text{flat}} = \frac{0.01944 \dfrac{\text{Btu}}{\text{hr-ft-°F}}}{\dfrac{2}{12} \text{ ft}}(0.092)(0.689)^{1/3}(13,152)^{0.675} = 5.72 \text{ Btu/hr-ft}^2\text{-°F}$$

$$\bar{h}_{\text{diamond}} = \frac{0.01944}{\frac{2}{12}\sqrt{2}} (0.222)(0.689)^{1/3}(18,599)^{0.588} = 5.24 \text{ Btu/hr-ft}^2\text{-°F}$$

so that

$$\bar{h} = \frac{5.72 + 5.24}{2} = 5.48 \text{ Btu/hr-ft}^2\text{-°F}$$

Then, if P is the perimeter of the shaft, the heat transfer is given by

$$\frac{q}{L} = \bar{h}P(T_s - T_\infty)$$

$$= \left(5.48 \frac{\text{Btu}}{\text{hr-ft}^2\text{-°F}}\right)\left(\frac{8}{12} \text{ ft}\right)[(440 - 80) \text{ °F}] = 1315 \text{ Btu/hr-ft}$$

7.17.   Water droplets at 180 °F in a cooling tower have an average diameter of 0.060 in.   The airstream, which moves at a velocity of 3 ft/sec relative to the water drops, is at 60 °F.   Determine the heat transfer coefficient.

At the film temperature, $T_f = (T_s + T_\infty)/2 = 120$ °F, the necessary properties are taken from Table B-4:

$$\nu_f = 19.32 \times 10^{-5} \text{ ft}^2/\text{sec} \qquad k_f = 0.01613 \text{ Btu/hr-ft-°F}$$

The Reynolds number is

$$\mathbf{Re}_{Df} = \frac{V_\infty D}{\nu_f} = \frac{\left(3 \frac{\text{ft}}{\text{sec}}\right)\left(\frac{0.060}{12} \text{ ft}\right)}{19.32 \times 10^{-5} \text{ ft}^2/\text{sec}} = 77.64$$

If we assume the interference from adjacent water droplets is negligible, (7.52) is valid; hence,

$$\bar{h} = \frac{k_f}{D}(0.37)(\mathbf{Re}_{Df})^{0.6} = \frac{0.01613 \frac{\text{Btu}}{\text{hr-ft-°F}}}{\frac{0.060}{12} \text{ ft}}(0.37)(77.64)^{0.6} = 16.25 \frac{\text{Btu}}{\text{hr-ft}^2\text{-°F}}$$

7.18.   An in-line tube bundle consists of 19 rows of 1-in.-o.d. tubes with 12 tubes in each row (in the direction of flow).   The tube spacing is 1.5 in. in the direction normal to the flow and 2.0 in. parallel to it.   The tube surfaces are maintained at 260 °F.   Air at 80 °F and 14.7 psia flows through the bundle with a maximum velocity of 30 ft/sec.   Calculate the total heat transfer from the bundle per foot of length.

At the film temperature, $T_f = (T_\infty + T_s)/2 = 170$ °F, the fluid properties are (from Table B-4)

$$\nu_f = 22.38 \times 10^{-5} \text{ ft}^2/\text{sec} \qquad k_f = 0.01735 \text{ Btu/hr-ft-°F} \qquad \mathbf{Pr}_f = 0.697$$

giving a maximum Reynolds number of

$$\mathbf{Re}_{\max} = \frac{V_{\max}D}{\nu_f} = \frac{\left(30 \frac{\text{ft}}{\text{sec}}\right)\left(\frac{1}{12} \text{ ft}\right)}{22.38 \times 10^{-5} \text{ ft}^2/\text{sec}} = 11,171$$

In terms of Fig. 7-8, the geometric configuration is:

$$\frac{a}{D} = \frac{1.5}{1} = 1.5 \qquad \frac{b}{D} = \frac{2}{1} = 2$$

Table 7-7 gives $C_1 = 0.299$ and $n = 0.602$, and (7.54) gives

$$\bar{h} = \frac{k_f}{D} C_1 (\mathbf{Re}_{max})^n$$

$$= \frac{0.01735 \dfrac{\text{Btu}}{\text{hr-ft-°F}}}{\dfrac{1}{12}\text{ ft}} (0.299)(11,171)^{0.602} = 17.03 \text{ Btu/hr-ft}^2\text{-°F}$$

The total heat transfer per unit length is $q/L = \bar{h}\pi DN(T_s - T_\infty)$, where N is the total number of tubes.

$$\frac{q}{L} = \left(17.03 \frac{\text{Btu}}{\text{hr-ft}^2\text{-°F}}\right) \pi \left(\frac{1}{12}\text{ ft}\right) [(19)(12)][(260 - 80) \text{ °F}] = 1.83 \times 10^5 \text{ Btu/hr-ft}$$

7.19. Water at 60 °F flows across a 12-in.-wide staggered tube bank (Fig. 7-11) carrying combustion gases which keep the tube surfaces at 580 °F. For each 1-ft length of tube bank, water is supplied in a 6-in.-i.d. pipe, flowing at a velocity of 5 ft/sec. Estimate the temperature of the water after passing over the tube bank.

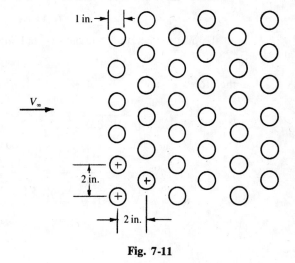

Fig. 7-11

At the film temperature, $T_f = (T_\infty + T_s)/2 = 320$ °F, the fluid parameters, from Table B-3, are:

$$\nu_f = 0.204 \times 10^{-5} \text{ ft}^2/\text{sec}$$
$$k_f = 0.393 \text{ Btu/hr-ft-°F}$$
$$\mathbf{Pr}_f = 1.099$$

The flow area per foot of length in the tube bank is equal to the free space between the tubes; therefore,

$$A = [12 \text{ in.} - 6(1 \text{ in.})](12 \text{ in.}) = 72 \text{ in}^2$$

and the maximum velocity in the tube bank is given by

$$V_{max}A = V_{pipe}A_{pipe}$$

$$V_{max} = \left(5 \frac{\text{ft}}{\text{sec}}\right)\left(\frac{\pi}{4}\right)\frac{(6 \text{ in.})^2}{72 \text{ in}^2} = 1.96 \text{ ft/sec}$$

This gives the maximum Reynolds number, i.e.

$$\mathbf{Re}_{max} = \frac{V_{max}D}{\nu_f} = \frac{\left(1.96 \dfrac{\text{ft}}{\text{sec}}\right)\left(\dfrac{1}{12}\text{ ft}\right)}{0.204 \times 10^{-5} \text{ ft}^2/\text{sec}} = 80,208$$

From the tube bank's geometry,

$$\frac{a}{D} = \frac{b}{D} = 2$$

which gives $C_1 = 0.482$ and $n = 0.556$, from Table 7-7. Equation (7.54) gives

$$\bar{h}_{10} = \frac{k_f}{D} C_1 (\mathbf{Re}_{max})^n = \frac{0.393 \dfrac{\text{Btu}}{\text{hr-ft-}^\circ\text{F}}}{\dfrac{1}{12} \text{ ft}} (0.482)(80,208)^{0.556} = 1212 \text{ Btu/hr-ft}^2\text{-}^\circ\text{F}$$

This value must be modified by the appropriate factor from Table 7-8 to account for the fact that the heat transfer coefficient has not reached its constant value produced by passing over a minimum of 10 tubes; therefore,

$$\bar{h} = (0.95)(1212) = 1151 \text{ Btu/hr-ft}^2\text{-}^\circ\text{F}$$

For $N = 36$ tubes, the heat transferred per foot of length to the water is given by

$$q = \bar{h}AN(T_\infty - T_s)$$

$$= 1151 \frac{\text{Btu}}{\text{hr-ft}^2\,^\circ\text{F}} \left(\pi \frac{1}{12} \text{ ft} \right) (1 \text{ ft})(36)[(580 - 60)\,^\circ\text{F}] = 5.64 \times 10^6 \text{ Btu/hr}$$

Since the water gains this heat, a heat balance on it gives its exit temperature.

$$q = \dot{m}c_p(T_{out} - T_{in}) = (\rho A V)_{in} c_p(T_{out} - T_{in})$$

At the inlet temperature, the density and specific heat are

$$\rho_{in} = 62.48 \text{ lbm/ft}^3 \qquad c_{p,in} = 1.0007 \text{ Btu/lbm-}^\circ\text{F}$$

therefore,

$$T_{out} = \frac{q}{(\rho A V)_{in} c_{p,in}} + T_{in}$$

$$= \frac{\left(5.64 \times 10^6 \dfrac{\text{Btu}}{\text{hr}} \right) \left(\dfrac{\text{hr}}{3600 \text{ sec}} \right)}{\left(62.48 \dfrac{\text{lbm}}{\text{ft}^3} \right) (\pi/4)(0.5 \text{ ft})^2 \left(5 \dfrac{\text{ft}}{\text{sec}} \right) \left(1.0007 \dfrac{\text{Btu}}{\text{lbm-}^\circ\text{F}} \right)} + 60\,^\circ\text{F} = 85.52\,^\circ\text{F}$$

7.20. Liquid mercury flows through a 20-mm-i.d. copper tube at the rate of 1 kg/s. The mercury enters at 12 °C and is heated to 28 °C as it passes through the tube. For a constant heat flux at the wall, which is at an average temperature of 40 °C, determine the length of tube required.

At the mean bulk temperature,

$$T_b = \frac{T_i + T_o}{2} = \frac{12 + 28}{2} = 20\,^\circ\text{C}$$

the fluid properties, from Table B-3, are:

$$\rho = (847.71)(16.02) = 13,580 \text{ kg/m}^3 \qquad\qquad k = (5.02)(1.7296) = 8.683 \text{ W/m-K}$$

$$c_p = (0.0333)(4184) = 139.33 \text{ J/kg-K} \qquad\quad \mathbf{Pr} = 0.0249$$

$$\nu = (0.123 \times 10^{-5})(0.0929) = 1.143 \times 10^{-7} \text{ m}^2/\text{s}$$

The heat gained by the mercury is

$$q = \dot{m}c_p(T_o - T_i) = \left(1 \frac{\text{kg}}{\text{s}} \right) \left(139.33 \frac{\text{J}}{\text{kg-K}} \right) [(28 - 12) \text{ K}]$$

$$= 2229 \text{ J/s} = 2229 \text{ W}$$

Equation (7.58) is valid for the average heat-transfer coefficient. It requires the Reynolds number–Prandtl number product; therefore,

$$\textbf{Re}_D = \frac{VD}{\nu} = \frac{\dot{m}D}{\rho A \nu} = \frac{4\dot{m}}{\pi D \rho \nu}$$

$$= \frac{4(1 \text{ kg/s})}{\pi(0.020 \text{ m})(13,580 \text{ kg/m}^3)(1.143 \times 10^{-7} \text{ m}^2/\text{s})} = 41,014$$

and

$$\bar{h} = \frac{k_b}{D}[4.82 + (0.0185)(\textbf{Re}_D \textbf{ Pr})^{0.827}]$$

$$= \frac{8.683 \text{ W/m-K}}{0.020 \text{ m}}\{4.82 + (0.0185)[(41,014)(0.0249)]^{0.827}\} = 4566 \text{ W/m}^2\text{-K}$$

The heat gained by the fluid results from the convective process, i.e. $q = \bar{h}\pi DL(T_s - T_b)$, or

$$L = \frac{q}{\bar{h}\pi D(T_s - T_b)}$$

$$= \frac{2229 \text{ W}}{(4566 \text{ W/m}^2\text{-K})\pi(0.020 \text{ m})[(40 - 20) \text{ K}]} = 0.3884 \text{ m}$$

Supplementary Problems

7.21.   Parallel flat plates are spaced 3 inches apart.   Estimate the length required for the boundary layers to meet when air at 80 °F flows between them at 100 ft/sec.

   Ans.   7.003 ft, assuming the laminar leading edge effect is negligible

7.22.   Air at 80 °F and 1 atm flows at 100 ft/sec over a 2-ft-long flat plate, which is maintained at 260 °F.   What is the heat transfer per unit width from one side of the plate, assuming $\textbf{Re}_c = 500,000$?

   Ans.   3435 Btu/hr-ft

7.23.   A 20-mph wind blows parallel to a 100-ft-long by 20-ft-high wall.   Determine the heat transfer from the wall, which is held at 70 °F, to the ambient air, which is at 30 °F.   *Ans.* 2.48×10^5 Btu/hr

7.24.   Ethylene glycol at 32 °F flows at the rate of 75 ft/sec parallel to a 2-ft-square, thin flat plate at 104 °F, which is suspended from a balance.   Assume the fluid flows over both sides of the plate and that the critical Reynolds number is 500,000.   (*a*) What drag should be indicated by the balance?   (*b*) What is the heat transfer rate from the plate to the fluid?   *Ans.*   (*a*) 123.7 lbf; (*b*) 2.25×10^5 Btu/hr

7.25.   Determine the heat transfer coefficient for the flow of liquid ammonia at a bulk temperature of 20 °C in a long, 30-mm-dia. tube which is at 40 °C.   The velocity is 6 m/s.   Are the effects of viscosity significant?   If the fluid were water, would the effects of viscosity be significant?

   Ans. $\bar{h} = 18.612$ kW/m^2-K; +1.48%; +6.22%

7.26.   A 3-in.-o.d. steam pipe without insulation is exposed to a 30-mph wind blowing normal to it.   The surface temperature of the pipe is 200 °F and the air is at 40 °F.   Find the heat loss per foot of pipe.

   Ans.   1291 Btu/hr-ft

7.27. Atmospheric air at 70 °F flows normal to a tube bank consisting of 15 transverse and 12 longitudinal rows arranged in-line with $a = 0.38$ in., $b = 0.31$ in. The tubes, which are 0.25-in.-o.d., are kept at 200 °F. The maximum air velocity is 4 ft/sec. What is the average film coefficient of the bundle?

Ans.   8.467 Btu/hr-ft²-°F

7.28. Determine the average heat-transfer coefficient in a staggered tube bundle having 12 20-mm-o.d. tubes per row spaced 40 mm apart in both directions. Water at 20 °C flows over the 100 °C tubes at a free-volume velocity of 0.20 m/s.   *Ans.*   3.497 kW/m²-K

Chapter 8

Natural Convection

The past three chapters dealt with heat transfer in fluids whose motion was caused primarily by changes in fluid pressure produced by external work. In this chapter we shall consider *natural* or *free convection* in which fluid moves under the influence of buoyant forces arising from changes in density.

In natural convection the velocity is zero at the heated body (no-slip boundary condition), increases rapidly in a thin boundary layer adjacent to the body, and becomes zero again far from the body. In practice, natural convection and forced convection commonly occur simultaneously. The analysis which we make in a given case must then be determined by which is predominant. If both natural and forced convection are approximately of equal importance, both must be accounted for in our analysis.

8.1 VERTICAL FLAT PLATE

One of the most common, and simplest, natural convective problems occurs when a vertical plane surface is subjected to a cooler (or warmer) surrounding fluid. Figure 8-1 shows the boundary layer adjacent to a heated vertical plane surface. For convenience, the hydrodynamic and thermal boundary layers are shown as coincident, which occurs only when the Prandtl number is unity (as was the case with forced convection). When the vertical plate is cooler than the surrounding fluid, the physical problem is inverted, but the mathematical treatment is unchanged.

At the outset the boundary layer is laminar; but at some distance from the leading edge, depending upon fluid properties and the thermal gradient, transition to turbulent flow occurs. As a rule of thumb, transition from laminar to turbulent flow occurs when the product of the Prandtl number and the *Grashof number*, where

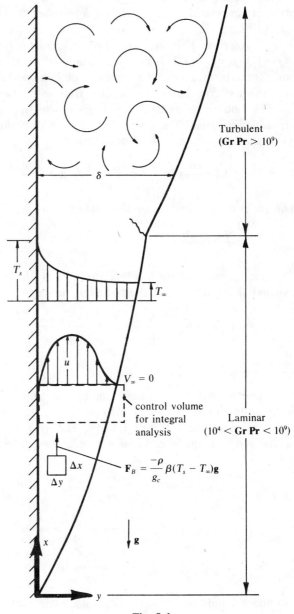

Fig. 8-1

$$\mathbf{Gr} \equiv \frac{g\beta(T_s - T_\infty)L^3}{\nu^2} \propto \frac{\text{buoyancy force}}{\text{viscous force}} \tag{8.1}$$

equals approximately 10^9.   To determine **Pr** and **Gr**, consider all properties constant at either their film-temperature or their *reference-temperature* values [see (8.16) below].   In (8.1), β is the *coefficient of volume expansion*, defined by

$$\beta \equiv \frac{1}{v}\left(\frac{\partial v}{\partial T}\right)_p \tag{8.2}$$

For an ideal gas having the equation of state $p = \rho RT$, $\beta = 1/T$.   The Grashof–Prandtl number product is sometimes called *Rayleigh number*, $\mathbf{Ra} = \mathbf{Gr} \cdot \mathbf{Pr}$.

Resulting as it does from density variations, free convective flow is compressible flow.   However, *if the temperature difference between the plate and fluid is small*, an analysis for $u(x, y)$, $v(x, y)$, and $T(x, y)$ can be made in which the density is treated as constant, except in the body force term, $\rho g/g_c$, where ρ must be considered a function of temperature.   (It is the variation of ρ in this term that is responsible for the buoyant force.)

The boundary layer assumptions of Section 6.1 are equally applicable in this quasi-incompressible analysis.   The derivation of the x-direction equation of motion is identical to Problem 6.1 when the body force is included, giving

$$\frac{\rho}{g_c}\left(u\frac{\partial u}{\partial x} + v\frac{\partial u}{\partial y}\right) = -\frac{\partial p}{\partial x} - \frac{\rho}{g_c}g + \mu_f\frac{\partial^2 u}{\partial y^2} \tag{8.3}$$

The pressure gradient along the plate results from the change in elevation; hence,

$$\frac{\partial p}{\partial x} = -\frac{\rho_\infty}{g_c}g \tag{8.4}$$

To relate ρ on the right-hand side of (8.3) to temperature, we introduce the volume coefficient of expansion written across the boundary layer:

$$\beta \approx \frac{1}{v_\infty}\left(\frac{v - v_\infty}{T - T_\infty}\right) = \rho_\infty\left(\frac{1/\rho - 1/\rho_\infty}{T - T_\infty}\right)$$

or

$$\rho_\infty - \rho \approx \rho\beta(T - T_\infty) \tag{8.5}$$

The approximation becomes more accurate as the temperature differential approaches zero.   Substituting (8.4) and (8.5) into (8.3), we have the free convective x-momentum equation for a vertical flat plate.   Noting that the energy equation (6.18) and the continuity equation (6.3) are unchanged from the forced convective, incompressible boundary layer case, the governing equations are:

x-momentum: $\qquad u\dfrac{\partial u}{\partial x} + v\dfrac{\partial u}{\partial y} = g\beta(T - T_\infty) + \nu\dfrac{\partial^2 u}{\partial y^2}$ $\qquad\qquad$ (8.6)

energy: $\qquad\qquad u\dfrac{\partial T}{\partial x} + v\dfrac{\partial T}{\partial y} = \alpha\dfrac{\partial^2 T}{\partial y^2}$ $\qquad\qquad$ (8.7)

continuity: $\qquad\quad \dfrac{\partial u}{\partial x} + \dfrac{\partial v}{\partial y} = 0$ $\qquad\qquad$ (8.8)

Boundary conditions for an isothermal plate are:

at $y = 0$: $\quad u = 0$, $v = 0$, $T = T_s$

at $y = \infty$: $\quad u = 0$, $T = T_\infty$, $\dfrac{\partial u}{\partial y} = 0$, $\dfrac{\partial T}{\partial y} = 0$

Note that the momentum and energy equations are coupled in temperature. In forced convection, the hydrodynamic problem could be solved independent of the thermal problem; this is not possible in free convection.

Similarity Solution: Isothermal Plate

Pohlhausen noted that the free convective velocity and temperature profiles exhibit a similarity property, analogous to that observed by Blasius for the forced convective problem. Using Pohlhausen's similarity parameter

$$\eta \equiv \frac{y}{x}\left[\frac{\mathbf{Gr}_x}{4}\right]^{1/4} \tag{8.9}$$

where $\mathbf{Gr}_x$ is given by (8.1) with L replaced by x, and the dimensionless temperature

$$\theta \equiv \frac{T - T_\infty}{T_s - T_\infty} \tag{8.10}$$

S. Ostrach solved the governing equations for a wide range of Prandtl numbers. His results are shown in Figs. 8-2 and 8-3. Problem 8.1 outlines the reduction of the three partial differential equations to two ordinary differential equations.

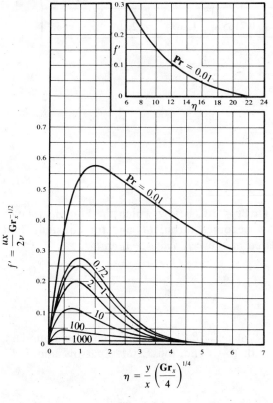

Fig. 8-2

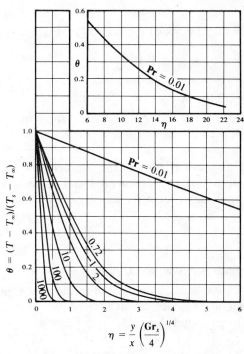

Fig. 8-3

Knowing the temperature distribution permits determination of the local heat flux. At the wall, heat transfer takes place by conduction only, since the fluid velocity is zero; therefore,

$$\frac{q}{A} = -k\left.\frac{\partial T}{\partial y}\right|_{y=0} = -\frac{k}{x}(T_s - T_\infty)\left(\frac{\mathbf{Gr}_x}{4}\right)^{1/4}\left.\frac{d\theta}{d\eta}\right|_{\eta=0} \tag{8.11}$$

This energy must be convected away; hence,

$$\frac{q}{A} = h(T_s - T_\infty) \tag{8.12}$$

Combining (8.11) and (8.12), the local Nusselt number is

$$\mathbf{Nu}_x \equiv \frac{hx}{k} = -\left(\frac{\mathbf{Gr}_x}{4}\right)^{1/4} \frac{d\theta}{d\eta}\bigg|_{\eta=0} \tag{8.13}$$

It is obvious from Fig. 8-3 that the slope of the temperature profile at the wall depends upon the Prandtl number of the fluid. With the slope at the wall written as a function of Prandtl number, $-F_1(\mathbf{Pr})$, (8.13) becomes

$$\mathbf{Nu}_x = F_1(\mathbf{Pr})\left(\frac{\mathbf{Gr}_x}{4}\right)^{1/4} \tag{8.14}$$

Values of F_1 are given in Table 8-1.

Table 8-1

Pr	0.01	0.72	0.733	1.0	2.0	10.0	100.0	1000.0
$F_1(\mathbf{Pr})$	0.0812	0.5046	0.5080	0.5671	0.7165	1.1694	2.191	3.966

By the usual method of averaging,

$$\bar{h} = \frac{1}{L}\int_0^L h(x)\,dx$$

resulting in

$$\overline{\mathbf{Nu}} \equiv \frac{\bar{h}L}{k} = \frac{4}{3}F_1(\mathbf{Pr})\left(\frac{\mathbf{Gr}_L}{4}\right)^{1/4} \tag{8.15}$$

This result is valid in the laminar flow regime $10^4 < \mathbf{Gr}\,\mathbf{Pr} < 10^9$, below which the boundary layer approximations are invalid and above which the flow becomes turbulent, and for constant fluid properties except for density. Properties are evaluated at the *reference temperature*

$$T_{\text{ref}} = T_s + (0.38)(T_\infty - T_s) \tag{8.16}$$

Similarity Solution: Constant Heat Flux

For a uniform heat flux along the plate, (8.6), (8.7) and (8.8) are valid, but the boundary condition which must be used is $q/A =$ constant. With this condition E. M. Sparrow and J. L. Gregg found that

$$\overline{\mathbf{Nu}} \equiv \frac{\bar{h}L}{k} = F_2(\mathbf{Pr})\left(\frac{\mathbf{Gr}_L}{4}\right)^{1/4} \tag{8.17}$$

Values of F_2 are given in Table 8-2. Comparing Tables 8-1 and 8-2, we note that

$$\frac{4}{3}F_1(\mathbf{Pr}) \approx (0.95)F_2(\mathbf{Pr})$$

which suggests that a single correlation equation might be applicable to a variety of cases.

Table 8-2

Pr	0.1	1.0	10.0	100.0
$F_2(\mathbf{Pr})$	0.335	0.811	1.656	3.083

Integral Solution: Isothermal Plate

By equating the sum of the external forces in the x-direction to the net x-momentum flux out of a control volume which extends through the boundary layer (see Fig. 8-1), we get the integral momentum equation

$$\frac{d}{dx} \int_0^\delta u^2 \, dy = \int_0^\delta g\beta(T - T_\infty) \, dy - \nu \frac{\partial u}{\partial y}\bigg|_{y=0} \tag{8.18}$$

in which the thermal and hydrodynamic boundary layers have been assumed of equal thickness, as shown in Fig. 8-1. The integral energy equation is obtained as in Problem 6.14:

$$\alpha \frac{\partial T}{\partial y}\bigg|_{y=0} = \frac{d}{dx}\left[\int_0^\delta (T_\infty - T)u \, dy\right] \tag{8.19}$$

These equations, which are coupled because of the gravitational body force, may be solved simultaneously when the velocity and temperature profiles are known or can be closely approximated. Logical assumptions for the profiles are:

$$\theta \equiv \frac{T - T_\infty}{T_s - T_\infty} = \left(1 - \frac{y}{\delta}\right)^2 \tag{8.20}$$

$$\frac{u}{V} = \frac{y}{\delta}\left(1 - \frac{y}{\delta}\right)^2 \tag{8.21}$$

In (8.21) the fictitious velocity V, chosen to nondimensionalize the expression, is an **unknown** function of x, as is δ. These parameters may be intuitively expressed as exponential functions:

$$V = C_1 x^a \qquad \delta = C_2 x^b$$

Substituting (8.20) and (8.21) into (8.18) and (8.19) and integrating, we get

$$\frac{1}{105}\frac{d}{dx}(V^2\delta) = \frac{1}{3}g\beta(T_s - T_\infty)\delta - \nu\frac{V}{\delta}$$

$$2\alpha\frac{T_s - T_\infty}{\delta} = \frac{1}{30}(T_s - T_\infty)\frac{d}{dx}(V\delta)$$

which can be further simplified by substituting the expressions for V and δ, and equating exponents to get $a = 1/2$ and $b = 1/4$. Upon simplification, the results are

$$\frac{\delta}{x} = (3.93)\left(\frac{0.952 + \mathbf{Pr}}{\mathbf{Gr}_x \mathbf{Pr}^2}\right)^{1/4} \tag{8.22}$$

$$V = (5.17)\frac{\nu}{x}\left(\frac{\mathbf{Gr}_x}{0.952 + \mathbf{Pr}}\right)^{1/2} \tag{8.23}$$

At the wall the heat balance is

$$\frac{q}{A} = -k\frac{\partial T}{\partial y}\bigg|_{y=0} = h(T_s - T_\infty)$$

from which $h = 2k/\delta$, upon using the temperature profile, (8.20). This, together with (8.22), gives the local Nusselt number as

$$\blacksquare \qquad \mathbf{Nu}_x \equiv \frac{hx}{k} = \frac{2x}{\delta} = (0.508)\left[\frac{\mathbf{Gr}_x\,\mathbf{Pr}^2}{0.952 + \mathbf{Pr}}\right]^{1/4} \qquad (8.24)$$

Whenever $\mathbf{Nu}_x \propto \mathbf{Gr}_x^{1/4}$, as in (8.14) and (8.24), the average heat-transfer coefficient is given by

$$\overline{\mathbf{Nu}} \equiv \frac{\bar{h}L}{k} = \frac{4}{3}\mathbf{Nu}_L \qquad (8.25)$$

In (8.24) and (8.25), as in the exact solutions (8.14) and (8.15), fluid properties are evaluated at the reference temperature (8.16).

When $\mathbf{Gr\,Pr} > 10^9$, the flow becomes turbulent. Assuming the one-seventh-power law for the velocity and temperature profiles, E.R.G. Eckert and T. W. Jackson obtained a *turbulent free-convective equation*

$$\mathbf{Nu}_x \equiv \frac{hx}{k} = (0.0295)\left[\frac{\mathbf{Gr}_x\,\mathbf{Pr}^{7/6}}{1 + (0.494)\mathbf{Pr}^{2/3}}\right]^{2/5} \qquad (8.26)$$

where the fluid properties are evaluated at the mean film temperature, $T_f = (T_s + T_\infty)/2$. Obtained by an integral analysis, this result can be approximated by a simplified equation and integrated to give an average Nusselt number

$$\overline{\mathbf{Nu}} \equiv \frac{\bar{h}L}{k} = (0.0210)(\mathbf{Gr}_L\,\mathbf{Pr})^{2/5} \qquad (8.27)$$

8.2 EMPIRICAL CORRELATIONS: ISOTHERMAL SURFACES

In the preceding section we have seen that the Nusselt number for free convection is a function of the Grashof and Prandtl numbers. Engineering data, for both the laminar and turbulent regimes, correlate well for many simple geometric configurations in a single equation,

$$\blacksquare \qquad \frac{\bar{h}L}{k} = \overline{\mathbf{Nu}} = C(\mathbf{Gr}_L\,\mathbf{Pr})^a \qquad (8.28)$$

where L is a characteristic length appropriate for the configuration. As a rule of thumb, the exponent is usually 1/4 for laminar and 1/3 for turbulent flow. All fluid properties are evaluated at the mean film temperature, $T_f = (T_s + T_\infty)/2$.

Table 8-3 gives the constants in (8.28) for common geometric configurations. The characteristic lengths L_v and L_h in the table refer to vertical and horizontal dimensions, respectively. "Large" cylinders are those with radii which are large compared to the boundary layer thickness δ. The values cited for horizontal plates are for square configurations; however, they are sufficiently accurate for engineering calculations on rectangular plates when the length-to-width ratios are small. For horizontal circular disks of diameter D, the constants pertaining to horizontal plates may be used, together with $L = (0.9)D$.

8.3 FREE CONVECTION IN ENCLOSED SPACES

Free convection is influenced by other surfaces or objects being near the surface which generates the convective currents. Two cases, horizontal and vertical fluid layers, commonly occur. In both we define an average heat-transfer coefficient $\bar{h}$ by

$$q = \bar{h}A(T_1 - T_2) \qquad (8.29)$$

where T_1 and T_2 are the temperatures of the opposing surfaces, each of which has area A. Fluid

Table 8-3

Configuration	Gr_L Pr	Characteristic Length, L	C	a
Vertical Plates and Large Cylinders				
Laminar	10^{-1} to 10^4	L_v	See Fig. 8-4	
Laminar	10^4 to 10^9	L_v	0.59	1/4
Turbulent	10^9 to 10^{12}	L_v	0.13	1/3
Small Vertical Cylinders (Wires)				
	10^{-14} to 10^{-1}	D	See Fig. 8-5	
Horizontal Plates				
Laminar (heated surface up or cooled surface down)	10^5 to 2×10^7	$L = (L_1 + L_2)/2$	0.54	1/4
Turbulent (heated surface up or cooled surface down)	2×10^7 to 3×10^{10}	$L = (L_1 + L_2)/2$	0.14	1/3
Laminar (heated surface down or cooled surface up)	3×10^5 to 3×10^{10}	$L = (L_1 + L_2)/2$	0.27	1/4
Inclined Plates (small θ) Multiply Grashof number by $\cos \theta$, where θ is the angle of inclination from the vertical, and use vertical plate constants				
Long Horizontal Cylinders (0.002 in. $< D <$ 12 in.)				
Laminar	$<10^4$	D	See Fig. 8-6	
Laminar	10^4 to 10^9	D	0.53	1/4
Turbulent	10^9 to 10^{12}	D	0.13	1/3
Fine Horizontal Wires ($D <$ 0.002 in.)				
Laminar		D	0.4	0
Miscellaneous Solid Shapes (spheres, short cylinders, blocks)				
Laminar	10^{-4} to 10^4	$\dfrac{1}{L} = \dfrac{1}{L_v} + \dfrac{1}{L_h}$	See Fig. 8-7	
Laminar	10^4 to 10^9		0.60	1/4

properties are evaluated at the arithmetic mean of the two surface temperatures, $(T_1 + T_2)/2$. And the characteristic length in the Grashof number is the distance between the surfaces, b; i.e. if $T_1 > T_2$,

$$\mathbf{Gr}_b \equiv \frac{g\beta(T_1 - T_2)b^3}{\nu^2} \qquad (8.30)$$

Horizontal Air Layers—Isothermal Walls

For horizontal air layers there are two cases: (1) if the upper plate is at the higher temperature, no convection occurs (since the lighter fluid is above the more dense fluid); (2) if the lower plate is warmer, an unstable condition exists, and convective motion occurs.

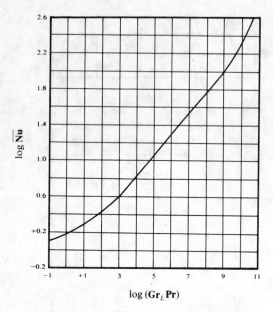

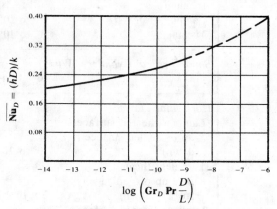

Fig. 8-4. Correlation for heated vertical plates.
(Adapted from W. H. McAdams, *Heat Transmission*, 3d ed., p. 173. Copyright 1954. McGraw-Hill Book Company. Used by permission.)

Fig. 8-5. Correlation for small vertical cylinders. (Adapted from J. R. Kyte, A. J. Madden, and E. L. Piret, *Chem. Eng. Progr.*, **49**:657 (1953), by permission.)

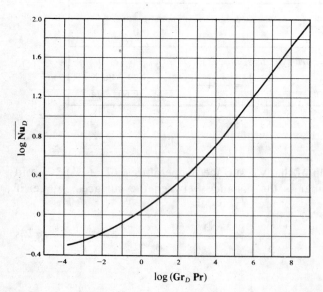

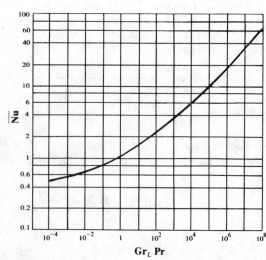

Fig. 8-6. Correlation for horizontal cylinders. (Adapted from W. H. McAdams, *Heat Transmission*, 3d ed. Copyright 1954. McGraw-Hill Book Company. Used by permission.)

Fig. 8-7. Correlation for miscellaneous solid shapes.

Upper plate warmer.   As there is no fluid motion, the heat transfer is by conduction, and the heat transfer coefficient in (8.29) can be evaluated by

$$q = \bar{h}A(T_1 - T_2) = kA\left(\frac{T_1 - T_2}{b}\right)$$

or

$$\overline{\mathbf{Nu}_b} \equiv \frac{\bar{h}b}{k} = 1.0 \tag{8.31}$$

Lower plate warmer.   The fluid motion depends upon the Grashof number $\mathbf{Gr}_b$.   At Grashof numbers less than about 2000, the convective velocities are very low; heat transfer is primarily by conduction, and (8.31) applies.   As the Grashof number increases, the conduction and convection modes coexist, until convection predominates at $\mathbf{Gr}_b > 10^4$, and the following correlations apply:

$$\overline{\mathbf{Nu}_b} = (0.195)\mathbf{Gr}_b^{1/4} \qquad (8.32)$$
$$10^4 < \mathbf{Gr}_b < 4 \times 10^5$$

$$\overline{\mathbf{Nu}_b} = (0.068)\mathbf{Gr}_b^{1/3} \qquad (8.33)$$
$$\mathbf{Gr}_b > 4 \times 10^5$$

The lower range, (8.32), corresponds to a well-ordered process in which the fluid circulates in small hexagonal cells, called *Benard cells*, rising in the center of each cell and descending at the boundaries (Fig. 8-8).

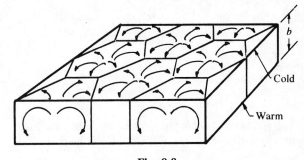

Fig. 8-8

Horizontal Liquid Layers—Isothermal Walls

S. Globe and D. Dropkin recommended the following correlation for mercury, water and silicone oils, over the wide range $0.02 < \mathbf{Pr} < 8750$:

$$\overline{\mathbf{Nu}_b} = (0.069)\mathbf{Gr}_b^{1/3}\,\mathbf{Pr}^{0.407} \qquad 3 \times 10^5 < \mathbf{Gr}_b\,\mathbf{Pr} < 7 \times 10^9 \tag{8.34}$$

Fluid properties are evaluated at the average of the two surface temperatures.

Vertical Enclosed Spaces

The effects of geometry are more complicated in vertical spaces than for horizontal layers.   Both plate height L and spacing b are important, in addition to the Rayleigh number $\mathbf{Ra}_b = \mathbf{Gr}_b\,\mathbf{Pr}$.   R. K. MacGregor and A. F. Emery characterize the behavior in regimes, as shown schematically in Fig. 8-9.

Vertical air layers—isothermal walls.   The following results have ranges of applicability approximating the regimes shown in Fig. 8-9.

$$\overline{\mathbf{Nu}_b} = 1 \qquad\qquad\qquad \mathbf{Gr}_b < 2000 \tag{8.35}$$

$$\overline{\mathbf{Nu}_b} = (0.18)\mathbf{Gr}_b^{1/4}\left(\frac{L}{b}\right)^{-1/9} \qquad 2 \times 10^4 < \mathbf{Gr}_b < 2 \times 10^5 \tag{8.36}$$

$$\overline{\mathbf{Nu}_b} = (0.065)\mathbf{Gr}_b^{1/3}\left(\frac{L}{b}\right)^{-1/9} \qquad 2 \times 10^5 < \mathbf{Gr}_b < 1.1 \times 10^7 \tag{8.37}$$

Fluid properties are evaluated at $(T_1 + T_2)/2$.   These equations are valid for $L/b > 3$; for smaller values, the correlation for a single vertical plate should be applied to each surface.

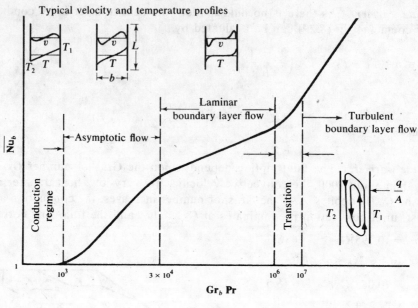

Fig. 8-9

Vertical liquid layers—constant heat flux. The heat transfer coefficients for a number of liquids in vertical enclosures with constant heat flux are given by the following relations.

$$\overline{\mathbf{Nu}_b} = (0.42)(\mathbf{Gr}_b\,\mathbf{Pr})^{1/4}\,\mathbf{Pr}^{0.012}\left(\frac{L}{b}\right)^{-0.30} \tag{8.38}$$

for

$$10^4 < \mathbf{Gr}_b\,\mathbf{Pr} < 10^7 \qquad 1 < \mathbf{Pr} < 2\times10^4 \qquad 10 < L/b < 40$$

$$\overline{\mathbf{Nu}_b} = (0.046)(\mathbf{Gr}_b\,\mathbf{Pr})^{1/3} \tag{8.39}$$

for

$$10^6 < \mathbf{Gr}_b\,\mathbf{Pr} < 10^9 \qquad 1 < \mathbf{Pr} < 20 \qquad 1 < L/b < 40$$

Other Geometrical Configurations

For air enclosed between *two concentric spheres* E. N. Bishop, L. R. Mack and J. A. Scanlan give the convective parameter in terms of the *effective thermal conductivity, k_e,*

$$\frac{k_e}{k} = 0.106\,\mathbf{Gr}_b^{0.276} \tag{8.40}$$

for

$$2\times10^4 < \mathbf{Gr}_b < 3.6\times10^6 \qquad 0.25 < b/r_i < 1.50$$

where b is the difference between the sphere radii, $b = r_o - r_i$. The heat transfer is then given by the steady-state conduction equation for a spherical shell,

$$q = \frac{4\pi k_e r_i r_o}{r_o - r_i}(T_1 - T_2) \tag{8.41}$$

L. B. Evans and N. E. Stefany showed that transient free-convective heating or cooling in *closed vertical or horizontal cylindrical enclosures* is correlated by

$$\overline{\mathbf{Nu}_D} \equiv \frac{\bar{h}D}{k} = (0.55)(\mathbf{Gr}_L \,\mathbf{Pr})^{1/4} \qquad 0.75 < L/D < 2.0 \qquad (8.42)$$

where the Grashof number is formed with the cylinder length, and the fluid properties are evaluated at $T_f = (T_s + T_\infty)/2$.

Free-convective coefficients inside *spherical cavities* of diameter D are given by F. Kreith:

$$\overline{\mathbf{Nu}_D} = (0.59)(\mathbf{Gr}_D \,\mathbf{Pr})^{1/4} \qquad 10^4 < \mathbf{Gr}_D \,\mathbf{Pr} < 10^9 \qquad (8.43)$$

$$\overline{\mathbf{Nu}_D} = (0.13)(\mathbf{Gr}_D \,\mathbf{Pr})^{1/3} \qquad 10^9 < \mathbf{Gr}_D \,\mathbf{Pr} < 10^{12} \qquad (8.44)$$

In these equations the Grashof number is based on the cavity diameter, and fluid properties are evaluated at the mean film temperature.

8.4 MIXED FREE AND FORCED CONVECTION

Throughout Chapters 6 and 7, on forced convection, the effects of buoyancy were neglected, a valid approach in moderate- to high-velocity fluids. Free convection may be significant, however, when low-velocity fluids flow over heated (or cooled) surfaces. A measure of the influence of free convection is provided by the ratio

$$\frac{\mathbf{Gr}}{\mathbf{Re}^2} \propto \frac{\text{buoyancy force}}{\text{inertia force}} \qquad (8.45)$$

whose significance may be verified by combining (5.16) and (8.1). *For $\mathbf{Gr}/\mathbf{Re}^2 > 1.0$ free convection is important.* The regimes of convection are:

1. Free convection: $\mathbf{Gr} \gg \mathbf{Re}^2$ (Sections 8.1, 8.2, and 8.3)
2. Forced convection: $\mathbf{Gr} \ll \mathbf{Re}^2$ (Chapters 6 and 7)
3. Mixed free and forced convection: $\mathbf{Gr} \approx \mathbf{Re}^2$ (this section)

Combining these with the two hydrodynamic domains—laminar and turbulent—yields $3 \times 2 = 6$ subregimes. These are mapped in Fig. 8-10 for flow in vertical tubes and in Fig. 8-11 for flow in horizontal tubes. These figures may be used to ascertain whether superimposed free convection is of significance for

$$10^{-2} < \mathbf{Pr}\,\frac{D}{L} < 1.0$$

The Grashof number is formed with the tube diameter as the characteristic length and with the difference between the tube wall and the bulk fluid temperatures.

In the correlation equations given in Fig. 8-11, $\mathbf{Gz}$ is the *Graetz number*, defined as

$$\mathbf{Gz} \equiv \mathbf{Re}_D \,\mathbf{Pr}\,\frac{D}{L} \qquad (8.46)$$

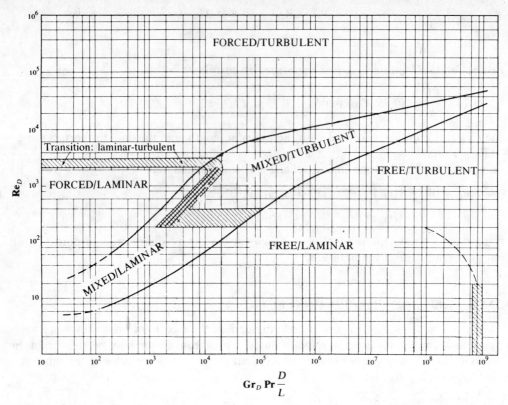

Fig. 8-10. Regimes of flow in vertical tubes. [Adapted from B. Metais and E.R.G. Eckert, *Trans. ASME*, Ser. C, *J. Heat Transfer*, **86** : 295 (1964).]

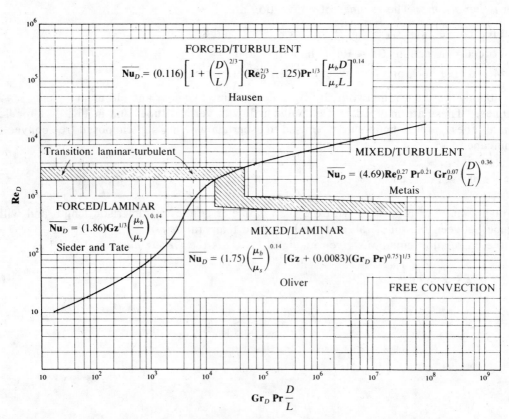

Fig. 8-11. Regimes of flow in horizontal tubes. [Adapted from B. Metais and E.R.G. Eckert, *Trans. ASME*, Ser. C, *J. Heat Transfer*, **86** : 295 (1964).]

Solved Problems

8.1. Using Pohlhausen's similarity parameter,

$$\eta = \frac{y}{x}\left(\frac{\mathbf{Gr}_x}{4}\right)^{1/4}$$

and a stream function $\psi(x, y)$ that involves an unknown function $f(\eta)$,

$$\psi(x, y) = f(\eta)\left[4\nu\left(\frac{\mathbf{Gr}_x}{4}\right)^{1/4}\right]$$

outline the reduction of the three partial differential equations (8.6), (8.7) and (8.8) to two ordinary differential equations. Cite the boundary conditions.

From the definition of a stream function ψ,

$$u = \frac{\partial\psi}{\partial y}\quad\text{and}\quad v = -\frac{\partial\psi}{\partial x}$$

so that (8.8) is satisfied. Expressing the velocity components in terms of x and η, we get

$$u = \frac{\partial\psi}{\partial y} = \frac{\partial\psi}{\partial\eta}\frac{\partial\eta}{\partial y} = f'(\eta)\left[4\nu\left(\frac{\mathbf{Gr}_x}{4}\right)^{1/4}\right]\frac{1}{x}\left(\frac{\mathbf{Gr}_x}{4}\right)^{1/4} = \frac{2\nu}{x}\mathbf{Gr}_x^{1/2}f'$$

$$-v = \frac{\partial\psi}{\partial x} = f'(\eta)\frac{\partial\eta}{\partial x}\left[4\nu\left(\frac{\mathbf{Gr}_x}{4}\right)^{1/4}\right] + f(\eta)\frac{d}{dx}\left[4\nu\left(\frac{\mathbf{Gr}_x}{4}\right)^{1/4}\right]$$

$$= -\frac{\nu}{x}\left(\frac{\mathbf{Gr}_x}{4}\right)^{1/4}(\eta f' - 3f)$$

Defining a dimensionless temperature,

$$\theta \equiv \frac{T - T_\infty}{T_s - T_\infty}$$

taking the respective derivatives and substituting into (8.6) and (8.7), we get

$$f''' + 3ff'' - 2(f')^2 + \theta = 0$$
$$\theta'' + 3\,\mathbf{Pr}\,f\theta' = 0$$

where the primes indicate differentiation with respect to η. Although θ can be eliminated between these two equations, it is simpler to consider them as simultaneous equations coupled through the function f. Each solution must be for a particular Prandtl number, since $\mathbf{Pr}$ appears as a parameter. The boundary conditions are:

at $\eta = 0$: $f = 0$, $f' = 0$, $\theta = 1$
as $\eta \to \infty$: $f' \to 0$, $\theta \to 0$

Figures 8-2 and 8-3 give the solution for a wide range of Prandtl numbers.

8.2. Water is heated by a 6-in. by 6-in. vertical flat plate which is maintained at 126 °F. Using the similarity solution of Section 8.1, find the heat transfer rate when the water is at 68 °F.

At the reference temperature, $T_{\text{ref}} = T_s + (0.38)(T_\infty - T_s) = 104$ °F, the pertinent parameters from Table B-3 are:

$$\rho = 62.09\text{ lbm/ft}^3 \qquad k = 0.363\text{ Btu/hr-ft-°F}$$
$$\nu = 0.708 \times 10^{-5}\text{ ft}^2/\text{sec} \qquad \mathbf{Pr} = 4.34$$

From Fig. 8-3, the temperature gradient at the wall is, for $\mathbf{Pr} \approx 4$,

$$\left.\frac{d\theta}{d\eta}\right|_{\eta=0} \approx -0.9 = -F_1(\mathbf{Pr})$$

and (8.15) gives the average heat-transfer coefficient as

$$\bar{h} = \frac{4}{3}(0.9)\frac{k}{L}\left(\frac{\mathbf{Gr}_L}{4}\right)^{1/4}$$

In order to get the Grashof number, the coefficient of volume expansion β may be obtained from (8.5), where ρ_∞ is taken as the density at a temperature only slightly removed from the reference or film temperature. Choosing this as 100 °F,

$$\rho_\infty = \rho_{100} = 62.13 \text{ lbm/ft}^3$$

by interpolation in Table B-3; therefore,

$$\beta_{104} \approx \rho_{100}\left[\frac{\frac{1}{\rho_{104}} - \frac{1}{\rho_{100}}}{104 - 100}\right] = (62.13)\left[\frac{\frac{1}{62.09} - \frac{1}{62.13}}{4}\right] = 1.611 \times 10^{-4} \text{ °F}^{-1}$$

Then

$$\mathbf{Gr}_L = \frac{g\beta(T_s - T_\infty)L^3}{\nu^2}$$

$$= \frac{(32.2 \text{ ft/sec}^2)(1.611 \times 10^{-4} \text{ °F}^{-1})[(126 - 68) \text{ °F}](0.5 \text{ ft})^3}{(0.708 \times 10^{-5})^2 \text{ ft}^4/\text{sec}^2} = 7.50 \times 10^8$$

and

$$\bar{h} = \frac{4}{3}(0.9)\left[\frac{0.363 \frac{\text{Btu}}{\text{hr-ft-°F}}}{0.5 \text{ ft}}\right]\left[\frac{7.50 \times 10^8}{4}\right]^{1/4} = 101.95 \text{ Btu/hr-ft}^2\text{-°F}$$

The heat transfer is then

$$q = \bar{h}A(T_s - T_\infty) = \left(101.95 \frac{\text{Btu}}{\text{hr-ft}^2\text{-°F}}\right)(0.25 \text{ ft}^2)[(126 - 68) \text{ °F}] = 1478 \text{ Btu/hr}$$

8.3.   Solve Problem 8.2 using the empirical equation (8.28).

At the mean film temperature, $T_f = (T_s + T_\infty)/2 = 97.0$ °F, the pertinent parameters are:

$$k = 0.360 \text{ Btu/hr-ft-°F} \qquad \mathbf{Pr} = 4.86 \qquad \mathbf{Gr}_L = 6.16 \times 10^8 \qquad \mathbf{Ra} = \mathbf{Gr}_L\,\mathbf{Pr} = 2.994 \times 10^9$$

The flow is turbulent; Table 8-3 and (8.28) give

$$\frac{\bar{h}L}{k} = (0.13)(\mathbf{Gr}_L\,\mathbf{Pr})^{1/3}$$

$$\bar{h} = \frac{0.36 \frac{\text{Btu}}{\text{hr-ft-°F}}}{0.5 \text{ ft}}(0.13)(2.994 \times 10^9)^{1/3} = 134.9 \text{ Btu/hr-ft}^2\text{-°F}$$

This gives a larger value for the heat transfer than that obtained in Problem 8.2, i.e.

$$q_{\text{emp}} = \frac{134.9}{101.95}(1478) = 1956 \text{ Btu/hr}$$

It is worth noting here that this value is perhaps more reliable than that of Problem 8.2, since the similarity solution required an estimation of the slope in Fig. 8-3.

8.4.   Assume that the heating element of Problem 8.2 is a 2-in. by 18-in. rectangle rather than being square. With the two inch dimension vertical, estimate the heat transfer.

This changes the Grashof number, viz.

$$\mathbf{Gr}_L = \left(\frac{2}{6}\right)^3 (6.16 \times 10^8) = 2.281 \times 10^7$$

and

$$\mathbf{Gr}_L\,\mathbf{Pr} = (2.281 \times 10^7)(4.86) = 1.108 \times 10^8$$

The laminar constants of Table 8-3 give

$$\frac{\bar{h}L}{k} = (0.59)(\mathbf{Gr}_L\,\mathbf{Pr})^{1/4}$$

$$\bar{h} = \frac{0.360\ \text{Btu/hr-ft-°F}}{\dfrac{2}{12}\ \text{ft}}(0.59)(1.108 \times 10^8)^{1/4} = 130.75\ \text{Btu/hr-ft}^2\text{-°F}$$

This gives the heat transfer:

$$q = \bar{h}A(T_s - T_\infty) = (130.75)(0.25)(126 - 68) = 1896\ \text{Btu/hr}$$

Comparing this result with the square-plate empirical result of Problem 8.3, we conclude that the square plate is better.

8.5. Estimate the heat loss from a vertical wall exposed to nitrogen at one atmosphere and 40 °F. The wall is 6 ft high and 8 ft wide. It is maintained at 120 °F.

From Table B-4, for nitrogen at a mean film temperature of $T_f = (T_s + T_\infty)/2 = 80$ °F,

$$\rho = 0.0713\ \text{lbm/ft}^3 \qquad k = 0.01514\ \text{Btu/hr-ft-°F}$$
$$\nu = 16.82 \times 10^{-5}\ \text{ft}^2/\text{sec} \qquad \mathbf{Pr} = 0.713$$

For the gas, $\beta = 1/T = 1/540 = 1.852 \times 10^{-3}$ °R^{-1}, giving a Grashof number of

$$\mathbf{Gr}_L = \frac{g\beta(T_s - T_\infty)L^3}{\nu^2}$$

$$= \frac{(32.2\ \text{ft/sec}^2)(1.852 \times 10^{-3}\ \text{°R}^{-1})[(120 - 40)\ \text{°R}](6\ \text{ft})^3}{(16.82 \times 10^{-5})^2\ \text{ft}^4/\text{sec}^2} = 3.64 \times 10^{10}$$

and

$$\mathbf{Gr}_L\,\mathbf{Pr} = (3.64 \times 10^{10})(0.713) = 2.59 \times 10^{10}$$

The flow is turbulent; (8.28) with the appropriate constants from Table 8-3 gives

$$\frac{\bar{h}L}{k} = (0.13)(\mathbf{Gr}_L\,\mathbf{Pr})^{1/3}$$

$$\bar{h} = \frac{0.01514\ \dfrac{\text{Btu}}{\text{hr-ft-°F}}}{6\ \text{ft}}(0.13)(2.59 \times 10^{10})^{1/3} = 0.9714\ \text{Btu/hr-ft}^2\text{-°F}$$

and the heat loss is given by

$$q = \bar{h}A(T_s - T_\infty) = \left(0.9714\ \frac{\text{Btu}}{\text{hr-ft}^2\text{-°F}}\right)(48\ \text{ft}^2)[(120 - 40)\ \text{°F}] = 3730\ \text{Btu/hr}$$

8.6. What is the maximum vertical velocity in the boundary layer of Problem 8.5 at $x = 3$ ft (from the bottom of the wall)?

Using (8.22) and (8.23) and the parameters from Problem 8.5, we get

$$\mathbf{Gr}_x = \left(\frac{3}{6}\right)^3 \mathbf{Gr}_L = \frac{3.64 \times 10^{10}}{8} = 4.55 \times 10^9$$

$$\delta = (3.93)x \left[\frac{0.952 + \mathbf{Pr}}{\mathbf{Gr}_x \, \mathbf{Pr}^2}\right]^{1/4} = (3.93)(3 \text{ ft}) \left[\frac{0.952 + 0.713}{(4.55 \times 10^9)(0.713)^2}\right]^{1/4} = 0.061 \text{ ft}$$

$$V = (5.17)\frac{\nu}{x}\left[\frac{\mathbf{Gr}_x}{0.952 + \mathbf{Pr}}\right]^{1/2} = (5.17)\left(\frac{16.82 \times 10^{-5} \frac{\text{ft}^2}{\text{sec}}}{3 \text{ ft}}\right)\left(\frac{4.55 \times 10^9}{0.952 + 0.713}\right)^{1/2} = 15.15 \frac{\text{ft}}{\text{sec}}$$

Hence, from (8.21),

$$\frac{u}{V} = \frac{y}{\delta}\left(1 - \frac{y}{\delta}\right)^2$$

$$u = \frac{15.15}{0.061} y \left(1 - \frac{y}{0.061}\right)^2 = (248.4)y(1 - 16.39\,y)^2 \tag{1}$$

Setting du/dy equal to zero to maximize, we get

$$(805.89)y^2 - (65.56)y + 1 = 0$$

$$y = \frac{65.56 \pm \sqrt{4298.1 - 3223.6}}{2(805.89)} = \frac{65.56 \pm 32.78}{2(805.89)} = 0.061 \text{ ft} \quad \text{or} \quad 0.020 \text{ ft}$$

The root $y = 0.061$ ft corresponds to $u = 0$, the boundary layer edge. Thus, $y = 0.020$, and, from (1),

$$u_{max} = (248.4)(0.020)[1 - (16.39)(0.020)]^2 = 2.245 \text{ ft/sec}$$

8.7.   In Problem 8.5, what is the mass-flow rate of nitrogen past the station $x = 3$ ft?

Assuming the density is constant, the mass-flow rate is given by (W = width):

$$\dot{m} = \rho W \int_0^\delta u \, dy$$

$$= (248.4)\rho W \int_0^\delta (y - 32.78\,y^2 + 268.63\,y^3)\, dy$$

$$= (248.4)\rho W \left[\frac{y^2}{2} - 10.93\,y^3 + 67.16\,y^4\right]_0^{\delta = 0.061}$$

$$= (248.4)\left(0.0713\,\frac{\text{lbm}}{\text{ft}^3}\right)(8 \text{ ft})\left[\frac{(0.061)^2}{2} - (10.93)(0.061)^3 + (67.16)(0.061)^4\right]\frac{\text{ft}^2}{\text{sec}}$$

$$= 0.044 \text{ lbm/sec}$$

8.8.   Estimate the heat transfer from a 40-W incandescent bulb at 127 °C to 27 °C quiescent air.   Approximate the bulb as a 50-mm-dia. sphere.   What percent of the power is lost by free convection?

From Table B-4 the required parameters, evaluated at $T_f = (T_s + T_\infty)/2 = 77$ °C, are:

$$\nu = (22.38 \times 10^{-5})(0.0929) = 2.079 \times 10^{-5} \text{ m}^2/\text{s} \qquad \mathbf{Pr} = 0.697$$

$$k = (0.01735)(1.729) = 0.0300 \text{ W/m-K} \qquad \beta = 1/T = 1/350 = 2.857 \times 10^{-3} \text{ K}^{-1}$$

Evaluating the Grashof number, where the characteristic length L is the diameter of the sphere, gives

$$\mathbf{Gr}_L = \frac{g\beta(T_s - T_\infty)D^3}{\nu^2}$$

$$= \frac{(9.80 \text{ m/s}^2)(2.857 \times 10^{-3} \text{ K}^{-1})(100 \text{ K})(0.050 \text{ m})^3}{(2.079 \times 10^{-5})^2 \text{ m}^4/\text{s}^2} = 8.1 \times 10^5$$

Equation (8.28) governs when the appropriate constants are chosen from Table 8-3; hence,

$$\frac{\bar{h}D}{k} = (0.60)(\mathbf{Gr}_L \, \mathbf{Pr})^{1/4}$$

$$\bar{h} = \frac{0.0300 \text{ W/m-K}}{0.050 \text{ m}}(0.60)[(8.1 \times 10^5)(0.697)]^{1/4} = 9.87 \text{ W/m}^2\text{-K}$$

The heat transfer is

$$q = \bar{h}A(T_s - T_\infty) = \left(9.87 \, \frac{\text{W}}{\text{m}^2\text{-K}}\right)\pi(0.050 \text{ m})^2(100 \text{ K}) = 7.75 \text{ W}$$

The percentage lost by free convection is, therefore,

$$\frac{7.75}{40}(100\%) = 19.37\%$$

This result is of the same order of magnitude as that obtained for the same configuration in forced convection (Problem 7.15). In such cases, both free and forced convection should be considered, as in Section 8.4.

8.9. What heat load is generated in a restaurant by a 1.0-m by 0.8-m grill which is maintained at 134 °C? The room temperature is 20 °C.

The appropriate parameters from Table B-4, evaluated at the mean film temperature,

$$T_f = \frac{T_s + T_\infty}{2} = 77 \text{ °C}$$

are:

$$\nu = (22.38 \times 10^{-5})(9.29 \times 10^{-2}) = 2.08 \times 10^{-5} \text{ m}^2/\text{s}$$
$$k = (0.01735)(1.729) = 0.0300 \text{ W/m-K}$$
$$\mathbf{Pr} = 0.697$$

For air, $\beta = 1/T = 1/350 = 2.857 \times 10^{-3} \text{ K}^{-1}$, and the Grashof number is

$$\mathbf{Gr}_L = \frac{g\beta(T_s - T_\infty)L^3}{\nu^2}$$

where $L = 0.9$ m, the average of the lengths of the two edges.

$$\mathbf{Gr}_L = \frac{(9.80 \text{ m/s}^2)(2.857 \times 10^{-3} \text{ K}^{-1})(114 \text{ K})(0.9 \text{ m})^3}{(2.08 \times 10^{-5})^2 \text{ m}^4/\text{s}^2} = 5.38 \times 10^9$$

Using (8.28), with the constants from Table 8-3 for a heated plate facing up, we get

$$\frac{\bar{h}L}{k} = (0.14)(\mathbf{Gr}_L \, \mathbf{Pr})^{1/3}$$

$$\bar{h} = \frac{0.0300 \text{ W/m-K}}{0.9 \text{ m}}(0.14)[(5.38 \times 10^9)(0.697)]^{1/3} = 7.25 \text{ W/m}^2\text{-K}$$

This gives a heat transfer of

$$q = \bar{h}A(T_s - T_\infty) = \left(7.25 \, \frac{\text{W}}{\text{m}^2\text{-K}}\right)[(1.0 \times 0.8) \text{ m}^2](114 \text{ K}) = 661.2 \text{ W}$$

8.10. What electrical power is required to maintain a 0.003-in.-diameter, 2-ft-long vertical wire at 260 °F in an atmosphere of quiescent air at 80 °F? The wire's resistance is 0.0036 ohms per foot.

At a mean film temperature $T_f = (T_s + T_\infty)/2 = 170$ °F, fluid properties from Table B-4 are:

$$\nu = 22.38 \times 10^{-5} \text{ ft}^2/\text{sec} \qquad \mathbf{Pr} = 0.697$$
$$k = 0.01735 \text{ Btu/hr-ft-°F} \qquad \beta = 1/T = 1/630 = 1.587 \times 10^{-3} \text{ °R}^{-1}$$

The characteristic length for the wire is its diameter, giving

$$\mathbf{Gr}_D = \frac{g\beta(T_s - T_\infty)D^3}{\nu^2}$$

$$= \frac{\left(32.2 \dfrac{\text{ft}}{\text{sec}^2}\right)(1.587 \times 10^{-3} \text{ °R}^{-1})[(260 - 80) \text{ °R}]\left(\dfrac{0.003}{12} \text{ ft}\right)^3}{(22.38 \times 10^{-5})^2 \text{ ft}^4/\text{sec}^2} = 2.869 \times 10^{-3}$$

The parameter required for using Fig. 8-5 is

$$\mathbf{Gr}_D \, \mathbf{Pr} \, \frac{D}{L} = (2.869 \times 10^{-3})(0.697)\left(\frac{0.003}{24}\right) = 2.50 \times 10^{-7}$$

$$\log\left(\mathbf{Gr}_D \, \mathbf{Pr} \, \frac{D}{L}\right) = -6.60$$

From Fig. 8-5,

$$\overline{\mathbf{Nu}_D} = \frac{\bar{h}D}{k} \approx 0.37$$

$$\bar{h} \approx (0.37)\left(\frac{0.01735 \dfrac{\text{Btu}}{\text{hr-ft-°F}}}{\dfrac{0.003}{12} \text{ ft}}\right) = 25.68 \text{ Btu/hr-ft}^2\text{-°F}$$

The ohmic power loss is given by $I^2R = q = \bar{h}A(T_s - T_\infty)$. Thus

$$I^2R = \left(25.68 \frac{\text{Btu}}{\text{hr-ft}^2\text{-°F}}\right)\left[\pi\left(\frac{0.003}{12} \text{ ft}\right)(2 \text{ ft})\right][(260 - 80) \text{ °F}]$$

$$= \left(7.26 \frac{\text{Btu}}{\text{hr}}\right)\left(\frac{\text{W}}{3.412 \text{ Btu/hr}}\right) = 2.13 \text{ W}$$

8.11. Atmospheric air is between two parallel, vertical plates separated by 1 in. The plates, which are 6 ft high and 4 ft wide, are at temperatures of 120 °F and 40 °F. Estimate the heat transfer across the air space.

Evaluating the fluid properties at the average temperature of the two plates, 80 °F, we have, using Table B-4,

$$\nu = 16.88 \times 10^{-5} \text{ ft}^2/\text{sec} \qquad k = 0.01516 \text{ Btu/hr-ft-°F}$$

and

$$\beta = 1/T = 1/540 = 1.852 \times 10^{-3} \text{ °R}^{-1}$$

The Grashof number is based upon the thickness between the plates:

$$\mathbf{Gr}_b = \frac{g\beta(T_1 - T_2)b^3}{\nu^2}$$

$$= \frac{\left(32.2 \, \frac{\text{ft}}{\text{sec}^2}\right)(1.852 \times 10^{-3} \, °\text{R}^{-1})[(120 - 40) \, °\text{R}]\left(\frac{1}{12} \, \text{ft}\right)^3}{(16.88 \times 10^{-5} \, \text{ft}^2/\text{sec})^2} = 9.69 \times 10^4$$

and (*8.36*) gives the heat transfer coefficient, i.e.

$$\overline{\mathbf{Nu}_b} \equiv \frac{\bar{h}b}{k} = (0.18)\mathbf{Gr}_b^{1/4}\left(\frac{L}{b}\right)^{-1/9}$$

$$\bar{h} = \frac{0.01516 \, \dfrac{\text{Btu}}{\text{hr-ft-°F}}}{\dfrac{1}{12} \, \text{ft}} (0.18)(9.69 \times 10^4)^{1/4}\left[\frac{6}{1/12}\right]^{-1/9} = 0.359 \, \text{Btu/hr-ft}^2\text{-°F}$$

The heat transfer is given by (*8.29*):

$$q = \bar{h}A(T_1 - T_2) = \left(0.359 \, \frac{\text{Btu}}{\text{hr-ft}^2\text{-°F}}\right)(24 \, \text{ft}^2)[(120 - 40) \, °\text{F}] = 689.3 \, \text{Btu/hr}$$

8.12. Air at 50 psia is contained between two concentric spheres having radii of 4 in. and 3 in. Estimate the heat transfer when the inner sphere is at 120 °F and the outer sphere is at 40 °F.

The required parameters may be determined using the data from Table B-4 when the properties are evaluated at the average temperature, 80 °F.

$$\mu_m = 1.241 \times 10^{-5} \, \text{lbm/sec-ft} \qquad k = 0.01516 \, \text{Btu/hr-ft-°F}$$

Since the air is above atmospheric pressure, the density is given by

$$\rho = \frac{p}{RT} = \frac{50(144) \, \text{lbf/ft}^2}{\left(53.3 \, \dfrac{\text{ft-lbf}}{\text{lbm-°R}}\right)(540 \, °\text{R})} = 0.250 \, \text{lbm/ft}^3$$

which gives a kinematic viscosity of

$$\nu = \frac{\mu_m}{\rho} = \frac{1.241 \times 10^{-5} \, \text{lbm/sec-ft}}{0.250 \, \text{lbm/ft}^3} = 4.96 \times 10^{-5} \, \text{ft}^2/\text{sec}$$

The coefficient of volume expansion is the reciprocal of absolute temperature, i.e.

$$\beta = 1/T = 1/540 = 1.852 \times 10^{-3} \, °\text{R}^{-1}$$

For this configuration the Grashof number is based upon the gap spacing, $b = r_o - r_i$:

$$\mathbf{Gr}_b = \frac{g\beta(T_1 - T_2)b^3}{\nu^2}$$

$$= \frac{\left(32.2 \, \frac{\text{ft}}{\text{sec}^2}\right)(1.852 \times 10^{-3} \, °\text{R}^{-1})[(120 - 40) \, °\text{R}]\left(\frac{1}{12} \, \text{ft}\right)^3}{(4.96 \times 10^{-5} \, \text{ft}^2/\text{sec})^2} = 1.12 \times 10^6$$

The effective thermal conductivity, k_e, is determined from (*8.40*) ($b/r_i = 0.33$).

$$k_e = k(0.106)\mathbf{Gr}_b^{0.276}$$

$$= \left(0.01516 \, \frac{\text{Btu}}{\text{hr-ft-°F}}\right)(0.106)(1.12 \times 10^6)^{0.276} = 0.0751 \, \text{Btu/hr-ft-°F}$$

The heat transfer is then given by (8.41):

$$q = \frac{4\pi k_e r_i r_o}{r_o - r_i}(T_1 - T_2)$$

$$= \frac{4\pi \left(0.0751 \dfrac{\text{Btu}}{\text{hr-ft-}^\circ\text{F}}\right)\left(\dfrac{3}{12}\text{ ft}\right)\left(\dfrac{4}{12}\text{ ft}\right)}{\dfrac{4-3}{12}\text{ ft}}[(120 - 40)\,^\circ\text{F}] = 75.50 \text{ Btu/hr}$$

8.13.   Atmospheric air passes through a 20-mm-diameter horizontal tube at an average velocity of 30 mm/s. The tube is maintained at 127 °C, and the bulk temperature of the air is 27 °C. Estimate the heat transfer if the tube is 1 m long.

For the bulk temperature, we get the fluid properties from Table B-4.

$$\nu = (16.88 \times 10^{-5})(0.0929) = 1.568 \times 10^{-5} \text{ m}^2/\text{s} \qquad \mathbf{Pr} = 0.708$$
$$k = (0.01516)(1.729) = 0.0262 \text{ W/m-K} \qquad\qquad \mu = (1.241 \times 10^{-5})(1.488)$$
$$= 1.847 \times 10^{-5} \text{ kg/m-s}$$

and

$$\beta = 1/T = 1/300 = 3.33 \times 10^{-3} \text{ K}^{-1}$$

At the wall the viscosity is

$$\mu_s = (1.536 \times 10^{-5})(1.488) = 2.286 \times 10^{-5} \text{ kg/m-s}$$

In order to determine the regime of interest, we need the Reynolds and Grashof numbers.

$$\mathbf{Re}_D = \frac{VD}{\nu} = \frac{(0.030 \text{ m/s})(0.020 \text{ m})}{1.568 \times 10^{-5} \text{ m}^2/\text{s}} = 38.28$$

$$\mathbf{Gr}_D = \frac{g\beta(T_s - T_b)D^3}{\nu^2}$$

$$= \frac{(9.80 \text{ m/s}^2)(3.33 \times 10^{-3} \text{ K}^{-1})[(127 - 27)\text{ K}](0.020 \text{ m})^3}{(1.568 \times 10^{-5} \text{ m}^2/\text{s})^2} = 1.06 \times 10^5$$

and

$$\mathbf{Gr}_D\,\mathbf{Pr}\,\frac{D}{L} = (1.06 \times 10^5)(0.708)\left(\frac{0.020}{1}\right) = 1504$$

Using these parameters, we see from Fig. 8-11 that the flow is laminar, mixed convection; Oliver's equation will be used to evaluate the Nusselt number. The Graetz number is, from (8.46),

$$\mathbf{Gz} = \mathbf{Re}_D\,\mathbf{Pr}\,\frac{D}{L} = (38.28)(0.708)\left(\frac{0.020}{1}\right) = 0.542$$

Thus,

$$\overline{\mathbf{Nu}_D} = \frac{\bar{h}D}{k} = (1.75)\left[\frac{1.847 \times 10^{-5}}{2.286 \times 10^{-5}}\right]^{0.14}\{0.542 + (0.0083)[(1.06 \times 10^5)(0.708)]^{0.75}\}^{1/3} = 5.72$$

and

$$\bar{h} = \frac{k}{D}(5.72) = \frac{0.0262 \text{ W/m-K}}{0.020 \text{ m}}(5.72) = 7.49 \text{ W/m}^2\text{-K}$$

The heat transfer for a length of 1 m is given by

$$q = \bar{h}A(T_s - T_b) = \left(7.49 \frac{\text{W}}{\text{m}^2\text{-K}}\right)[\pi(0.020 \text{ m})(1 \text{ m})][(127 - 27)\text{ K}] = 47.1 \text{ W}$$

As a matter of interest, we might compare the result for $\bar{h}$ with that obtained from the equation given in Fig. 8-11 for the forced convection, laminar regime.

$$\overline{\mathbf{Nu}_D} = \frac{\bar{h}D}{k} = (1.86)\mathbf{Gz}^{1/3}\left(\frac{\mu_b}{\mu_s}\right)^{0.14}$$

$$\bar{h} = \frac{0.0262 \text{ W/m-K}}{0.020 \text{ m}}(1.86)(0.542)^{1/3}\left(\frac{1.847 \times 10^{-5}}{2.286 \times 10^{-5}}\right)^{0.14} = 1.93 \text{ W/m}^2\text{-K}$$

Therefore, if the calculation were made on the basis of forced convection, laminar flow alone, the result would be highly erroneous.

Supplementary Problems

8.14. For an ideal gas having the equation of state $p = \rho RT$, show that the coefficient of volume expansion is $1/T$.

8.15. Using the empirical equation (*8.28*) for turbulent flow on a vertical flat plate, show that the heat transfer coefficient is independent of plate height L.

8.16. For film properties evaluated at 127 °C, estimate the temperature difference $(T_s - T_\infty)$ required to lose 4 W by natural convection from a 120-mm-diameter sphere to surrounding air.   *Ans.*   17.75 °C

8.17. The front panel of a dishwasher is at 95 °F during the drying cycle.   What is the rate of heat gain by the room, which is maintained at 65 °F?   The panel is 2.5 ft square.
Ans.   108.9 Btu/hr using (*8.28*)

8.18. An uninsulated 10.75-in.-o.d. steam line passes through a 9-ft-high room in which the air is at 120 °F.   What is the heat loss per foot of pipe if its surface is at 760 °F when (*a*) horizontal, (*b*) vertical?   *Ans.*   (*a*) 2193 Btu/hr-ft; (*b*) 2394 Btu/hr-ft using (*8.28*)

Chapter 9

Boiling and Condensation

This chapter deals with the two most common *phase-change processes*: vaporization and its inverse, condensation. As in simple convection, a heat transfer coefficient h is used to relate the heat flux to the temperature differential between the heating surface and the saturated liquid.

$$q = hA(T_s - T_{sat}) \tag{9.1}$$

However, since phase-change processes involve changes in density, viscosity, specific heat and thermal conductivity of the fluid, while the fluid's latent heat is either liberated (condensation) or absorbed (vaporization), the heat transfer coefficient for boiling and condensation is much more complicated than that for single-phase convective processes. Because of this, most engineering calculations involving boiling and condensation are made from empirical correlations.

9.1 BOILING PHENOMENA

Consider a pool of fluid being heated from below, e.g. by a submerged wire. For low rates of heat addition, vapor will be formed at the free surface. As the heat flux increases, bubbles form at the heater surface and change in size while rising through the fluid, in addition to the free-surface vaporization. This bubble formation, with its attendant agitation, is called *boiling,* or *ebullition.*

The behavior of a fluid during boiling is highly dependent upon the *excess temperature,* $\Delta T = T_s - T_{sat}$, measured from the boiling point of the fluid. Figure 9-1 indicates six different regimes for typical *pool boiling*; the heat flux curve is commonly called the *boiling curve.*

Regime I.　Heat is transferred by free convection, described in detail in Chapter 8.

Regime II.　Bubbles begin to appear at the heating surface and rise to the free surface individually.

Regime III.　The boiling action becomes so vigorous that individual bubbles combine with others very rapidly to form a vapor bubble column reaching to the free surface.

Regime IV.　Bubbles form so rapidly that they blanket the heating surface, preventing fresh fluid from moving in to take their place. The increased resistance of this film reduces the heat flux, and the heat transfer decreases with increasing temperature differential. Because the film intermittently collapses and reappears, this regime is very unstable.

Regime V.　The film on the heater surface becomes stable. As ΔT reaches about 1000 °F, radiant heat transfer comes into play—in fact, becomes predominant—and the heat flux again rises with increasing ΔT.

The peak heat flux, point B, is called the *burnout point.* It is the condition at which the increased heat flux produced by a rise in ΔT is offset by the increased resistance of the vapor blanket around the heater. The two effects balance, producing what is sometimes called the *boiling crisis, burnout,* or *departure from nucleate boiling.* For many common fluids, the temperature at D is above the melting point of most heater materials, and failure of the heater occurs before reaching it. If the heater does not melt, the boiling curve continues to rise beyond point D.

216

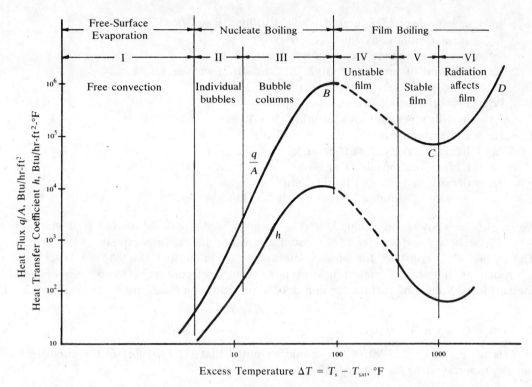

Fig. 9-1.   Horizontal, Chromel C 0.040-in.-diameter heating wire in water at 1 atm.

As boiling is predominantly a local phenomenon, the heat transfer coefficient h is normally given without the overbar, as in (9.1). Most applications, however, require an average heat flux. Since burnout of heating elements is a common problem in boiling, and since the largest heat flux is a local quantity for a given regime, the local value is the one that should be used in design, being the conservative value.

9.2   POOL BOILING

Free Convection (Regime I)

Using the general equation (8.28), the heat transfer in this regime is given by

$$\frac{q}{A} = C\frac{k}{L}(\mathbf{Gr}_L \, \mathbf{Pr})^a (T_s - T_b) \tag{9.2}$$

where T_b is the bulk temperature, and where the constants a and C are taken from Table 8-3. Since $\mathbf{Gr}_L \equiv g\beta(T_s - T_b)L^3/\nu^2$ and since the exponent a is usually 1/4 for laminar flow and 1/3 for turbulent flow, the heat transfer in this regime varies with ΔT to the 5/4 power for laminar flow, 5/3 for turbulent.

Nucleate Boiling (Regimes II and III)

The most commonly accepted general correlation for heat transfer in the nucleate boiling regimes is that due to W. M. Rohsenow.

$$\frac{q}{A} = \mu_l h_{\mathrm{fg}} \sqrt{\frac{g(\rho_l - \rho_v)}{g_c \sigma}} \left[\frac{c_l(T_s - T_{\mathrm{sat}})}{h_{\mathrm{fg}} \, \mathbf{Pr}_l^{1.7} \, C_{sf}} \right]^3 \tag{9.3}$$

where c_l = specific heat of saturated liquid, Btu/lbm-°F or J/kg-K
       C_{sf} = surface–fluid constant (Table 9-1)
        g = local gravitational acceleration, ft/sec^2 or m/s^2
       g_c = constant of proportionality, 32.17 lbm-ft/lbf-sec^2 or 1.0 kg-m/N-s^2
      h_{fg} = enthalpy of vaporization, Btu/lbm or J/kg
      $\mathbf{Pr}_l$ = Prandtl number of saturated liquid
      q/A = heat flux per unit area, Btu/hr-ft^2 or W/m^2
$T_s - T_{sat}$ = excess temperature, °F or K
       μ_l = liquid viscosity, lbm/ft-hr or kg/m-s
        σ = surface tension, lbf/ft or N/m
       ρ_l = density of saturated liquid, lbm/ft^3 or kg/m^3
       ρ_v = density of saturated vapor, lbm/ft^3 or kg/m^3

Note: In the above, and throughout this chapter, μ denotes the mass-based viscosity co-efficient.   Subscripts l and v refer to the liquid and vapor states, respectively.

The surface–fluid constant, for which some values are given in Table 9-1, is a function of the surface roughness (number of nucleating sites) and the angle of contact between the bubble and the heating surface.   Values of surface tension σ of some common fluids are given in Fig. 9-2.   For water,

$$\sigma = (0.00528)(1 - 0.0013\ T) \tag{9.4}$$

where T is in °F and σ is in lbf/ft.   It should be noted that *the heat flux in the nucleate boiling regimes is proportional to the cube of* ΔT.

Peak heat flux.   At the point of maximum heat transfer (point B of Fig. 9-1), the recommended correlation is

$$\frac{q}{A}\bigg|_{max} = (0.18)\rho_v h_{fg}\left[\frac{\sigma(\rho_l - \rho_v)gg_c}{\rho_v^2}\right]^{1/4}\left[\frac{\rho_l}{\rho_l + \rho_v}\right]^{1/2} \tag{9.5}$$

Observe that the peak heat flux is independent of the heating element.

Table 9-1

Surface–Fluid Combination	C_{sf}
Water–brass	0.006
Water–copper	0.013
Water–nickel	0.006
Water–platinum	0.013
CCl$_4$–copper	0.013
Benzene–chromium	0.010
n-Pentane–chromium	0.015
Ethyl alcohol–chromium	0.0027
Isopropyl alcohol–copper	0.0025
35% K$_2$CO$_3$–copper	0.0054
50% K$_2$CO$_3$–copper	0.0027
n-Butyl alcohol–copper	0.0030

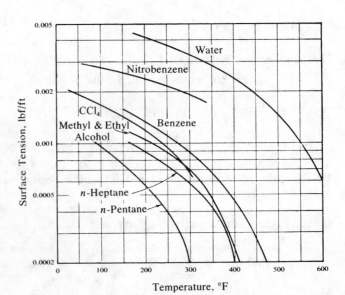

Fig. 9-2

Film Boiling (Regimes IV, V and VI)

Horizontal tube. From a study of conduction through the film on a heated tube and radiation from the tube, L. A. Bromley proposed that the boiling heat-transfer coefficient in these regimes be given by

$$h = h_c \left(\frac{h_c}{h}\right)^{1/3} + h_r \tag{9.6}$$

$$h_c = (0.62)\left[\frac{k_{vf}^3 \rho_{vf}(\rho_l - \rho_{vf})g(h_{fg} + 0.4\, c_{pvf}\, \Delta T)}{D\mu_{vf}\, \Delta T}\right]^{1/4} \tag{9.7}$$

$$h_r = \frac{\sigma\epsilon(T_s^4 - T_{sat}^4)}{T_s - T_{sat}} \tag{9.8}$$

In (9.8) σ is the Stefan-Boltzmann constant and ϵ is the emissivity of the surface. In (9.7), D is the outside diameter of the tube, and, as indicated by the additional subscript f, vapor properties are taken at the mean film temperature, $T_f = (T_s + T_{sat})/2$.

Equation (9.6) is difficult to use since h is in it implicitly. The following explicit equations are much simpler when the inherent approximation errors and ranges are acceptable, which is the case in most problems of engineering interest.

$$\pm 0.3 \text{ percent:} \quad h = h_c + h_r\left[\frac{3}{4} + \frac{1}{4}\frac{h_r}{h_c}\left(\frac{1}{2.62 + \dfrac{h_r}{h_c}}\right)\right] \quad \left(0 < \frac{h_r}{h_c} < 10\right)$$

$$\tag{9.9}$$

$$\pm 5 \text{ percent:} \quad h = h_c + \frac{3}{4}h_r \quad \left(\frac{h_r}{h_c} < 1\right)$$

Vertical tube. For vertical tubes, Y. Y. Hsu and J. W. Westwater proposed the correlation

$$h = (0.0020)\text{Re}^{0.6}\left[\frac{g\rho_v(\rho_l - \rho_v)k_v^3}{\mu_v^2}\right]^{1/3} \tag{9.10}$$

where

$$\text{Re} \equiv \frac{4\dot{m}}{\pi D\mu_v} \tag{9.11}$$

and $\dot{m}$ is the vapor mass-flow rate at the upper end of the tube. For like conditions, the rate of heat transfer is greater for vertical than for horizontal tubes.

Horizontal plane. Verified for boiling in pentane, carbon tetrachloride, benzene and ethyl alcohol, the following correlation is due to P. Berenson:

$$■ \quad h = (0.425)\left[\frac{k_v^3 \rho_v(\rho_l - \rho_v)g(h_{fg} + 0.4\, c_{pv}\, \Delta T)}{\mu_v\, \Delta T\, \sqrt{\sigma g_c/g(\rho_l - \rho_v)}}\right]^{1/4} \tag{9.12}$$

where σ once again denotes the surface tension. The similarity between this result and (9.7) should be noted.

Minimum heat flux. Using the hydrodynamic instability of the liquid-vapor boundary, N. Zuber and M. Tribus found the following equation for the minimum heat flux in film boiling (point C of Fig. 9-1).

$$■ \quad \left.\frac{q}{A}\right|_{min} = (0.09)\rho_{vf}h_{fg}\left[\frac{g(\rho_l - \rho_{vf})}{\rho_l + \rho_{vf}}\right]^{1/2}\left[\frac{g_c\sigma}{g(\rho_l - \rho_{vf})}\right]^{1/4} \tag{9.13}$$

Based upon the Zuber-Tribus analysis, Berenson found an expression for the excess temperature ΔT at the point of minimum heat flux (point C), viz.

$$\Delta T_C = 0.127 \frac{\rho_{vf} h_{fg}}{k_{vf}} \left[\frac{g(\rho_l - \rho_v)}{\rho_l + \rho_v} \right]^{2/3} \left[\frac{g_c \sigma}{g(\rho_l - \rho_v)} \right]^{1/2} \left[\frac{\mu_f}{g_0(\rho_l - \rho_v)} \right]^{1/3} \qquad (9.14)$$

Properties designated by a subscript f in (9.13) and (9.14) are evaluated at the mean film temperature, $T_f = (T_s + T_{sat})/2$; g_0 is the earth's standard gravitation, 32.17 ft/sec² or 9.81 m/s².

Simplified Relations for Water

Since water is the most common fluid used in boiling processes, some simplified relations for boiling with water at atmospheric pressure are given in Table 9-2. The constants in these formulas are not dimensionless; hence, care must be taken to express the quantities in the units given in the table.

The heat transfer coefficient at pressure p is given by the relation

$$h_p = h_a \left(\frac{p}{p_a} \right)^{0.4} \qquad (9.15)$$

where p_a is standard atmospheric pressure and h_a is taken from Table 9-2.

Table 9-2

Configuration	h_a, Btu/hr-ft²-°F	q/A, Btu/hr-ft²
Horizontal surface (in wide vessel)	$h_a = 151(\Delta T)^{1/3}$ $h_a = (0.168)(\Delta T)^3$	$q/A < 5000$ $5000 < q/A < 75,000$
Vertical surface (in wide vessel)	$h_a = 87(\Delta T)^{1/7}$ $h_a = (0.24)(\Delta T)^3$	$q/A < 1000$ $1000 < q/A < 20,000$
Vertical tube (interior)	$h_a \approx 189(\Delta T)^{1/3}$ $h_a \approx (0.21)(\Delta T)^3$	$q/A < 5000$ $5000 < q/A < 75,000$

9.3 FLOW (CONVECTION) BOILING

Flow Boiling Characteristics

Flow boiling may occur when a liquid flows through a passage or over a surface which is maintained at a higher temperature than the saturation temperature of the liquid. The flow is a two-phase mixture of the liquid and its vapor.

Figure 9-3(a) shows a vertical-tube evaporator. A subcooled liquid enters the evaporator, passing over the hotter wall where local nucleate boiling occurs. The flow is bubbly when there is less than about 10 percent vapor. With an increase in bubble agitation, there is an increase in the heat transfer coefficient. At higher qualities the flow becomes annular, with a thin liquid layer on the wall and a vapor core. The velocity of the vapor is much higher than that of the liquid. Heat is transferred through the film by conduction, and vaporization takes place at the liquid-vapor interface primarily, although some bubbles still form at the wall.

In the transition from annular to vapor (also called *mist* or *fog*) flow, the heat transfer coefficient drops sharply. Burnout sometimes occurs at this transition because a liquid film of high thermal conductivity is replaced by a low-thermal-conductivity vapor at the wall. Vapor flow continues until the quality reaches 100 percent, after which the heat transfer coefficient may be determined by the appropriate equations for forced convection, Chapters 6 and 7, using vapor properties.

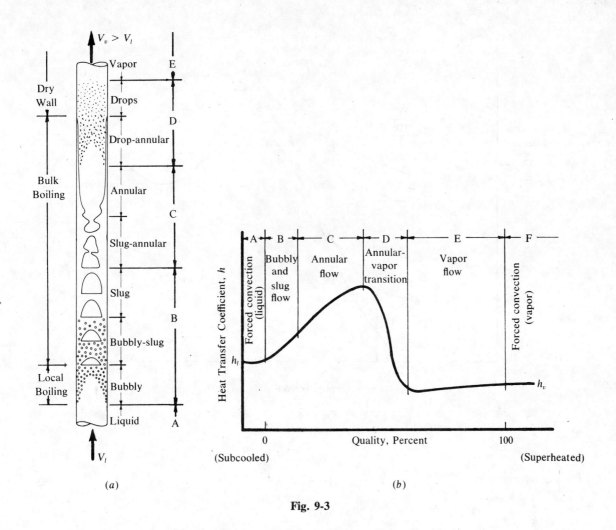

Fig. 9-3

A simplified approach for determination of the heat transfer in flow boiling is to sum the convective effect, either forced or natural convection without boiling, and the boiling effect.

$$\frac{q}{A} = \frac{q}{A}\bigg|_{conv} + \frac{q}{A}\bigg|_{bo} \tag{9.16}$$

Here $(q/A)_{conv} = \bar{h}(T_s - T_b)$, where T_b is the bulk temperature, and $\bar{h}$ is given by the appropriate relations of Chapters 6 and 7. The heat transfer due to boiling, $(q/A)_{bo}$, is given by the relations of Section 9.2, where convection is absent.

Nucleate Boiling

Figure 9-4 shows the boiling curve for a submerged heating wire, taken from Fig. 9-1, with the effects of convection superimposed upon it in the nucleate boiling regimes. The fully-developed flow boiling curve tends to become asymptotic to a projection of the nonflow boiling curve. The forced convection effects, as shown in the figure, were determined by the use of (7.51) for normal flow across a single cylinder.

In the fully-developed boiling region, little heat is transferred by forced convection. In this region, the heat flux for subcooled water inside tubes is given by (British units only):

$$\frac{q}{A} = (0.074)(\Delta T)^{3.86} \qquad (30 < p < 90 \text{ psia}) \tag{9.17}$$

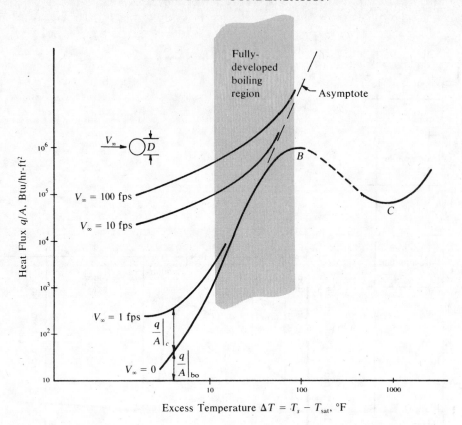

Fig. 9-4

or, for higher pressures,

$$\frac{q}{A} = \frac{p^{4/3}}{495}(\Delta T)^3 \qquad (100 < p < 2000 \text{ psia}) \tag{9.18}$$

Peak heat flux. For water flowing in a pipe of diameter D and length L, a simplified correlation for burnout heat flux was developed by W. K. Lowdermilk, C. D. Lanzo and B. L. Siegel. Their result, valid for the full range of quality and for inlet subcooling from 0 to 140 °F, is:

low velocity, high quality: $\left.\dfrac{q}{A}\right|_{\text{max}} = \dfrac{270(\rho V)^{0.85}}{D^{0.2}(L/D)^{0.85}} \qquad \left(1 < \dfrac{\rho V}{(L/D)^2} < 150\right)$

$$\tag{9.19}$$

high velocity, low quality: $\left.\dfrac{q}{A}\right|_{\text{max}} = \dfrac{1400(\rho V)^{0.50}}{D^{0.2}(L/D)^{0.15}} \qquad \left(150 < \dfrac{\rho V}{(L/D)^2} < 10,000\right)$

with q/A in Btu/hr-ft², ρ in lbm/ft³, V in fps, L in in. and D in in. The individual parameters in (9.19) are restricted to the ranges

$$14.7 < p < 100 \text{ psia} \qquad 25 < L/D < 250$$
$$0.1 < V < 98 \text{ fps} \qquad 0.051 < D < 0.188 \text{ in.}$$

Film Boiling

For forced convection of a liquid flowing normally across horizontal tubes, formulas (9.6), (9.7) and (9.8) may be used when $V_\infty < \sqrt{gD}$. If $V_\infty > 2\sqrt{gD}$, (9.7) is replaced by

$$h_c = (2.7) \left[\frac{V_\infty k_{vf} \rho_{vf} (h_{fg} + 0.4 \, c_{pf} \, \Delta T)}{D \, \Delta T} \right]^{1/2} \tag{9.20}$$

These laminar flow relations have been verified for: benzene, carbon tetrachloride, ethanol, n-heptane; tube diameters from 0.387 to 0.637 in.; and free-stream velocities from 0 to 14 fps.

9.4 CONDENSATION

Condensation, the inverse of boiling, occurs when a saturated vapor comes in contact with a surface at a lower temperature. The liquid collects on the surface, from which it drains under the influence of gravity or is carried off by the drag of the moving vapor. If the motion of the condensate is laminar, which is generally the case, heat is transferred from the vapor-liquid interface to the surface by conduction. The rate of heat transfer depends upon the thickness of the film, depicted for a vertical surface in Fig. 9-5. The film thickness depends upon the rate of condensation and the rate of removal of the condensate. For an inclined plate the drainage rate is lower, which increases the film thickness and decreases the rate of heat transfer.

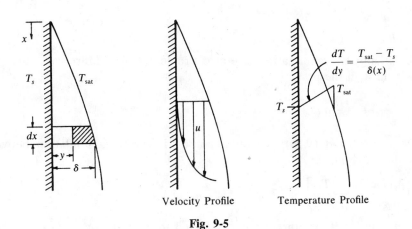

Velocity Profile   Temperature Profile

Fig. 9-5

Two distinct modes of condensation may occur, or they may occur together. The more common mode, *film condensation*, is characterized by a thin liquid film forming over the entire surface. This occurs on clean wettable surfaces in contact with noncontaminated vapors. *Dropwise condensation* occurs on nonwettable surfaces, such as Teflon in the presence of water vapor. In this case minute drops of condensate form, growing in size until they are carried away by gravity or vapor motion. In dropwise condensation, a portion of the condenser surface is exposed to the vapor, making heat transfer rates much larger than those in film condensation.

Only film condensation will be treated in this book.

Laminar Film Condensation

Since most condenser surfaces are short and the film velocity is small, condensation is usually laminar film condensation. For this case the Nusselt analysis provides an insight into the mechanism of condensation.

Making a force balance on a unit depth of the shaded element of Fig. 9-5, neglecting inertia terms (low velocity), we get

$$\frac{1}{g_c} \mu_l \frac{du}{dy} dx = \frac{1}{g_c} g(\rho_l - \rho_v)(\delta - y) \, dx \tag{9.21}$$

The term on the left is the viscous shear force at y, and the term on the right is the difference between the weight and buoyancy forces. The underlying assumptions are: (1) linear temperature gradient in the film; (2) uniform surface temperature T_s; (3) pure vapor at the saturation temperature T_{sat}; and (4) negligible shear at the liquid-vapor interface, i.e. low velocity. For the no-slip boundary condition, $u = 0$ at $y = 0$, (9.21) integrates to give

$$u = \frac{g(\rho_l - \rho_v)}{\mu_l}\left(y\delta - \frac{y^2}{2}\right) \tag{9.22}$$

The condensate mass-flow rate per unit depth, $\dot{m}'$, at any elevation x is given by

$$\dot{m}' = \int_0^\delta \rho_l u\, dy = \frac{\rho_l g(\rho_l - \rho_v)\delta^3}{3\mu_l} \tag{9.23}$$

from which the rate of change of mass flow with respect to condensate thickness is

$$\frac{d\dot{m}'}{d\delta} = \frac{\rho_l g(\rho_l - \rho_v)\delta^2}{\mu_l} \tag{9.24}$$

The excess mass flow rate $d\dot{m}'$ must come from the condensation at the interface, which is given by

$$d\dot{m}' = \frac{dq'}{h_{\text{fg}}} \tag{9.25}$$

where h_{fg} is the enthalpy of vaporization (see also Problem 9.16). Moreover, since the liberated heat is conducted through the film,

$$dq' = \frac{k_l(T_{\text{sat}} - T_s)}{\delta}\, dx \tag{9.26}$$

Combining (9.24), (9.25) and (9.26), we obtain, upon integration, a relation between the film thickness δ and the elevation x:

$$\delta = \left[\frac{4\mu_l k_l x(T_{\text{sat}} - T_s)}{\rho_l g(\rho_l - \rho_v)h_{\text{fg}}}\right]^{1/4} \tag{9.27}$$

Since the heat convected into the film is conducted through it to the plate

$$h_x(T_{\text{sat}} - T_s) = \frac{k_l(T_{\text{sat}} - T_s)}{\delta} \qquad \text{or} \qquad h_x = \frac{k_l}{\delta} \tag{9.28}$$

which, combined with (9.27), gives

$$\mathbf{Nu}_x \equiv \frac{h_x x}{k_l} = \left[\frac{\rho_l g(\rho_l - \rho_v)h_{\text{fg}}x^3}{4\mu_l k_l(T_{\text{sat}} - T_s)}\right]^{1/4} \tag{9.29}$$

Examination of this result shows that the thickening of the condensate film is analogous to the growth of a boundary layer over a flat plate in single-phase convection. Contrary to simple convection, however, an increase in temperature differential, being accompanied by an increase in film thickness, *reduces* the surface conductance.

By integrating the local value of conductance over the entire height L of the plate, we get the average heat-transfer coefficient

$$\bar{h} = \frac{4}{3}h_L = (0.943)\left[\frac{\rho_l g(\rho_l - \rho_v)h_{\text{fg}}k_l^3}{\mu_l L(T_{\text{sat}} - T_s)}\right]^{1/4}$$

More generally for a plate inclined at an angle ϕ with the horizontal, the result is

$$\bar{h} = (0.943)\left[\frac{\rho_l g(\rho_l - \rho_v)h_{\text{fg}}k_l^3}{\mu_l L(T_{\text{sat}} - T_s)}\sin\phi\right]^{1/4}$$

Experimental results have shown that this equation is conservative, yielding results approximately 20 percent lower than measured values. Therefore, the recommended relation for inclined (including vertical) plates is

$$\bar{h} = (1.13) \left[\frac{\rho_l g (\rho_l - \rho_v) h_{fg} k_l^3}{\mu_l L (T_{sat} - T_s)} \sin \phi \right]^{1/4} \tag{9.30}$$

Vertical tubes. Equation (9.30), with $\sin \phi = 1$, is also valid for the inside and outside surfaces of vertical tubes if the tubes are large in diameter D compared to the film thickness δ. However, (9.30) is not valid for inclined tubes, since the film flow would not be parallel to the axis of the tube.

Horizontal tubes. A Nusselt-type analysis for external condensation yields

$$\bar{h} = (0.725) \left[\frac{\rho_l g (\rho_l - \rho_v) h_{fg} k_l^3}{\mu_l D (T_{sat} - T_s)} \right]^{1/4} \tag{9.31}$$

When condensation takes place in a bank of n horizontal tubes arranged in a vertical tier, the condensate from an upper tube flows onto lower tubes affecting the heat transfer rate. In this case, an estimate can be made of the heat transfer, in the absence of empirical relations which account for splashing and other effects, by replacing D in (9.31) by nD.

Turbulent Film Condensation

When a liquid film is vigorous enough, heat is transferred not only by conduction but also by eddy diffusion, a characteristic of turbulence. This may occur on tall vertical surfaces or in banks of horizontal tubes. When such behavior occurs the laminar relations are no longer valid. This change occurs when the film Reynolds number, $\mathbf{Re}_f$, defined by

$$\mathbf{Re}_f \equiv \frac{V D_h \rho_l}{\mu_l} = \frac{4 \rho_l A V}{P \mu_l} \tag{9.32}$$

is approximately 1800; here the hydraulic diameter, $D_h \equiv 4A/P$, is the characteristic length. In this relation, A is the area over which the condensate flows and P is the wetted perimeter. For inclined surfaces of width W, $A/P = LW/W = L$; for vertical tubes, $A/P = \pi DL/\pi D = L$; and, for horizontal tubes, $A/P = \pi DL/L = \pi D$. It should be noted at this point that the transition Reynolds number for a horizontal tube is 3600 rather than 1800, since the film flows down two sides of the tube. This is academic, however, since turbulent flow rarely occurs on a horizontal tube, because of its small vertical dimension.

Noting that $\dot{m} = \rho_l A V$ and that $\dot{m}' = \dot{m}/P$, the film Reynolds number may be expressed as

$$\mathbf{Re}_f = \frac{4 \dot{m}'}{\mu_l} \tag{9.33}$$

where $\dot{m}'$ is the condensate mass flow per unit width for surfaces or per unit length for tubes. Its maximum value occurs at the lower edge of the surface.

Turbulent film condensation on vertical surfaces. The average heat-transfer coefficient, developed by C. G. Kirkbride, is

$$\bar{h} = (0.0076) \mathbf{Re}_f^{0.4} \left[\frac{\rho_l g (\rho_l - \rho_v) k_l^3}{\mu_l^2} \right]^{1/3} \tag{9.34}$$

which is valid for $\mathbf{Re}_f > 1800$.

Determination of the Film Flow Regime

Since the condensate velocity V is unknown in (9.32), a trial-and-error approach is necessary. Expressing the mass-flow rate in terms of the heat transfer, we get

$$\dot{m} = \frac{q}{h_{\text{fg}}} = \frac{\bar{h}A(T_{\text{sat}} - T_s)}{h_{\text{fg}}}$$

which, substituted into (9.33), gives

$$\mathbf{Re}_f = \frac{4\bar{h}A(T_{\text{sat}} - T_s)}{P\mu_l h_{\text{fg}}} \tag{9.35}$$

Since the transition film Reynolds number is known ($\mathbf{Re}_f|_{\text{crit}} \approx 1800$), the process can be simplified by eliminating the heat transfer coefficient for the initial calculation of the film Reynolds number.

From (9.30) and (9.35), *flow on vertical or inclined plates and vertical tubes is laminar if*

$$(4.52)\left[\frac{\rho_l g(\rho_l - \rho_v)k_l^3(T_{\text{sat}} - T_s)^3}{\mu_l^5 h_{\text{fg}}^3} L^3 \sin\phi\right]^{1/4} < 1800 \tag{9.36}$$

From (9.31) and (9.35), *flow on a bank of n horizontal tubes is laminar if*

$$(9.11)\left[\frac{\rho_l g(\rho_l - \rho_v)k_l^3(T_{\text{sat}} - T_s)^3}{\mu_l^5 h_{\text{fg}}^3} n^3 D^3\right]^{1/4} < 3600 \tag{9.37}$$

From (9.34) and (9.35), *flow on vertical surfaces is turbulent if*

$$(0.00296)\left[\frac{\rho_l g(\rho_l - \rho_v)k_l^3(T_{\text{sat}} - T_s)^3}{\mu_l^5 h_{\text{fg}}^3} L^3\right]^{5/9} > 1800 \tag{9.38}$$

If condition (9.36), (9.37) or (9.38) is satisfied, the left side gives $\mathbf{Re}_f$.

The condensation mechanism is somewhat different if the condensing vapor is superheated rather than saturated. Experimental results have shown that in most cases the effect of superheat may be ignored, and the equations for saturated vapors may be used with negligible error. It should be emphasized, however, that $(T_{\text{sat}} - T_s)$ is still the driving differential, and that the actual superheated vapor temperature does not enter the calculations. *In all the condensation relations, condensate properties are evaluated at the mean film temperature*, $T_f = (T_{\text{sat}} + T_s)/2$; *vapor properties are evaluated at the saturation temperature*; *and h_{fg} is that at the saturated vapor temperature.*

Solved Problems

9.1.   Using Fig. 9-1, estimate the excess temperature for a 0.040-in.-diameter, horizontal, 6-in.-long, Chromel C wire submerged in water at atmospheric pressure. The voltage drop in the wire is 14.7 V and the current is 42.8 A.

An energy balance gives

$$q = EI = hA\,\Delta T$$

Since 1 W = 1 V-A,

$$q = [(14.7)(42.8)\text{ W}]\left(3.413\,\frac{\text{Btu}}{\text{hr-W}}\right) = 2147.32\text{ Btu/hr}$$

The surface area of the wire is

$$A = \pi DL = \pi\left(\frac{0.040}{12}\text{ ft}\right)\left(\frac{6}{12}\text{ ft}\right) = 0.005236\text{ ft}^2$$

therefore

$$\frac{q}{A} = \frac{2147.32}{0.005236} = 4.1 \times 10^5 \text{ Btu/hr-ft}^2$$

From Fig. 9-1, $\Delta T \approx 50$ °F.

9.2. A 1.0-mm-diameter wire, 150 mm long, is submerged horizontally in water at atmospheric pressure. The wire has a steady-state applied voltage drop of 10.1 V and a current of 52.3 A. Determine the heat flux in W/m² and the approximate wire temperature in degrees Celsius.

The electrical energy input rate is

$$q = EI = (10.1)(52.3) = 528.23 \text{ W}$$

The wire surface area is

$$A = \pi DL = \pi(1.0 \times 10^{-3} \text{ m})(150 \times 10^{-3} \text{ m}) = 4.7124 \times 10^{-4} \text{ m}^2$$

The boiling energy flux is

$$\frac{q}{A} = \frac{528.23 \text{ W}}{4.7124 \times 10^{-4} \text{ m}^2} = 1.121 \times 10^6 \text{ W/m}^2$$

To approximate the excess temperature using Fig. 9-1, convert q/A to British Engineering units.

$$\frac{q}{A} = (1.121 \times 10^6 \text{ W/m}^2) \left(\frac{\text{Btu/hr-ft}^2}{3.15248 \text{ W/m}^2} \right) = 355,573 \frac{\text{Btu}}{\text{hr-ft}^2}$$

From Fig. 9-1,

$$\Delta T \approx (40 \text{ °F}) \left(\frac{5 \text{ °C}}{9 \text{ °F}} \right) = 22 \text{ °C}$$

and

$$T_s = 100 + 22 = 122 \text{ °C}.$$

9.3. A 6-in.-long, 0.040-in.-diameter nickel wire submerged horizontally in water at 100 psig requires 131.8 A at 2.18 V to maintain the wire at 350.08 °F. What is the heat transfer coefficient?

From *Steam Tables*, the saturation temperature at $p = 100 + 14.7 = 114.7$ psia is 337.92 °F; therefore $\Delta T = 350.08 - 337.92 = 12.16$ °F. A heat balance on the wire gives

$$q = EI = hA\,\Delta T$$

whence

$$h = \frac{EI}{A\,\Delta T} = \frac{[(2.18)(131.8) \text{ W}](3.413 \text{ Btu/hr-W})}{\pi \left(\dfrac{0.040}{12} \text{ ft} \right) \left(\dfrac{6}{12} \text{ ft} \right) (12.16 \text{ °F})} = 1.54 \times 10^4 \text{ Btu/hr-ft}^2\text{-°F}$$

9.4. In a laboratory experiment, a current of 193 A burns out a 12-in.-long, 0.040-in.-diameter nickel wire which is submerged horizontally in water at atmospheric pressure. What was the voltage at burnout?

The peak heat flux is given by (*9.5*):

$$\left. \frac{q}{A} \right|_{\text{max}} = (0.18)\rho_v h_{\text{fg}} \left[\frac{\sigma(\rho_l - \rho_v)gg_c}{\rho_v^2} \right]^{1/4} \left[\frac{\rho_l}{\rho_l + \rho_v} \right]^{1/2}$$

Taking the pertinent parameters from Table B-3 and from *Steam Tables*, we get

$$\frac{q}{A}\bigg|_{max} = (0.18)\left(0.0373\ \frac{lbm}{ft^3}\right)\left(970.4\ \frac{Btu}{lbm}\right)$$

$$\times \left[\frac{\left(0.004\ \frac{lbf}{ft}\right)(59.97 - 0.0373)\ \frac{lbm}{ft^3}\left(32.17\ \frac{ft}{sec^2}\right)\left(32.17\ \frac{lbm\text{-}ft}{lbf\text{-}sec^2}\right)}{\left(0.0373\ \frac{lbm}{ft^3}\right)^2}\right]^{1/4}$$

$$\times \left[\frac{59.97}{59.97 + 0.0373}\right]^{1/2}$$

$$= \left(133.84\ \frac{Btu}{sec\text{-}ft^2}\right)\left(3600\ \frac{sec}{hr}\right) = 4.82 \times 10^5\ Btu/hr\text{-}ft^2$$

The burnout voltage, E_b, must satisfy $E_b I = q_{max}$. Thus,

$$E_b = \frac{q_{max}}{I} = \frac{A}{I}\frac{q}{A}\bigg|_{max} = \frac{\pi\left(\frac{0.040}{12}\ ft\right)(1\ ft)}{193\ A}\left(4.82 \times 10^5\ \frac{Btu}{hr\text{-}ft^2}\right)\left(\frac{1\ V\text{-}A}{3.413\ Btu/hr}\right)$$

$$= 7.66\ V$$

9.5. Estimate the peak heat flux, in W/m^2, for boiling water at normal atmospheric pressure.

Equation (9.5) may be used with the parameters in SI units.

$$\frac{q}{A}\bigg|_{max} = (0.18)\rho_v h_{fg}\left[\frac{\sigma(\rho_l - \rho_v)gg_c}{\rho_v^2}\right]^{1/4}\left[\frac{\rho_l}{\rho_l + \rho_v}\right]^{1/2}$$

$\rho_l = (1/0.016715)(16.02) = 958.42\ kg/m^3$, from *Steam Tables*

$\rho_v = (1/26.8)(16.02) = 0.60\ kg/m^3$, from *Steam Tables*

$$h_{fg} = \left(970.3\ \frac{Btu}{lbm}\right)\left(\frac{lbm}{0.454\ kg}\right)\left(1054.8\ \frac{J}{Btu}\right)$$

$$= 2.25 \times 10^6\ J/kg = 2.25\ MW\text{-}s/kg$$

$$\sigma = \left(0.004\ \frac{lbf}{ft}\right)\left(4.448\ \frac{N}{lbf}\right)\left(\frac{1}{0.3048}\ \frac{ft}{m}\right) = 0.0584\ N/m,\ \text{from Fig. 9-2}$$

$g_c = 1.0\ kg\text{-}m/N\text{-}s^2$

$g = 9.80\ m/s^2$

$$\frac{q}{A}\bigg|_{max} = (0.18)\left(0.60\ \frac{kg}{m^3}\right)\left(2.25\ \frac{MW\text{-}s}{kg}\right)$$

$$\times \left[\frac{(0.0584\ N/m)(958.42 - 0.60)kg/m^3(9.8\ m/s^2)(1.0\ kg\text{-}m/N\text{-}s^2)}{(0.60\ kg/m^3)^2}\right]^{1/4}$$

$$\times \left[\frac{958.42}{958.42 + 0.60}\right]^{1/2}$$

$$= 1.517\ MW/m^2$$

9.6. A heated nickel plate at 222 °F is submerged horizontally in water at atmospheric pressure. What is the heat transfer per unit area?

For an excess temperature

$$\Delta T = T_s - T_{sat} = 222 - 212 = 10 \,°F$$

Fig. 9-1 indicates that the boiling is most likely nucleate, with (9.3) being valid.

$$\frac{q}{A} = \mu_l h_{fg} \sqrt{\frac{g(\rho_l - \rho_v)}{g_c \sigma}} \left[\frac{c_l(T_s - T_{sat})}{h_{fg} \, \mathbf{Pr}_l^{1.7} \, C_{sf}} \right]^3$$

The required parameters, from Table B-3 except where noted, are:

$h_{fg} = 970.4$ Btu/lbm, from *Steam Tables* $\sigma = 0.004$ lbf/ft, from Fig. 9-2

$\rho_l = 59.97$ lbm/ft³ $c_l = 1.007$ Btu/lbm-°F

$\rho_v = 0.0373$ lbm/ft³, from *Steam Tables* $\mathbf{Pr}_l = 1.74$

$\mu_l = 0.682$ lbm/ft-hr $C_{sf} = 0.006$, from Table 9-1

The heat transfer is

$$\frac{q}{A} = \left(0.682 \, \frac{\text{lbm}}{\text{ft-hr}}\right) \left(970.4 \, \frac{\text{Btu}}{\text{lbm}}\right) \sqrt{\frac{\left(32.2 \, \frac{\text{ft}}{\text{sec}^2}\right) \left[(59.97 - 0.0373) \, \frac{\text{lbm}}{\text{ft}^3}\right]}{\left(32.2 \, \frac{\text{ft-lbm}}{\text{lbf-sec}^2}\right) \left(0.004 \, \frac{\text{lbf}}{\text{ft}}\right)}}$$

$$\times \left[\frac{\left(1.007 \, \frac{\text{Btu}}{\text{lbm-°F}}\right)(10 \,°F)}{\left(970.4 \, \frac{\text{Btu}}{\text{lbm}}\right)(1.74)^{1.7}(0.006)} \right]^3$$

$$= 2.48 \times 10^4 \text{ Btu/hr-ft}^2$$

The peak heat flux for water at 1 atm was found in Problem 9.4 to be

$$\left. \frac{q}{A} \right|_{max} = 4.82 \times 10^5 \text{ Btu/hr-ft}^2$$

Since $q/A < (q/A)_{max}$, nucleate boiling does in fact occur, and the heat transfer is as calculated.

9.7. How does the result of Problem 9.6 compare with that obtained from Table 9-2?

From Table 9-2, $h_a = 151(\Delta T)^{1/3}$, so that

$$\frac{q}{A} = h_a \, \Delta T = 151(\Delta T)^{4/3} = 151(10)^{4/3} = 3.25 \times 10^3 \text{ Btu/hr-ft}^2$$

which deviates by a considerable amount, illustrating the possible errors involved in boiling heat-transfer correlations.   In this case an engineer would be more comfortable accepting the result from the Rohsenow correlation.

9.8. If the plate of Problem 9.6 were copper, what would be the heat transfer rate?

We note from the Rohsenow equation, (9.3), that all parameters are identical to those in Problem 9.6 except the surface–fluid constant C_{sf}; therefore, if the boiling remains nucleate,

$$\left. \frac{q}{A} \right|_{copper} = \left[\frac{C_{sf \text{ nickel}}}{C_{sf \text{ copper}}} \right]^3 \times \left. \frac{q}{A} \right|_{nickel}$$

From Table 9-1, $C_{sf \text{ copper}} = 0.013$; hence

$$\left. \frac{q}{A} \right|_{copper} = \left(\frac{0.006}{0.013} \right)^3 \left(2.48 \times 10^4 \, \frac{\text{Btu}}{\text{hr-ft}^2} \right) = 2438 \text{ Btu/hr-ft}^2$$

The boiling is certainly nucleate, since the peak heat flux, which is independent of the heater material, has the same value as in Problem 9.6.

9.9. A brass plate which is submerged horizontally in water at atmospheric pressure is heated at the rate of 0.7 MW/m². At what temperature, in °C, must the plate be held?

Assume that nucleate boiling occurs. Except for the excess temperature $\Delta T = T_s - T_{sat}$ and the surface–fluid constant C_{sf}, the parameters are identical to those in Problem 9.6. We may, then, use the Rohsenow equation, (9.3), to simplify the computation, viz.

$$\frac{(q/A)_{brass}}{(q/A)_{nickel}} = \left[\frac{(\Delta T)_{brass}}{(\Delta T)_{nickel}} \frac{C_{sf\ nickel}}{C_{sf\ brass}}\right]^3$$

From Table 9-1, $C_{sf\ brass} = 0.006$, and from Problem 9.6

$$(q/A)_{nickel} = (2.48 \times 10^4 \text{ Btu/hr-ft}^2)\left(3.1525 \frac{\text{W/m}^2}{\text{Btu/hr-ft}^2}\right) = 7.82 \times 10^4 \text{ W/m}^2$$

$$(\Delta T)_{nickel} = (10\ ^\circ\text{F})(5/9) = 5.56\ ^\circ\text{C}$$

Using these values in the above relation, we get

$$\frac{0.7 \times 10^6 \text{ W/m}^2}{7.82 \times 10^4 \text{ W/m}^2} = \left[\frac{(\Delta T)_{brass}}{5.56\ ^\circ\text{C}}\left(\frac{0.006}{0.006}\right)\right]^3$$

Solving,

$$(\Delta T)_{brass} = (5.56\ ^\circ\text{C})\left(\frac{0.7 \times 10^6}{7.82 \times 10^4}\right)^{1/3} = 11.54\ ^\circ\text{C}$$

and

$$(T_s)_{brass} = (\Delta T)_{brass} + T_{sat} = 11.54 + 100 = 111.54\ ^\circ\text{C}$$

Since the excess temperature $\Delta T = 11.54(9/5) = 20.78\ ^\circ\text{F}$, the assumption of nucleate boiling appears to be reasonable from Fig. 9-1.

9.10. Determine the dimensional constant for use in the first equation of Table 9-2 when SI units are used.

The constant 151 has units

$$\frac{[h]}{[\Delta T]^{1/3}} = \frac{\text{Btu/hr-ft}^2\text{-}^\circ\text{F}}{^\circ\text{F}^{1/3}}$$

Thus

$$h_a = \left(151 \frac{\text{Btu/hr-ft}^2\text{-}^\circ\text{F}}{^\circ\text{F}^{1/3}}\right)(\Delta T)^{1/3}$$

$$= \left[151 \frac{\text{Btu/hr-ft}^2\text{-}^\circ\text{F}}{^\circ\text{F}^{1/3}} \times \frac{5.6783 \text{ W/m}^2\text{-K}}{\text{Btu/hr-ft}^2\text{-}^\circ\text{F}} \times \left(\frac{9\ ^\circ\text{F}}{5\ \text{K}}\right)^{1/3}\right](\Delta T)^{1/3}$$

$$= \left(1043 \frac{\text{W/m}^2\text{-K}}{\text{K}^{1/3}}\right)(\Delta T)^{1/3}$$

where now h_a is in W/m²-K and ΔT is in K. This relation is valid for

$$\frac{q}{A} < (5000 \text{ Btu/hr-ft}^2)\left(\frac{3.1537 \text{ W/m}^2}{\text{Btu/hr-ft}^2}\right) = 15,769 \text{ W/m}^2 \approx 16 \text{ kW/m}^2$$

9.11. A 2-in.-diameter polished copper bar is submerged horizontally in a pool of water at atmospheric pressure and 68 °F. The bar is maintained at 300 °F. Estimate the heat transfer rate per foot of the bar.

The excess temperature is $\Delta T = T_s - T_{\text{sat}} = 300 - 212 = 88$ °F, which may be in the nucleate boiling regime III as shown in Fig. 9-1; therefore, (9.3) will be tried and compared with the peak heat flux, (9.5).

The required parameters, taken from Table B-3 except where noted, are given below. Note that the liquid parameters are evaluated at the saturation condition at 212 °F, since the temperature of the pool of water has little effect on the heat transfer. *Except for a slight effect due to subcooling, which is usually negligible in engineering calculations, the same heat is transferred whether the pool is at 68 °F, 150 °F, or 212 °F.*

$h_{\text{fg}} = 970.4$ Btu/lbm, from *Steam Tables*

$\rho_l = 59.97$ lbm/ft³

$\rho_v = 0.0373$ lbm/ft³, from *Steam Tables*

$\mu_l = \nu_l \rho_l = (0.316 \times 10^{-5} \text{ ft}^2/\text{sec})(59.97 \text{ lbm/ft}^3)$
$\qquad = 1.895 \times 10^{-4}$ lbm/ft-sec $= 0.682$ lbm/ft-hr

$\sigma = 0.004$ lbf/ft, from Fig. 9-2

$c_l = 1.007$ Btu/lbm-°F

$\mathbf{Pr}_l = 1.74$

$C_{sf} = 0.013$, from Table 9-1

Substituting these values into (9.3), we get

$$\frac{q}{A} = \left(0.682 \, \frac{\text{lbm}}{\text{ft-hr}}\right)\left(970.4 \, \frac{\text{Btu}}{\text{lbm}}\right)$$
$$\times \sqrt{\frac{\left(32.2 \, \dfrac{\text{ft}}{\text{sec}^2}\right)\left[(59.97 - 0.0373) \, \dfrac{\text{lbm}}{\text{ft}^3}\right]}{\left(32.2 \, \dfrac{\text{ft-lbm}}{\text{lbf-sec}^2}\right)\left(0.004 \, \dfrac{\text{lbf}}{\text{ft}}\right)}} \left[\frac{\left(1.007 \, \dfrac{\text{Btu}}{\text{lbm-°F}}\right)(88 \text{ °F})}{\left(970.4 \, \dfrac{\text{Btu}}{\text{lbm}}\right)(1.74)^{1.7}(0.013)}\right]^3$$
$$= 1.666 \times 10^6 \text{ Btu/hr-ft}^2$$

The maximum heat flux is as found in Problem 9.4:

$$\left.\frac{q}{A}\right|_{\text{max}} = 4.82 \times 10^5 \text{ Btu/hr-ft}^2$$

Since $(q/A)_{\text{max}}$ is less than that given by the Rohsenow equation, film boiling exists, and (9.6), (9.7) and (9.8) will be used to evaluate. Four additional vapor properties are required, evaluated at the mean film temperature,

$$T_f = \frac{T_s + T_{\text{sat}}}{2} = \frac{300 + 212}{2} = 256 \text{ °F}$$

hence, from Table B-4:

$k_{vf} = 0.0150$ Btu/hr-ft-°F

$c_{pvf} = 0.482$ Btu/lbm-°F

$\rho_{vf} = 0.0348$ lbm/ft³

$\mu_{vf} = 8.98 \times 10^{-6}$ lbm/ft-sec $= 0.0323$ lbm/ft-hr

and

$\epsilon = 0.023$, from Table B-6

Using these values, the appropriate heat transfer coefficients are:

$$h_c = (0.62)\left[\frac{k_{vf}^3\rho_{vf}(\rho_l - \rho_{vf})g(h_{fg} + 0.4\,c_{pvf}\,\Delta T)}{D\mu_{vf}\,\Delta T}\right]^{1/4}$$

$$= (0.62)\left[\frac{\left(0.0150\,\dfrac{\text{Btu}}{\text{hr-ft-}°\text{F}}\right)^3\left(0.0348\,\dfrac{\text{lbm}}{\text{ft}^3}\right)(59.97 - 0.0348)\dfrac{\text{lbm}}{\text{ft}^3}\left(32.2\,\dfrac{\text{ft}}{\text{sec}^2}\right)\left(\dfrac{3600\,\text{sec}}{\text{hr}}\right)^2}{\left(\dfrac{2}{12}\,\text{ft}\right)\left(0.0323\,\dfrac{\text{lbm}}{\text{ft-hr}}\right)(88\,°\text{F})}\right]^{1/4}$$

$$\times\left[970.4\,\frac{\text{Btu}}{\text{lbm}} + 0.4\left(0.482\,\frac{\text{Btu}}{\text{lbm-}°\text{F}}\right)\right]^{1/4}$$

$$= 30.84\ \text{Btu/hr-ft}^2\text{-}°\text{F}$$

$$h_r = \frac{\sigma\epsilon(T_s^4 - T_{\text{sat}}^4)}{T_s - T_{\text{sat}}}$$

$$= \frac{(0.1714)(0.023)\left[\left(\dfrac{760}{100}\right)^4 - \left(\dfrac{672}{100}\right)^4\right]}{88} = 0.0581\ \text{Btu/hr-ft}^2\text{-}°\text{F}$$

The radiation heat-transfer coefficient is negligible, which could have been guessed from the outset since the surface temperature is relatively low. Therefore, $h = h_c$, and

$$\frac{q}{A} = h\,\Delta T = (30.84)(88) = 2713.9\ \text{Btu/hr-ft}^2$$

Since

$$A = \pi DL = \pi\left(\frac{2}{12}\,\text{ft}\right)L = (0.524)L$$

the heat transfer per length of bar is

$$\frac{q}{L} = (0.524)(2713.9) = 1422\ \text{Btu/hr-ft}$$

By determining the excess temperature for the minimum heat flux, using (9.14), the boiling curve can be sketched for this configuration.

$$\Delta T_C = (0.127)\frac{\left(0.0348\,\dfrac{\text{lbm}}{\text{ft}^3}\right)\left(970.4\,\dfrac{\text{Btu}}{\text{lbm}}\right)}{\left(0.0150\,\dfrac{\text{Btu}}{\text{hr-ft-}°\text{F}}\right)\left(\dfrac{\text{hr}}{3600\,\text{sec}}\right)}\left[\frac{\left(32.2\,\dfrac{\text{ft}}{\text{sec}^2}\right)(59.97 - 0.0373)\dfrac{\text{lbm}}{\text{ft}^3}}{(59.97 + 0.0373)\dfrac{\text{lbm}}{\text{ft}^3}}\right]^{2/3}$$

$$\times\left[\frac{\left(32.2\,\dfrac{\text{lbm-ft}}{\text{lbf-sec}^2}\right)\left(0.004\,\dfrac{\text{lbf}}{\text{ft}}\right)}{\left(32.2\,\dfrac{\text{ft}}{\text{sec}^2}\right)(59.97 - 0.0373)\dfrac{\text{lbm}}{\text{ft}^3}}\right]^{1/2}\left[\frac{\left(\dfrac{\text{hr}}{3600\,\text{sec}}\right)\left(0.555\,\dfrac{\text{lbm}}{\text{ft-hr}}\right)}{\left(32.2\,\dfrac{\text{ft}}{\text{sec}^2}\right)(59.97 - 0.0373)\dfrac{\text{lbm}}{\text{ft}^3}}\right]^{1/3}$$

$$= 366\ °\text{F}$$

The boiling curve is shown in Fig. 9-6.

9.12. What change in heat transfer would occur in Problem 9.1 if the water were flowing normal to the wire at 10 fps? Assume the water temperature to be 212 °F.

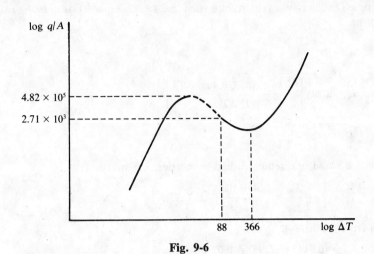

Fig. 9-6

The change is that due to convection, $(q/A)_{conv}$ of (9.16), which can be assessed by the use of (7.51),

$$\frac{\bar{h}D}{k_f} = C \, \mathbf{Pr}_f^{1/3} \, \mathbf{Re}_{Df}^n$$

where the constants C and n depend upon the flow regimes as given in Table 7-6.

Evaluating the fluid (vapor) properties at the film temperature

$$T_f = \frac{T_\infty + T_s}{2} = T_{sat} + \frac{\Delta T}{2} = 212 + \frac{50}{2} = 237 \, °F$$

we get the following values from Table B-4:

$$k_f = 0.0145 \text{ Btu/hr-ft-}°F \qquad \mathbf{Pr}_f = 1.0528 \qquad \nu = 2.431 \times 10^{-4} \text{ ft}^2/\text{sec}$$

Therefore

$$\mathbf{Re}_{Df} = \frac{VD}{\nu} = \frac{(10 \text{ ft/sec})[(0.040/12) \text{ ft}]}{2.431 \times 10^{-4} \text{ ft}^2/\text{sec}} = 137.12$$

The appropriate constants are $C = 0.683$, $n = 0.466$; hence,

$$\bar{h} = \frac{0.0148 \text{ Btu/hr-ft-}°F}{(0.040/12) \text{ ft}} (0.683)(1.0528)^{1/3}(137.12)^{0.466} = 30.56 \text{ Btu/hr-ft}^2\text{-}°F$$

and

$$\left. \frac{q}{A} \right|_{conv} = \bar{h}(T_s - T_\infty) = (30.56 \text{ Btu/hr-ft}^2\text{-}°F)[(262 - 212) \, °F] = 1528 \text{ Btu/hr-ft}^2$$

If we add this value to that obtained for boiling (Problem 9.1), we get

$$\frac{q}{A} = \left. \frac{q}{A} \right|_{conv} + \left. \frac{q}{A} \right|_{bo} = 1528 + (4.1 \times 10^5) = 4.12 \times 10^5 \text{ Btu/hr-ft}^2$$

which is approximately the value shown in Fig. 9-4 at $\Delta T \approx 50$ °F.

It should be noted that the "convective" contribution in this case is negligible, which is what the shaded portion of Fig. 9-4 emphasizes.

9.13. Film boiling occurs when water flows normal to a polished 15-mm-diameter copper tube at the rate of 3 m/s. Determine the boiling heat-transfer coefficient when the tube is maintained at 114 °C.

The tube size and fluid velocity are such that (9.20) applies. It must be used together with (9.6) and (9.8); i.e.

$$h = h_c \left(\frac{h_c}{h}\right)^{1/3} + h_r$$

$$h_c = (2.7) \left[\frac{V_\infty k_{vf} \rho_{vf}(h_{fg} + 0.4 \, c_{pf} \, \Delta T)}{D \, \Delta T}\right]^{1/2}$$

$$h_r = \frac{\sigma\epsilon(T_s^4 - T_{sat}^4)}{T_s - T_{sat}}$$

Except where noted, the required fluid properties, evaluated at

$$T_f = \frac{T_s + T_{sat}}{2} = \frac{114 + 100}{2} = 107 \, °C$$

are taken from Table B-4.

$$k_{vf} = (0.0142)(1.7296) = 0.02456 \text{ W/m-K}$$

$$\rho_{vf} = (0.0366)(16.02) = 0.5863 \text{ kg/m}^3$$

$$c_{pf} = (0.492)(4184) = 2058.5 \text{ J/kg-K}$$

$$h_{fg} = 2.25 \times 10^6 \text{ J/kg, from } Steam \, Tables$$

$$\sigma = 5.6697 \times 10^{-8} \text{ W/m}^2\text{-K}^4$$

$$\epsilon = 0.023, \text{ from Table B-6}$$

$$h_c = (2.7)\left[\frac{(3 \text{ m/s})(0.02456 \text{ W/m-K})(0.5863 \text{ kg/m}^3)\{(2.25 \times 10^6 \text{ W-s/kg}) + 0.4\,(2058.5 \text{ W-s/kg-K})(14 \text{ K})\}}{(0.015 \text{ m})(14 \text{ K})}\right]^{1/2}$$

$$= 1841.58 \text{ W/m}^2\text{-K}$$

$$h_r = \frac{(5.6697 \times 10^{-8} \text{ W/m}^2\text{-K}^4)(0.023)[(387)^4 - (373)^4]\text{K}^4}{(387 - 373) \text{ k}} = 0.2863 \text{ W/m}^2\text{-K}$$

Therefore,

$$h = (1841.58)\left(\frac{1841.58}{h}\right)^{1/3} + 0.2863$$

Solving by trial, $h \approx 1841.79$ W/m²-K.

9.14.   A wide vertical cooling fin, approximating a flat plate 0.3 m high, is exposed to steam at atmospheric pressure. The fin is maintained at 90 °C by cooling water. Determine the heat transfer and also the condensate mass-flow rate per unit width.

Assuming that the condensate film is laminar, (9.30) will be used to get the average heat-transfer coefficient, with condensate properties evaluated at the mean film temperature,

$$T_f = \frac{T_{sat} + T_s}{2} = \frac{100 + 90}{2} = 95 \, °C$$

Since $\sin \phi = 1$, the equation is:

$$\bar{h} = (1.13)\left[\frac{\rho_l g(\rho_l - \rho_v)h_{fg}k_l^3}{\mu_l L(T_{sat} - T_s)}\right]^{1/4}$$

where $\rho_l = (1/0.016654)(16.02) = 961.9$ kg/m³, from *Steam Tables*

$\rho_v = (1/26.8)(16.02) = 0.598$ kg/m³, from *Steam Tables* at T_{sat}

$h_{fg} = (976.0 \text{ Btu/lbm})(\text{lbm}/0.454 \text{ kg})(1054.8 \text{ J/Btu}) = 2.27 \times 10^6$ J/kg, from *Steam Tables* at T_{sat}

$$g = 9.8 \text{ m/s}^2$$
$$k_l = (0.3913)(1.7296) = 0.6767 \text{ W/m-K}$$
$$\mu_l = 3.0 \times 10^{-4} \text{ kg/m-s}$$

Thus,

$$\bar{h} = (1.13) \left[\frac{(961.9 \text{ kg/m}^3)(9.8 \text{ m/s}^2)(961.9 - 0.598)\text{kg/m}^3(2.27 \times 10^6 \text{ J/kg})(0.6767 \text{ W/m-K})^3}{(3.0 \times 10^{-4} \text{ kg/m-s})(0.3 \text{ m})(10 \text{ K})} \right]^{1/4}$$

$$= 1.04 \times 10^4 \text{ J/s-m}^2\text{-K} = 1.04 \times 10^4 \text{ W/m}^2\text{-K}$$

Checking the film Reynolds number with (9.35), we get

$$\mathbf{Re}_f = \frac{4\bar{h}L(T_{\text{sat}} - T_s)}{\mu_l h_{\text{fg}}} = \frac{4(1.04 \times 10^4 \text{ W/m}^2\text{-K})(0.3 \text{ m})(10 \text{ K})}{(3 \times 10^{-4} \text{ kg/m-s})(2.27 \times 10^6 \text{ W-s/kg})} = 183 < 1800$$

and the laminar assumption was correct. The heat transfer rate is, therefore,

$$\frac{q}{A} = \bar{h}(T_{\text{sat}} - T_s) = (1.04 \times 10^4 \text{ W/m}^2\text{-K})(10 \text{ K}) = 0.104 \text{ MW/m}^2$$

and, by (9.33),

$$\dot{m}' = \frac{\mu_l \mathbf{Re}_f}{4} = \frac{(3.0 \times 10^{-4} \text{ kg/s-m})(183)}{4} = 0.0137 \text{ kg/s-m}$$

9.15. A horizontal, 2-in.-o.d. tube is surrounded by saturated steam at 2.0 psia. The tube is maintained at 90 °F. What is the average heat-transfer coefficient?

The average heat-transfer coefficient is given by (9.31), which requires the following property data, taken from Table B-3 and *Steam Tables*. The liquid properties are evaluated at the mean film temperature, $T_f = (T_{\text{sat}} + T_s)/2 = (126 + 90)/2 = 108$ °F.

$$\rho_l = 62.03 \text{ lbm/ft}^3 \qquad\qquad k_l = 0.364 \text{ Btu/hr-ft-°F}$$
$$\rho_v = 0.00576 \text{ lbm/ft}^3 \text{ (at } T_{\text{sat}}) \qquad \mu_l = 4.26 \times 10^{-4} \text{ lbm/ft-sec}$$
$$h_{\text{fg}} = 1022.1 \text{ Btu/lbm (at } T_{\text{sat}})$$

Using these data, we get

$$\bar{h} = (0.725) \left[\frac{\rho_l g(\rho_l - \rho_v)h_{\text{fg}}k_l^3}{\mu_l D(T_{\text{sat}} - T_s)} \right]^{1/4}$$

$$= (0.725) \left[\frac{\left(62.03 \frac{\text{lbm}}{\text{ft}^3} \right) \left(32.2 \frac{\text{ft}}{\text{sec}^2} \right) (62.03 - 0.00576) \frac{\text{lbm}}{\text{ft}^3} \left(1022.1 \frac{\text{Btu}}{\text{lbm}} \right) \left(0.364 \frac{\text{Btu}}{\text{hr-ft-°F}} \right)^3}{\left(4.26 \times 10^{-4} \frac{\text{lbm}}{\text{ft-sec}} \right) \left(\frac{2}{12} \text{ ft} \right) (126 - 90) \text{ °F} \left(\frac{\text{hr}}{3600 \text{ sec}} \right)} \right]^{1/4}$$

$$= 1241.6 \text{ Btu/hr-ft}^2\text{-°F}$$

9.16. The relations given in Section 9.4 are somewhat simplified, although sufficiently accurate for most engineering calculations, since they do not account for the change of enthalpy from the saturation temperature, T_{sat}, to the film temperature, T, which varies across the film. This refinement can be taken into account by considering the total enthalpy change, Δh_{fg}, as the vapor condenses and cools to the film temperature. In equation form,

$$\Delta h_{\text{fg}} = \frac{1}{\dot{m}'} \int_0^\delta \rho_l u c_p (T_{\text{sat}} - T)\, dy \qquad\qquad (1)$$

Calculate Δh_{fg} for a linear temperature profile in the condensate film, i.e.

$$T = T_s + \frac{y}{\delta}(T_{sat} - T_s) \tag{2}$$

Using (9.22), (9.23) and (2) for u, $\dot{m}'$ and T, respectively, we get

$$\Delta h_{fg} = \frac{3\mu_l}{\rho_l g(\rho_l - \rho_v)\delta^3} \int_0^\delta \frac{\rho_l g(\rho_l - \rho_v)}{\mu_l} \left(y\delta - \frac{y^2}{2}\right) c_p \left[(T_{sat} - T_s) - \frac{y}{\delta}(T_{sat} - T_s)\right] dy$$

$$= \frac{3c_p(T_{sat} - T_s)}{\delta^3} \int_0^\delta \left(y\delta - \frac{3}{2}y^2 + \frac{y^3}{2\delta}\right) dy = \frac{3c_p(T_{sat} - T_s)}{\delta^3} \left[\frac{y^2\delta}{2} - \frac{y^3}{2} + \frac{y^4}{8\delta}\right]_0^\delta$$

$$= \frac{3}{8} c_p(T_{sat} - T_s) \tag{3}$$

For more accurate results in Section 9.4, h_{fg} should be replaced by $h_{fg} + \Delta h_{fg} = h'_{fg}$.

9.17. What effect does the refinement developed in Problem 9.16 have on the heat transfer coefficient of Problem 9.15?

At $T_f = 108\ °F$ the specific heat of water is 0.998 Btu/lbm-°F; therefore, using (3) of Problem 9.16,

$$h'_{fg} = h_{fg} + \Delta h_{fg} = 1022.1\ \frac{Btu}{lbm} + \frac{3}{8}\left(0.998\ \frac{Btu}{lbm\text{-}°F}\right)[(126 - 90)\ °F] = 1035.6\ Btu/lbm$$

Hence, a more accurate result for Problem 9.15 is

$$\bar{h} = (1241.6)\left(\frac{1035.6}{1022.1}\right)^{1/4} = 1245.7\ Btu/hr\text{-}ft^2\text{-}°F$$

giving an error of 0.33 percent, negligible in most engineering calculations.

9.18. A vertical plate 4 ft high is maintained at 140 °F in the presence of saturated steam at atmospheric pressure. Estimate the heat transfer per unit width and the condensation rate per unit width.

The parameters which will be required, with condensate properties evaluated at the mean film temperature,

$$T_f = \frac{T_{sat} + T_s}{2} = \frac{212 + 140}{2} = 176\ °F$$

are:

$\rho_l = 60.81\ lbm/ft^3$

$c_{pl} = 1.0023\ Btu/lbm\text{-}°F$

$\mu_l = 0.238 \times 10^{-3}\ lbm/ft\text{-}sec$

$k_l = 0.386\ Btu/hr\text{-}ft\text{-}°F$

$h_{fg} = 970.4\ Btu/lbm$ (from *Steam Tables* at T_{sat})

$\rho_v = 0.0373\ lbm/ft^3$ (from *Steam Tables* at T_{sat})

$A = LW = 4\ ft^2$

$P = W = 1\ ft$

Testing for laminar flow with (9.36):

$$(4.52) \left[\frac{\left(60.81\ \frac{lbm}{ft^3}\right)\left(32.2\ \frac{ft}{sec^2}\right)(60.81 - 0.0373)\ \frac{lbm}{ft^3}\left(0.386\ \frac{Btu}{hr\text{-}ft\text{-}°F}\right)^3 (212 - 140)^3\ °F^3}{\left(0.238 \times 10^{-3}\ \frac{lbm}{ft\text{-}sec}\right)^5 \left(970.4\ \frac{Btu}{lbm}\right)^3 \left(\frac{3600\ sec}{hr}\right)^3}(4\ ft)^3\right]^{1/4} = 1203 < 1800$$

Thus, the flow is laminar and $\mathbf{Re}_f = 1203$. We may now get the condensate rate from (9.33):

$$\dot{m}' = \frac{\mathbf{Re}_f\,\mu_l}{4} = \frac{1203\left(0.238 \times 10^{-3}\,\dfrac{\text{lbm}}{\text{ft-sec}}\right)}{4} = 0.0716 \text{ lbm/sec-ft}$$

or $\dot{m}' = 258$ lbm/hr-ft. The heat transfer per unit width is then

$$q' = \dot{m}'h_{\text{fg}} = \left(258\,\frac{\text{lbm}}{\text{hr-ft}}\right)\left(970.4\,\frac{\text{Btu}}{\text{lbm}}\right) = 250{,}363 \text{ Btu/hr-ft}$$

9.19. How would doubling the plate height of Problem 9.18 affect the heat transfer rate?

The fluid parameters are identical to those of Problem 9.18. The flow will likely be turbulent, however. Testing with (9.38),

$$(0.00296)\left[\frac{\left(60.81\,\dfrac{\text{lbm}}{\text{ft}^3}\right)\left(32.2\,\dfrac{\text{ft}}{\text{sec}^2}\right)(60.81 - 0.0373)\dfrac{\text{lbm}}{\text{ft}^3}\left(0.386\,\dfrac{\text{Btu}}{\text{hr-ft-}^\circ\text{F}}\right)^3 (212 - 140)^3\,{}^\circ\text{F}^3}{\left(0.238 \times 10^{-3}\,\dfrac{\text{lbm}}{\text{ft-sec}}\right)^5 \left(970.4\,\dfrac{\text{Btu}}{\text{lbm}}\right)^3 \left(\dfrac{3600\text{ sec}}{\text{hr}}\right)^3}(8\text{ ft})^3\right]^{5/9} = 2302 > 1800$$

The film is turbulent, as assumed, and $\mathbf{Re}_f = 2302$. Since $q' \propto \mathbf{Re}_f$, the ratio of the heat transfer rates is

$$\frac{q'_8}{q'_4} = \frac{2302}{1203} = 1.91$$

and the heat transfer rate is approximately doubled by doubling the plate height.

The heat transfer rate increases more rapidly, however, as the plate is heightened further. For example, increasing the plate height by a factor of 4 (to 16 ft) gives more than a sixfold increase in heat transfer. This illustrates the effect which eddy diffusion has on the heat transfer rate.

Supplementary Problems

9.20. For equilibrium of a spherical vapor bubble in a liquid, a balance of forces gives

$$\pi r^2(p_v - p_l) = 2\pi r\sigma$$

where r is the bubble radius, σ is the surface tension, and subscripts v and l represent the vapor and liquid, respectively. Combining this equilibrium condition with the *Clausius-Clapeyron equation*,

$$\frac{dp}{dT} = \frac{h_{\text{fg}}}{v_v T_v}$$

which relates the pressure and temperature between a saturated liquid and its saturated vapor, and with the ideal gas law, determine the relationship between the degree of superheat and bubble radius.

Ans. $T_v - T_{\text{sat}} = \dfrac{2R_v T_v^2 \sigma}{h_{\text{fg}} p_v r}$

9.21. A tungsten wire submerged horizontally in water at atmospheric pressure is electrically heated, maintaining the surface at 225 °F. The wire is 0.040-in.-diameter and 12 in. long; $E = 2$ V and $I = 20$ A. Determine the heat transfer coefficient.   *Ans.* $h = 1003$ Btu/hr-ft²-°F

9.22.   A heated brass plate is submerged vertically in water at atmospheric pressure. The plate is maintained at 232 °F. What is the heat transfer rate per unit area?   *Ans.* 1.99×10^5 Btu/hr-ft^2

9.23.   Using Table 9-2, estimate the heat transfer per unit area from a horizontal flat plate submerged in water at atmospheric pressure. The plate is held at 106 °C.   *Ans.*   11.37 kW/m^2

9.24.   Estimate the maximum nucleate boiling heat flux for water in a pressure cooker at 30 psia. *Ans.* 6.4×10^5 Btu/hr-ft^2

9.25.   For laminar film condensation, what is the ratio of heat transfer to a horizontal tube of large diameter to that to a vertical tube of the same size for the same temperature difference? *Ans.* $(0.64)(L/D)^{1/4}$

9.26.   What L/D ratio will produce the same laminar-film-condensation-controlled heat-transfer rate to a tube in both the vertical and horizontal orientations? Assume the tube diameter is large compared with the condensate thickness.   *Ans.*   5.89

9.27.   A 1-ft-high vertical plate is exposed to saturated steam at atmospheric pressure. If the plate is maintained at 188 °F, determine the heat transfer per unit width and the condensation rate per unit width.   *Ans.*   40,445 Btu/hr-ft; 41.7 lbm/hr-ft

Chapter 10

Heat Exchangers

A *heat exchanger* is any device that effects the transfer of thermal energy from one fluid to another. In the simplest exchangers the hot and cold fluids mix directly; more common are those in which the fluids are separated by a wall. This type, called a *recuperator*, may range from a simple plane wall between two flowing fluids to complex configurations involving multiple passes, fins, or baffles. In this case conductive and convective heat-transfer, and sometimes radiation, principles are required to describe the energy exchange process.

Many factors enter into the design of heat exchangers, including thermal analysis, size, weight, structural strength, pressure drop and cost. Our primary concern in this chapter shall be thermal analysis. Except for structural strength and cost, the remaining factors may be adequately evaluated utilizing the principles of earlier chapters. The ASME Code for Unfired Pressure Vessels sets structural design standards, while cost evaluation is obviously an optimization process dependent upon the other design parameters.

10.1 TYPES OF HEAT EXCHANGERS

Common heat exchangers include the *flat-plate, shell-and-tube* and *crossflow* types. A double-pipe exchanger, the simplest form of the shell-and-tube type, is shown in Fig. 10-1. If the fluids both flow in the same direction, as shown, it is referred to as a *parallel-flow* type; if they flow in opposite directions, a *counterflow* type. Figure 10-2 shows a shell-and-tube exchanger with several tubes, two passes and baffles.

In crossflow heat exchangers the fluids flow at right angles to each other, as illustrated in Fig. 10-3. If a fluid can move about freely while passing through the exchanger, the fluid is said to be *mixed*. Figure 10-4 shows a crossflow type with both fluids unmixed. Here the temperature distribution with distance is skewed because the fluid in a given flow path is subjected to a temperature difference unlike that experienced by the fluid in any other path at the same distance from the inlet.

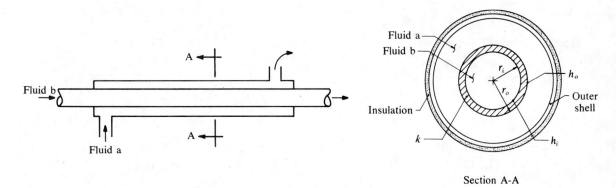

Section A-A

Fig. 10-1

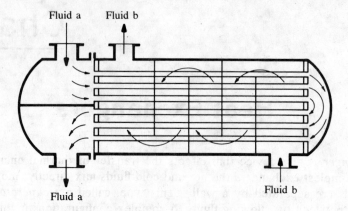

Fluid a　　Fluid b

Fluid a

Fluid b

Fig. 10-2

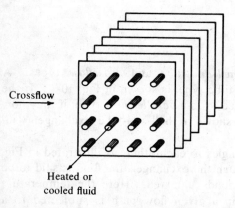

Crossflow →

Heated or
cooled fluid

(a) Both Fluids Unmixed

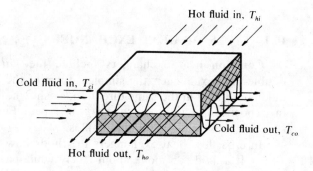

Hot fluid in, T_{hi}

Cold fluid in, T_{ci}

Cold fluid out, T_{co}

Hot fluid out, T_{ho}

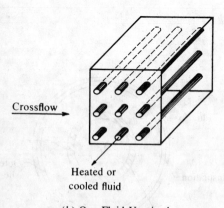

Crossflow →

Heated or
cooled fluid

(b) One Fluid Unmixed

Fig. 10-3

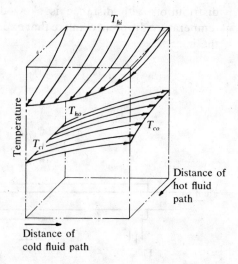

T_{hi}

T_{ho}

T_{co}

T_{ci}

Temperature

Distance of
hot fluid
path

Distance of
cold fluid path

Fig. 10-4

10.2 HEAT TRANSFER CALCULATIONS

The primary objective in the thermal design of heat exchangers is to determine the necessary surface area required to transfer heat at a given rate for given fluid temperatures and flow rates. This is facilitated by employing the overall heat-transfer coefficient, U, introduced in Section 2.7, in the fundamental heat transfer relation

■ $$q = UA\,\overline{\Delta T} \tag{10.1}$$

where $\overline{\Delta T}$ is an average effective temperature difference for the entire heat exchanger.

Overall Heat-Transfer Coefficient

Equation (*2.30*) shows that the overall heat-transfer coefficient U is proportional to the reciprocal of the sum of the thermal resistances. For the common configurations which we shall encounter,

plane wall: $$U = \frac{1}{1/h_o + L/k + 1/h_i} \tag{10.2}$$

cylindrical wall: $$U_o = \frac{1}{r_o/r_i h_i + [r_o \ln (r_o/r_i)]/k + 1/h_o} \tag{10.3}$$

$$U_i = \frac{1}{1/h_i + [r_i \ln (r_o/r_i)]/k + r_i/r_o h_o} \tag{10.4}$$

where subscripts i and o represent the inside and outside surfaces of the wall, respectively. It is important to note that the area for convection is not the same for both fluids in the case of a cylindrical wall; therefore, the overall heat-transfer coefficient and the surface area must be compatible, i.e. $q = U_o A_o \overline{\Delta T} = U_i A_i \overline{\Delta T}$.

For the preliminary design of heat exchangers it is advantageous to be able to estimate overall heat-transfer coefficients. Table 10-1 gives approximate values of U for some commonly encountered fluids. The wide range of values cited results from a diversity of heat exchanger materials (of different thermal conductivities, k) and flow conditions (influencing the film coefficients, h), as well as geometric configuration.

Log-Mean Temperature Difference

Before making heat transfer calculations it is necessary to define the remaining term in (*10.1*), $\overline{\Delta T}$. Consider, for instance, a parallel-flow flat-plate exchanger, whose temperature profiles are shown in Fig. 10-5. We shall assume that:

1. U is constant throughout the exchanger

2. the system is adiabatic; heat exchange takes place only between the two fluids

3. the temperatures of both fluids are constant over a given cross section and can be represented by bulk temperatures

4. the specific heats of the fluids are constant

Based upon these assumptions, the heat transfer between the hot and cold fluids for a differential

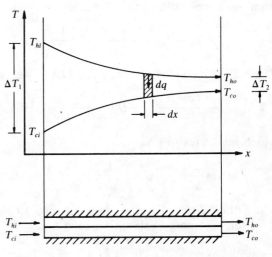

Fig. 10-5

Table 10-1

Fluid	U	
	Btu/hr-ft²-°F	W/m²-K
Oil to oil	30–55	170–312
Organics to organics	10–60	57–340
Steam to:		
Aqueous solutions	100–600	567–3400
Fuel oil, heavy	10–30	57–170
Light	30–60	170–340
Gases	5–50	28–284
Water	175–600	993–3400
Water to:		
Alcohol	50–150	284–850
Brine	100–200	567–1135
Compressed air	10–30	57–170
Condensing alcohol	45–120	255–680
Condensing ammonia	150–250	850–1420
Condensing Freon-12	80–150	454–850
Condensing oil	40–100	227–567
Gasoline	60–90	340–510
Lubricating oil	20–60	113–340
Organic solvents	50–150	284–850
Water	150–300	850–1700

length dx is

$$dq = U(T_h - T_c)\, dA \qquad\qquad (10.5)$$

since dA is the product of length dx and a constant width. The energy gained by the cold fluid is equal to that given up by the hot fluid, i.e.

$$dq = \dot{m}_c c_c\, dT_c = -\dot{m}_h c_h\, dT_h \qquad\qquad (10.6)$$

where $\dot{m}$ is the mass-flow rate and c is the specific heat. Solving for the temperature differentials from equations (10.6) and subtracting, we get

$$d(T_h - T_c) = -\left(\frac{1}{\dot{m}_h c_h} + \frac{1}{\dot{m}_c c_c}\right) dq \qquad\qquad (10.7)$$

Eliminating dq between (10.5) and (10.7) yields

$$\frac{d(T_h - T_c)}{(T_h - T_c)} = -U\left(\frac{1}{\dot{m}_h c_h} + \frac{1}{\dot{m}_c c_c}\right) dA \qquad\qquad (10.8)$$

which integrates to give

$$\ln\frac{\Delta T_2}{\Delta T_1} = -UA\left(\frac{1}{\dot{m}_h c_h} + \frac{1}{\dot{m}_c c_c}\right) \qquad\qquad (10.9)$$

where the ΔT terms are as shown in Fig. 10-5.

From an energy balance on each fluid,

$$\dot{m}_h c_h = \frac{q}{(T_{hi} - T_{ho})} \qquad \dot{m}_c c_c = \frac{q}{(T_{co} - T_{ci})}$$

and substitution of these expressions into (10.9) gives

$$\ln \frac{\Delta T_2}{\Delta T_1} = -UA \frac{(T_{hi} - T_{ho}) + (T_{co} - T_{ci})}{q}$$

or, in terms of the differences in end temperatures,

$$q = UA \frac{\Delta T_2 - \Delta T_1}{\ln (\Delta T_2/\Delta T_1)} \qquad (10.10)$$

Upon comparing this result with equation (10.1), we see that

$$\overline{\Delta T} = \frac{\Delta T_2 - \Delta T_1}{\ln (\Delta T_2/\Delta T_1)} \equiv \Delta T_{lm} \qquad (10.11)$$

This average effective temperature difference is called the *log-mean temperature difference* (LMTD). It can easily be shown that the subscripts 1 and 2 may be interchanged without changing the value of ΔT_{lm}; hence, the designation of ends for use in (10.10) and (10.11) is arbitrary.

Equations (10.10) and (10.11) can also be shown to hold for other single-pass exchangers such as the counterflow flat plate and the parallel flow or counterflow double-pipe configurations. Also, these equations are valid for single-pass parallel flow and counterflow evaporators and condensers where one of the fluids remains at a constant temperature.

Correction Factors for Complex Heat Exchangers

For more complex heat exchangers, such as those involving multiple tubes, several shell passes or crossflow, determination of the average effective temperature difference is so difficult that the usual practice is to modify (10.1) by a correction factor F, giving

$$q = UAF \, \Delta T_{lm} \qquad (10.12)$$

in which ΔT_{lm} is that for a counterflow double-pipe exchanger with the same fluid inlet and outlet temperatures as in the more complex design. Correction factors for several common configurations are given in Figs. 10-6 through 10-9. In these figures the notation (T, t) to denote the temperatures of the two fluid streams has been introduced, since it is immaterial whether the hot fluid flows through the shell or the tubes.

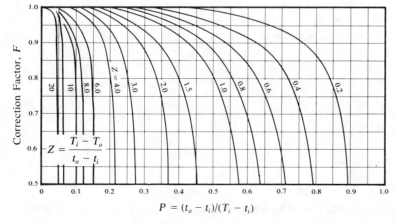
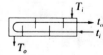

Fig. 10-6. One shell pass and an even number of tube passes.

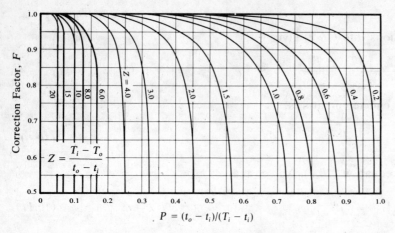
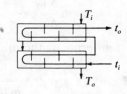

$$P = (t_o - t_i)/(T_i - t_i)$$

Fig. 10-7.   Two shell passes and twice an even number of tube passes.

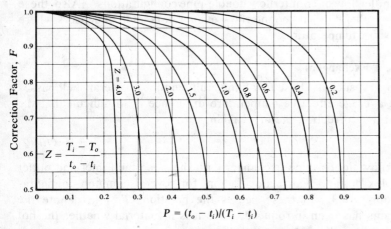
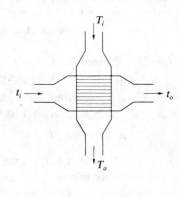

$$P = (t_o - t_i)/(T_i - t_i)$$

Fig. 10-8.   Crossflow with one fluid mixed.

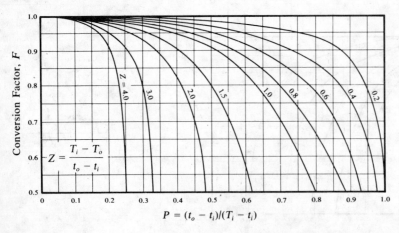
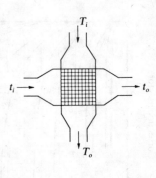

$$P = (t_o - t_i)/(T_i - t_i)$$

Fig. 10-9.   Crossflow with both fluids unmixed.

244

10.3 HEAT EXCHANGER EFFECTIVENESS (NTU METHOD)

If more than one of the inlet or outlet temperatures of the heat exchanger are unknown, the LMTD method of Section 10.2 is unwieldy, requiring a trial-and-error iterative approach.   Another approach introduces a definition of heat exchanger *effectiveness*:

$$\epsilon \equiv \frac{\text{actual heat transfer}}{\text{maximum possible heat transfer}} = \frac{q_{\text{actual}}}{q_{\text{max}}} \tag{10.13}$$

where the maximum possible heat transfer is that which would result if one fluid underwent a temperature change equal to the maximum temperature difference available—the temperature of the entering hot fluid minus the temperature of the entering cold fluid.   This method uses the effectiveness ϵ to eliminate the unknown discharge temperature and gives a solution for effectiveness in terms of other known parameters ($\dot{m}$, c, A and U).

Letting $C \equiv \dot{m}c$,

$$q_{\text{actual}} = C_h(T_{hi} - T_{ho}) = C_c(T_{co} - T_{ci}) \tag{10.14}$$

which indicates that the energy given up by the hot fluid is gained by the cold fluid.   The maximum possible heat transfer occurs when the fluid of smaller C undergoes the maximum temperature difference available, i.e.

$$q_{\text{max}} = C_{\text{min}}(T_{hi} - T_{ci}) \tag{10.15}$$

This transfer would be attained in a counterflow exchanger of infinite area.   Combining (10.13) and (10.15), we get the basic equation for determining the heat transfer in heat exchangers with unknown discharge temperatures:

■ $\qquad q_{\text{actual}} = \epsilon C_{\text{min}}(T_{hi} - T_{ci}) \tag{10.16}$

Parallel-Flow Heat Exchanger

Consider the simple parallel-flow heat exchanger of Fig. 10-5 under the same assumptions used in Section 10.2 to determine the log-mean temperature difference.

Combining (10.13), (10.14) and (10.15), we get two expressions for effectiveness, viz.

$$\epsilon = \frac{C_h(T_{hi} - T_{ho})}{C_{\text{min}}(T_{hi} - T_{ci})} = \frac{C_c(T_{co} - T_{ci})}{C_{\text{min}}(T_{hi} - T_{ci})} \tag{10.17}$$

Since either the hot or the cold fluid may have the minimum value of C, there are two possible values of effectiveness:

$$C_h < C_c: \quad \epsilon_h = \frac{T_{hi} - T_{ho}}{T_{hi} - T_{ci}}$$

$$\tag{10.18}$$

$$C_c < C_h: \quad \epsilon_c = \frac{T_{co} - T_{ci}}{T_{hi} - T_{ci}}$$

where subscripts on ϵ designate the fluid which has the minimum C.   Returning to (10.9), it may be written in terms of the C's to give

$$\ln \frac{T_{ho} - T_{co}}{T_{hi} - T_{ci}} = -UA \left(\frac{1}{C_h} + \frac{1}{C_c} \right) \tag{10.19}$$

or

$$\frac{T_{ho} - T_{co}}{T_{hi} - T_{ci}} = \exp \left[-\frac{UA}{C_h} \left(1 + \frac{C_h}{C_c} \right) \right] \tag{10.20}$$

From the energy balance equation (10.14),

$$T_{co} = T_{ci} + \frac{C_h}{C_c}(T_{hi} - T_{ho})$$

(10.21)

Combining (10.20) and (10.21) with the first equation (10.18), which assumes that the hotter fluid has the minimum value of C, we get

$$\epsilon_h = \frac{1 - \exp\left[-(UA/C_h)(1 + C_h/C_c)\right]}{1 + C_h/C_c}$$

(10.22)

If the colder fluid has the minimum value of C,

$$\epsilon_c = \frac{1 - \exp\left[-(UA/C_c)(1 + C_c/C_h)\right]}{1 + C_c/C_h}$$

(10.23)

Equations (10.22) and (10.23) may both be expressed as

■ $$\epsilon = \frac{1 - \exp\left[-(UA/C_{min})(1 + C_{min}/C_{max})\right]}{1 + C_{min}/C_{max}}$$

(10.24)

giving the effectiveness for a parallel-flow heat exchanger in terms of two dimensionless ratios. One of these, UA/C_{min}, is called the *number of transfer units*, i.e.

$$\mathbf{NTU} \equiv UA/C_{min}$$

(10.25)

Table 10-2

Exchanger Type	Effectiveness	See graph in
Parallel-flow: single-pass	$\epsilon = \dfrac{1 - \exp\left[-\mathbf{NTU}(1 + C)\right]}{1 + C}$	Fig. 10-10
Counterflow: single-pass	$\epsilon = \dfrac{1 - \exp\left[-\mathbf{NTU}(1 - C)\right]}{1 - C\exp\left[-\mathbf{NTU}(1 - C)\right]}$	Fig. 10-11
Shell-and-tube (one shell pass; 2, 4, 6, etc., tube passes)	$\epsilon_1 = 2\left[1 + C + \dfrac{1 + \exp\left[-\mathbf{NTU}(1 + C^2)^{1/2}\right]}{1 - \exp\left[-\mathbf{NTU}(1 + C^2)^{1/2}\right]}(1 + C^2)^{1/2}\right]^{-1}$	Fig. 10-12
Shell-and-tube (n shell passes; $2n$, $4n$, $6n$, etc., tube passes)	$\epsilon_n = \left[\left(\dfrac{1 - \epsilon_1 C}{1 - \epsilon_1}\right)^n - 1\right]\left[\left(\dfrac{1 - \epsilon_1 C}{1 - \epsilon_1}\right)^n - C\right]^{-1}$	Fig. 10-13 for $n = 2$
Crossflow (both streams unmixed)	$\epsilon \approx 1 - \exp\{C(\mathbf{NTU})^{0.22}\left[\exp\left[-C(\mathbf{NTU})^{0.78}\right] - 1\right]\}$	Fig. 10-14
Crossflow (both streams mixed)	$\epsilon = \mathbf{NTU}\left[\dfrac{\mathbf{NTU}}{1 - \exp(-\mathbf{NTU})} + \dfrac{(\mathbf{NTU})(C)}{1 - \exp\left[-(\mathbf{NTU})(C)\right]} - 1\right]^{-1}$	
Crossflow (stream C_{min} unmixed)	$\epsilon = C\{1 - \exp\left[-C\left[1 - \exp(-\mathbf{NTU})\right]\right]\}$	Fig. 10-15 (dashed curves)
Crossflow (stream C_{max} unmixed)	$\epsilon = 1 - \exp\{-C\left[1 - \exp\left[-(\mathbf{NTU})(C)\right]\right]\}$	Fig. 10-15 (solid curves)

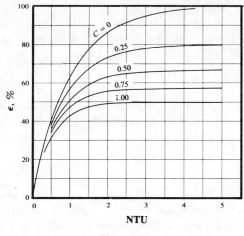

Fig. 10-10

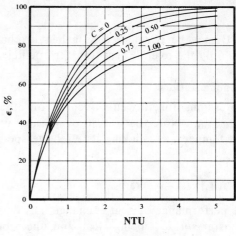

Fig. 10-11

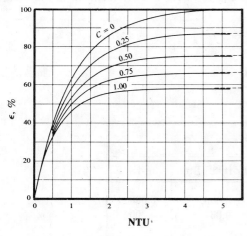

Fig. 10-12

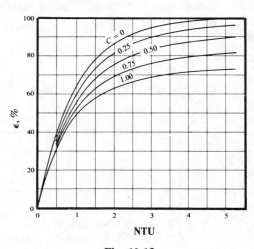

Fig. 10-13

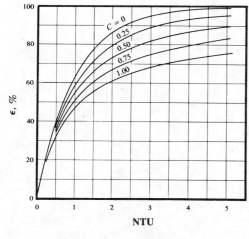

Fig. 10-14

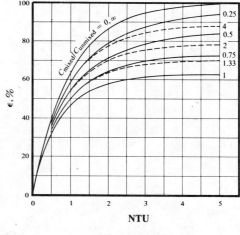

Fig. 10-15

The **NTU** may be considered as a heat exchanger size-factor. It should be noted that (*10.24*) contains only the overall heat-transfer coefficient, area, fluid properties and flow rates.

Other Configurations

Expressions for the effectiveness of other configurations are given in Table 10-2, where $C \equiv C_{min}/C_{max}$. Note that for an evaporator or condenser $C = 0$, because one fluid remains at a constant temperature, making its effective specific heat infinite.

10.4   FOULING FACTORS

The performance of heat exchangers as developed in the preceding sections depends upon the heat transfer surfaces being clean and uncorroded. Should surface deposits be present, thermal resistance increases, resulting in decreased performance. This added resistance is usually accounted for by a *fouling factor*, or *fouling resistance*, R_f, which must be included along with other thermal resistances when calculating the overall heat-transfer coefficient.

Fouling factors are determined experimentally by testing the heat exchanger in both the clean and dirty condition, being defined by

$$R_f = \frac{1}{U_{dirty}} - \frac{1}{U_{clean}}$$

(*10.26*)

Some typical values are given in Table 10-3.

Table 10-3

Fluid	R_f	
	hr-ft²-°F/Btu	m²-K/W
Seawater below 125 °F	0.0005	0.00009
Seawater above 125 °F	0.001	0.0002
Treated boiler feedwater above 125 °F	0.001	0.0002
Fuel oil	0.005	0.0009
Quenching oil	0.004	0.0007
Alcohol vapors	0.0005	0.00009
Steam, non-oil-bearing	0.0005	0.00009
Industrial air	0.002	0.0004
Refrigerating liquid	0.001	0.0002

Solved Problems

10.1. In a food processing plant a brine solution is heated from 10 °F to 20 °F in a double-pipe heat exchanger by water entering at 90 °F and leaving at 70 °F at the rate of 20 lbm/min. If the overall heat-transfer coefficient is 150 Btu/hr-ft²-°F, what heat exchanger area is required for (a) parallel flow and (b) counterflow?

The heat transfer from the water is given by

$$q = \dot{m} c_p \, \Delta T$$

where $c_p = 0.9985$ Btu/lbm-°F is taken from Table B-3; therefore,

$$q = \left(20 \, \frac{\text{lbm}}{\text{min}}\right)\left(\frac{60 \text{ min}}{\text{hr}}\right)\left(0.9985 \, \frac{\text{Btu}}{\text{lbm-°F}}\right) [(90 - 70) \, °\text{F}] = 23{,}964 \text{ Btu/hr}$$

(a) Figure 10-5 is a qualitative representation of the temperature distribution for the parallel-flow case. The log-mean temperature difference is given by (*10.11*),

$$\Delta T_{\text{lm}} = \frac{\Delta T_2 - \Delta T_1}{\ln (\Delta T_2 / \Delta T_1)}$$

for which $\Delta T_1 = 90 - 10 = 80$ °F and $\Delta T_2 = 70 - 20 = 50$ °F; hence,

$$\Delta T_{\text{lm}} = \frac{50 - 80}{\ln (50/80)} = 63.83 \text{ °F}$$

and

$$A = \frac{q}{U \, \Delta T_{\text{lm}}} = \frac{23{,}964 \text{ Btu/hr}}{\left(150 \, \dfrac{\text{Btu}}{\text{hr-ft}^2\text{-°F}}\right)(63.83 \text{ °F})} = 2.50 \text{ ft}^2$$

(b) The temperature distribution for the counterflow case is shown qualitatively in Fig. 10-16, from which

$$\Delta T_1 = 90 - 20 = 70 \text{ °F}$$

and

$$\Delta T_2 = 70 - 10 = 60 \text{ °F}$$

Equation (*10.11*) yields

$$\Delta T_{\text{lm}} = \frac{60 - 70}{\ln (60/70)} = 64.87 \text{ °F}$$

and

$$A = \frac{q}{U \, \Delta T_{\text{lm}}}$$

$$= \frac{23{,}964 \text{ Btu/hr}}{\left(150 \, \dfrac{\text{Btu}}{\text{hr-ft}^2\text{-°F}}\right)(64.87 \text{ °F})} = 2.46 \text{ ft}^2$$

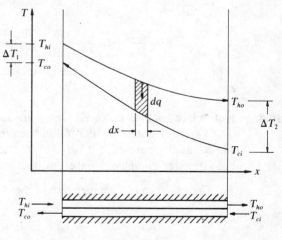

Fig. 10-16

10.2. Hot oil is used to heat water, flowing at the rate of 0.1 kg/s, from 40 °C to 80 °C in a counterflow double-pipe heat exchanger. For an overall heat-transfer coefficient of 300 W/m²-K, find the heat transfer area, if the oil enters at 105 °C and leaves at 70 °C.

The heat gained by the water is given by

$$q = \dot{m}_w c_w (T_{wo} - T_{wi})$$

From Table B-3, the specific heat of water at $(40 + 80)/2 = 60$ °C is

$$c_w = (0.9994)(4.184 \times 10^3) = 4181 \text{ J/kg-K}$$

This gives a heat transfer

$$q = (0.1 \text{ kg/s})(4181 \text{ W-s/kg-K})[(80 - 40) \text{ K}] = 1.67 \times 10^4 \text{ W}$$

Using (*10.10*), where $\Delta T_2 = 70 - 40 = 30$ K and $\Delta T_1 = 105 - 80 = 25$ K, we get

$$A = \frac{q}{U} \frac{\ln(\Delta T_2/\Delta T_1)}{\Delta T_2 - \Delta T_1} = \frac{1.67 \times 10^4 \text{ W}}{300 \text{ W/m}^2\text{-K}} \frac{\ln(30/25)}{(30 - 25) \text{ K}} = 2.03 \text{ m}^2$$

10.3. In a double-pipe counterflow heat exchanger, water at the rate of 60 lbm/min is heated from 65 °F to 95 °F by an oil having a specific heat of 0.36 Btu/lbm-°F. The oil enters the exchanger at 200 °F and leaves at 140 °F. Determine the heat exchanger area for an overall heat-transfer coefficient of 50 Btu/hr-ft²-°F.

The total heat transfer to the water is

$$q = \dot{m}c_p \, \Delta T = \left(60 \frac{\text{lbm}}{\text{min}}\right)\left(\frac{60 \text{ min}}{\text{hr}}\right)\left(1.0 \frac{\text{Btu}}{\text{lbm-°F}}\right)[(95 - 65) \text{ °F}] = 108{,}000 \text{ Btu/hr}$$

Referring to Fig. 10-16, the log-mean temperature difference is given by

$$\Delta T_{lm} = \frac{\Delta T_2 - \Delta T_1}{\ln(\Delta T_2/\Delta T_1)}$$

where $\Delta T_2 = 140 - 65 = 75$ °F and $\Delta T_1 = 200 - 95 = 105$ °F; hence,

$$\Delta T_{lm} = \frac{75 - 105}{\ln(75/105)} = \frac{-30}{-0.336} = 89.16 \text{ °F}$$

and the area is

$$A = \frac{q}{U \, \Delta T_{lm}} = \frac{108{,}000 \text{ Btu/hr}}{\left(50 \dfrac{\text{Btu}}{\text{hr-ft}^2\text{-°F}}\right)(89.16 \text{ °F})} = 24.23 \text{ ft}^2$$

10.4. Hot oil ($c_p = 2.09$ kJ/kg-K) flows through a counterflow heat exchanger at the rate of 0.63 kg/s. It enters at 193 °C and leaves at 65 °C. Cold oil ($c_p = 1.67$ kJ/kg-K) exits at 149 °C at the rate of 1.0 kg/s. What area is required to handle the load if the overall heat-transfer coefficient based on the inside area is 0.7 kW/m²-K?

The unknown inlet temperature of the cold oil may be found from an energy balance on the two fluids, i.e.

$$\dot{m}_c c_c (T_{co} - T_{ci}) = \dot{m}_h c_h (T_{hi} - T_{ho})$$

$$T_{ci} = T_{co} - \frac{\dot{m}_h c_h}{\dot{m}_c c_c}(T_{hi} - T_{ho})$$

$$= 149 \text{ °C} - \frac{(0.63 \text{ kg/s})(2.09 \text{ kJ/kg-K})}{(1.0 \text{ kg/s})(1.67 \text{ kJ/kg-K})}[(193 - 65) \text{ °C}] = 48.08 \text{ °C}$$

Knowing the inlet and exit temperatures for both fluids, we may now use (*10.10*) in conjunction with an energy balance on one fluid to determine the area. From an energy balance on the hot fluid, the heat transfer is given by

$$q = \dot{m}_h c_h (T_{hi} - T_{ho})$$

$$= (0.63 \text{ kg/s})(2.09 \text{ kJ/kg-K})[(193 - 65) \text{ K}] = 168.54 \text{ kJ/s} = 168.54 \text{ kW}$$

Referring to Fig. 10-16 and using (*10.10*), we have

$$A = \frac{q}{U} \frac{\ln(\Delta T_2/\Delta T_1)}{\Delta T_2 - \Delta T_1}$$

where

$$\Delta T_2 = T_{ho} - T_{ci} = 65 - 48.08 = 16.92 \text{ K}$$
$$\Delta T_1 = T_{hi} - T_{co} = 193 - 149 = 44 \text{ K}$$

Thus,

$$A = \frac{168.54 \text{ kW}}{0.700 \text{ kW/m}^2\text{-K}} \frac{\ln(16.92/44)}{(16.92 - 44) \text{ K}} = 8.5 \text{ m}^2$$

10.5. For the same parameters as in Problem 10.3, what area is required when using a shell-and-tube heat exchanger with the water making one shell pass and the oil making two tube passes?

In this case we need a correction factor F from Fig. 10-6 to use in the relation

$$q = UAF \Delta T_{lm}$$

From the nomenclature of Fig. 10-6: $T_i = 65$ °F, $T_o = 95$ °F, $t_i = 200$ °F, and $t_o = 140$ °F; therefore, the dimensionless parameters required to get the correction factor are:

$$P = \frac{t_o - t_i}{T_i - t_i} = \frac{140 - 200}{65 - 200} = 0.444 \qquad Z = \frac{T_i - T_o}{t_o - t_i} = \frac{65 - 95}{140 - 200} = 0.500$$

From Fig. 10-6, $F \approx 0.97$ and the area is

$$A = \frac{q}{UF \Delta T_{lm}} \approx \frac{108,000}{50(0.97)(89.16)} = 24.98 \text{ ft}^2$$

which is approximately 3% greater than that for the double-pipe counterflow exchanger.

10.6. Repeat Problem 10.5 for two shell passes and four tube passes.

The dimensionless parameters for this case are the same as those of Problem 10.5, but Fig. 10-7 must be used to obtain F, which approaches unity. Therefore, $A \approx 24.23 \text{ ft}^2$.

10.7. What area would be required for the conditions of Problem 10.2 if a shell-and-tube heat exchanger were substituted for the double-pipe heat exchanger? The water makes one shell pass, and the oil makes two tube passes.

Assuming that the overall heat-transfer coefficient remains at 300 W/m²-K, we must get a correction factor F from Fig. 10-6 to use in (*10.12*), viz.

$$q = UAF \Delta T_{lm}$$

The temperatures for use in the figure are:

$$T_i = 40 \text{ °C} \qquad T_o = 80 \text{ °C} \qquad t_i = 105 \text{ °C} \qquad t_o = 70 \text{ °C}$$

The dimensionless ratios are

$$P = \frac{t_o - t_i}{T_i - t_i} = \frac{70 - 105}{40 - 105} = 0.54 \qquad Z = \frac{T_i - T_o}{t_o - t_i} = \frac{40 - 80}{70 - 105} = 1.18$$

which gives a correction factor $F \approx 0.6$; therefore,

$$A = \frac{q}{UF \Delta T_{lm}} = \frac{q \ln(\Delta T_2/\Delta T_1)}{UF(\Delta T_2 - \Delta T_1)} \approx \frac{2.03 \text{ m}^2}{0.6} = 3.38 \text{ m}^2$$

10.8. A crossflow heat exchanger with both fluids unmixed is used to heat water (c_p = 4.181 kJ/kg-K) from 40 °C to 80 °C, flowing at the rate of 1.0 kg/s. What is the overall heat-transfer coefficient if hot engine oil (c_p = 1.9 kJ/kg-K), flowing at the rate of 2.6 kg/s, enters at 100 °C? The heat transfer area is 20 m^2.

The heat transfer is given by

$$q = \dot{m}_w c_w (T_{wo} - T_{wi}) = \dot{m}_e c_e (T_{ei} - T_{eo}) = UAF \frac{\Delta T_2 - \Delta T_1}{\ln (\Delta T_2 / \Delta T_1)}$$

The second equality will be used to get the exit oil temperature so that the log-mean temperature difference and the correction factor can be determined.

$$T_{eo} = T_{ei} - \frac{\dot{m}_e c_e}{\dot{m}_w c_w} (T_{wo} - T_{wi})$$

$$= 100 \text{ °C} - \frac{(2.6 \text{ kg/s})(1.9 \text{ kJ/kg-K})}{(1.0 \text{ kg/s})(4.181 \text{ kJ/kg-K})} [(80 - 40) \text{ °C}] = 52.74 \text{ °C}$$

This gives:

$$\Delta T_2 = T_{eo} - T_{wi} = 52.74 - 40 = 12.74 \text{ °C} \qquad \Delta T_1 = T_{ei} - T_{wo} = 100 - 80 = 20 \text{ °C}$$

In the nomenclature of Fig. 10-9,

$$T_i = 100 \text{ °C} \qquad T_o = 52.74 \text{ °C} \qquad t_i = 40 \text{ °C} \qquad t_o = 80 \text{ °C}$$

Evaluating the dimensionless parameters, we get

$$P = \frac{t_o - t_i}{T_i - t_i} = \frac{80 - 40}{100 - 40} = 0.67 \qquad Z = \frac{T_i - T_o}{t_o - t_i} = \frac{100 - 52.74}{80 - 40} = 1.18$$

and the correction factor is $F \approx 0.67$.

From the above heat transfer equations,

$$U = \frac{\dot{m}_w c_w (T_{wo} - T_{wi}) \ln (\Delta T_2 / \Delta T_1)}{AF (\Delta T_2 - \Delta T_1)}$$

$$\approx \frac{(1.0 \text{ kg/s})(4.181 \text{ kJ/kg-K})[(80 - 40) \text{ K}] \ln (12.74/20)}{(20 \text{ m}^2)(0.67)[(12.74 - 20) \text{ K}]} = 0.78 \text{ kW/m}^2\text{-K}$$

10.9. Estimate the surface area required in a crossflow heat exchanger with both fluids unmixed to cool 50,000 lbm/hr of air from 120 °F to 100 °F with water at 60 °F flowing at the rate of 115,000 lbm/hr. Assume that the average value of the overall heat-transfer coefficient is 30 Btu/hr-ft^2-°F.

The total heat transfer from the air is

$$q = (\dot{m} c_p \, \Delta T)_a = \left(50,000 \, \frac{\text{lbm}}{\text{hr}} \right) \left(0.24 \, \frac{\text{Btu}}{\text{lbm-°F}} \right) [(120 - 100) \text{ °F}] = 240,000 \text{ Btu/hr}$$

From an energy balance we can get the exit water temperature, i.e.

$$\dot{m}_a c_{pa} \, \Delta T_a = \dot{m}_w c_{pw} \, \Delta T_w = 240,000 \text{ Btu/hr}$$

$$\Delta T_w = \frac{240,000 \text{ Btu/hr}}{\left(115,000 \, \dfrac{\text{lbm}}{\text{hr}} \right) \left(1.0 \, \dfrac{\text{Btu}}{\text{lbm-°F}} \right)} = 2.09 \text{ °F}$$

and

$$T_{wo} = T_{wi} + \Delta T_w = 60 + 2.09 = 62.09 \text{ °F}$$

The surface area is given by (10.12) as

$$A = \frac{q}{UF\,\Delta T_{lm}}$$

where F is taken from Fig. 10-9 and ΔT_{lm} is the log-mean temperature difference for a counterflow double-pipe heat exchanger having the same fluid inlet and outlet temperatures. Referring to Fig. 10-16, $\Delta T_1 = 120 - 62.09 = 57.91$ °F and $\Delta T_2 = 100 - 60 = 40$ °F. Equation (10.11) gives the log-mean temperature difference as

$$\Delta T_{lm} = \frac{\Delta T_2 - \Delta T_1}{\ln(\Delta T_2/\Delta T_1)} = \frac{40 - 57.91}{\ln(40/57.91)} = 48.40 \text{ °F}$$

Letting the lowercase temperature nomenclature of Fig. 10-9 represent the air, the required dimensionless parameters are:

$$P = \frac{t_o - t_i}{T_i - t_i} = \frac{100 - 120}{60 - 120} = 0.33 \qquad Z = \frac{T_i - T_o}{t_o - t_i} = \frac{60 - 62.09}{100 - 120} = 0.10$$

The correction factor, from Fig. 10-9, is approximately unity; therefore,

$$A \approx \frac{240,000 \text{ Btu/hr}}{\left(30\ \dfrac{\text{Btu}}{\text{hr-ft}^2\text{-°F}}\right)(1.0)(48.40 \text{ °F})} = 165.29 \text{ ft}^2$$

10.10. What error would have been introduced in Problem 10.9 if the arithmetic-mean temperature difference, defined by $\overline{\Delta T_{am}} \equiv (\Delta T_2 + \Delta T_1)/2$, had been used rather than the log-mean temperature difference, ΔT_{lm}?

The arithmetic-mean temperature difference is

$$\overline{\Delta T_{am}} = \frac{40 + 57.91}{2} = 48.96 \text{ °F}$$

This gives an area of

$$A_{am} = \left(\frac{48.40}{48.96}\right)(165.29) = 163.42 \text{ ft}^2$$

which underspecifies the area by

$$\text{error} = \frac{165.29 - 163.42}{165.29} \times 100\% = 1.13\%$$

10.11. For what value of $\Delta T_2/\Delta T_1$ is the arithmetic-mean temperature difference,

$$\overline{\Delta T_{am}} \equiv (\Delta T_2 + \Delta T_1)/2$$

5 percent larger than the log-mean temperature difference, ΔT_{lm}?

We have

$$\frac{\overline{\Delta T_{am}}}{\Delta T_{lm}} = \frac{\frac{1}{2}(\Delta T_2 + \Delta T_1)}{(\Delta T_2 - \Delta T_1)/\ln(\Delta T_2/\Delta T_1)} = \frac{1}{2}\frac{(\Delta T_2/\Delta T_1) + 1}{(\Delta T_2/\Delta T_1) - 1}\ln(\Delta T_2/\Delta T_1)$$

For $\overline{\Delta T_{am}}/\Delta T_{lm} = 1.05$,

$$\frac{(\Delta T_2/\Delta T_1) + 1}{(\Delta T_2/\Delta T_1) - 1}\ln(\Delta T_2/\Delta T_1) = 2.10$$

Solving by trial, $\Delta T_2/\Delta T_1 = 2.2$.

It can be shown analytically that $\overline{\Delta T_{am}}/\Delta T_{lm}$ is a strictly increasing function of $\Delta T_2/\Delta T_1$ for $\Delta T_2/\Delta T_1 \geq 1$. Consequently, the simple arithmetic-mean temperature difference gives results to within 5 percent when the end temperature differences vary by no more than a factor of 2.2.

10.12. When new, a heat exchanger transfers 10 percent more heat than it does after being in service for six months. Assuming that it operates between the same temperature differentials and that there is insufficient scale buildup to change the effective surface area, determine the effective fouling factor in terms of its clean (new) overall heat-transfer coefficient.

The heat exchange ratio may be written as

$$\frac{q_{clean}}{q_{dirty}} = \frac{U_{clean} A \overline{\Delta T}}{U_{dirty} A \overline{\Delta T}} = 1.10 \quad \text{or} \quad U_{dirty} = \frac{U_{clean}}{1.10}$$

Substituting this into (10.26), we get

$$R_f = \frac{1}{U_{dirty}} - \frac{1}{U_{clean}} = \frac{1.10}{U_{clean}} - \frac{1}{U_{clean}} = \frac{0.10}{U_{clean}}$$

10.13. A double-pipe, parallel-flow heat exchanger uses oil ($c_p = 0.45$ Btu/lbm-°F) at an initial temperature of 400 °F to heat water, flowing at 500 lbm/hr, from 60 °F to 110 °F. The oil flow rate is 600 lbm/hr. (*a*) What heat exchanger area is required for an overall heat-transfer coefficient of 60 Btu/hr-ft²-°F? (*b*) Determine the number of transfer units (**NTU**). (*c*) Calculate the effectiveness of the heat exchanger.

(*a*) At an average water temperature of 85 °F, the specific heat is $c_{pw} = 0.998$ Btu/lbm-°F; therefore, an energy balance gives

$$(\dot{m}c_p \, \Delta T)_{oil} = (\dot{m}c_p \, \Delta T)_{water}$$

$$\left(600 \, \frac{lbm}{hr}\right)\left(0.45 \, \frac{Btu}{lbm\text{-}°F}\right)[(400 - T_o) \, °F] = \left(500 \, \frac{lbm}{hr}\right)\left(0.998 \, \frac{Btu}{lbm\text{-}°F}\right)[(110 - 60) \, °F]$$

$$T_{oil}\Big|_o = 400 - \frac{500(0.998)(50)}{600(0.45)} = 307.59 \, °F$$

This gives end temperature differences: $\Delta T_1 = 400 - 60 = 340 \, °F$ and $\Delta T_2 = 307.59 - 110 = 197.59 \, °F$, and the log-mean temperature difference is

$$\Delta T_{lm} = \frac{\Delta T_2 - \Delta T_1}{\ln(\Delta T_2/\Delta T_1)} = \frac{197.59 - 340}{\ln(197.59/340)} = 262.39 \, °F$$

The total heat transfer is $q = (\dot{m}c_p \, \Delta T)_{water} = 24{,}950$ Btu/hr and the area is given by

$$A = \frac{q}{U \, \Delta T_{lm}} = \frac{24{,}950 \, \text{Btu/hr}}{\left(60 \, \dfrac{Btu}{hr\text{-}ft^2\text{-}°F}\right)(262.39 \, °F)} = 1.58 \, ft^2$$

(*b*) The number of transfer units is given by **NTU** $= UA/C_{min}$. To determine C_{min} we must compare the products of the mass-flow rate and specific heat for the water and oil.

$$(\dot{m}c_p)_{water} = \left(500 \, \frac{lbm}{hr}\right)\left(0.998 \, \frac{Btu}{lbm\text{-}°F}\right) = 499 \, \frac{Btu}{hr\text{-}°F}$$

$$(\dot{m}c_p)_{oil} = \left(600 \, \frac{lbm}{hr}\right)\left(0.45 \, \frac{Btu}{lbm\text{-}°F}\right) = 270 \, \frac{Btu}{hr\text{-}°F}$$

Therefore, $C_{min} = 270$ Btu/hr-°F and

$$\text{NTU} = \frac{\left(60 \dfrac{\text{Btu}}{\text{hr-ft}^2\text{-}°\text{F}}\right)(1.58 \text{ ft}^2)}{270 \text{ Btu/hr-}°\text{F}} = 0.35$$

(c)   The effectiveness is given by (10.24).

$$\epsilon = \frac{1 - \exp\left[-\text{NTU}(1 + C_{\min}/C_{\max})\right]}{1 + C_{\min}/C_{\max}} = \frac{1 - \exp\left[-(0.35)\left(1 + \dfrac{270}{499}\right)\right]}{1 + \dfrac{270}{499}} = 27.0\%$$

which agrees very well with that shown graphically in Fig. 10-10.

10.14.   Water enters a counterflow, double-pipe heat exchanger at 100 °F, flowing at the rate of 100 lbm/min.   It is heated by oil ($c_p = 0.45$ Btu/lbm-°F) flowing at the rate of 200 lbm/min from an inlet temperature of 240 °F.   For an area of 140 ft² and an overall heat-transfer coefficient of 60 Btu/hr-ft²-°F, determine the total heat transfer.

Since the inlet temperatures are known and the ($\dot{m}c_p$)-products can be calculated, (10.16) may be used to determine the heat transfer after finding the heat exchanger effectiveness, ϵ.   Since the specific heat of water is approximately 1.0 Btu/lbm-°F,

$$(\dot{m}c_p)_{\text{water}} = \left(100 \frac{\text{lbm}}{\text{min}}\right)\left(1.0 \frac{\text{Btu}}{\text{lbm-}°\text{F}}\right) = 100 \text{ Btu/min-}°\text{F}$$

$$(\dot{m}c_p)_{\text{oil}} = \left(200 \frac{\text{lbm}}{\text{min}}\right)\left(0.45 \frac{\text{Btu}}{\text{lbm-}°\text{F}}\right) = 90 \text{ Btu/min-}°\text{F}$$

Therefore

$$\frac{C_{\min}}{C_{\max}} = \frac{90}{100} = 0.90$$

and

$$\text{NTU} = \frac{UA}{C_{\min}} = \frac{\left(60 \dfrac{\text{Btu}}{\text{hr-ft}^2\text{-}°\text{F}}\right)(140 \text{ ft}^2)}{\left(90 \dfrac{\text{Btu}}{\text{min-}°\text{F}}\right)\left(\dfrac{60 \text{ min}}{\text{hr}}\right)} = 1.56$$

Using these parameters with Fig. 10-11, we get $\epsilon \approx 0.62$; hence,

$$q = \epsilon C_{\min}(T_{hi} - T_{ci}) \approx (0.62)\left(90 \frac{\text{Btu}}{\text{min-}°\text{F}}\right)\left(\frac{60 \text{ min}}{\text{hr}}\right)[(240 - 100) °\text{F}] = 468{,}720 \text{ Btu/hr}$$

10.15.   Water enters a crossflow heat exchanger (both fluids unmixed) at 60 °F and flows at the rate of 60,000 lbm/hr to cool 80,000 lbm/hr of air from 250 °F.   For an overall heat-transfer coefficient of 40 Btu/hr-ft²-°F and an exchanger surface area of 2600 ft², what is the exit air temperature?

Taking the specific heats of the air and water to be constant at 0.24 and 1.0 Btu/lbm-°F, respectively, we get

$$(\dot{m}c_p)_{\text{air}} = \left(80{,}000 \frac{\text{lbm}}{\text{hr}}\right)\left(0.24 \frac{\text{Btu}}{\text{lbm-}°\text{F}}\right) = 19{,}200 \text{ Btu/hr-}°\text{F}$$

$$(\dot{m}c_p)_{\text{water}} = \left(60{,}000 \frac{\text{lbm}}{\text{hr}}\right)\left(1.0 \frac{\text{Btu}}{\text{lbm-}°\text{F}}\right) = 60{,}000 \text{ Btu/hr-}°\text{F}$$

which gives

$$\frac{C_{min}}{C_{max}} = \frac{C_{air}}{C_{water}} = \frac{19{,}200}{60{,}000} = 0.32$$

Also,

$$\textbf{NTU} = \frac{AU}{C_{min}} = \frac{(2600 \text{ ft}^2)\left(40 \dfrac{\text{Btu}}{\text{hr-ft}^2\text{-}°\text{F}}\right)}{19{,}200 \text{ Btu/hr-}°\text{F}} = 5.42$$

and from Fig. 10-14, $\epsilon \approx 0.94$.

The heat transfer is given by

$$q = \epsilon C_{min}(T_{hi} - T_{ci})$$

$$\approx (0.94)\left(19{,}200 \frac{\text{Btu}}{\text{hr-}°\text{F}}\right)[(250 - 60)\ °\text{F}] = 3.43 \times 10^6 \text{ Btu/hr}$$

An energy balance on the air then gives the exit temperature:

$$q = \dot{m}c_p(T_i - T_o)$$

$$T_o = -\frac{q}{\dot{m}c_p} + T_i$$

$$= -\frac{3.43 \times 10^6 \text{ Btu/hr}}{19{,}200 \text{ Btu/hr-}°\text{F}} + 250\ °\text{F} = 71.40\ °\text{F}$$

10.16. Solve Problem 10.15 for a one-shell-pass, 10-tube-pass heat exchanger.

The same dimensionless parameters hold, but the heat exchanger effectiveness is lower. From Fig. 10-12, $\epsilon \approx 0.83$; therefore,

$$q = \epsilon C_{min}(T_{hi} - T_{ci}) = [\dot{m}c_p(T_i - T_o)]_{air}$$

$$(0.83)(19{,}200)(250 - 60) = (80{,}000)(0.24)(250 - T_o)$$

$$T_o = 250 - 157.70 = 92.30\ °\text{F}$$

10.17. Hot water at 180 °F enters the tubes of a two-shell-pass, 8-tube-pass heat exchanger at the rate of 80 lbm/min, heating helium from 20 °F. The overall heat-transfer coefficient is 20 Btu/hr-ft²-°F, and the exchanger area is 110 ft². If the water exits at 120 °F, determine the exit temperature of the helium and its mass-flow rate.

Since the flow rate of the helium is unknown, there is no way to determine a priori the minimum heat-capacity rate. Assuming that the minimum fluid is the water,

$$C_w = \dot{m}c_p = \left(80 \frac{\text{lbm}}{\text{min}}\right)\left(1.0 \frac{\text{Btu}}{\text{lbm-}°\text{F}}\right) = 80 \text{ Btu/min-}°\text{F}$$

Based upon this assumption, the number of transfer units is

$$\textbf{NTU} = \frac{UA}{C_{min}} = \frac{\left(20 \dfrac{\text{Btu}}{\text{hr-ft}^2\text{-}°\text{F}}\right)(110 \text{ ft}^2)}{\left(80 \dfrac{\text{Btu}}{\text{min-}°\text{F}}\right)\left(\dfrac{60 \text{ min}}{\text{hr}}\right)} = 0.46$$

The top equation (10.18) holds, giving

$$\epsilon_h = \frac{T_{hi} - T_{ho}}{T_{hi} - T_{ci}} = \frac{180 - 120}{180 - 20} = 0.38$$

From Fig. 10-13 these parameters give $C = C_{min}/C_{max} \approx 0.2$, which validates the initial assumption of the water as the minimum fluid. Hence,

$$C_{He} \approx \frac{C_w}{0.2} = 400 \text{ Btu/min-°F} = (\dot{m}c_p)_{He}$$

In the temperature range being considered, the specific heat of helium is 1.242 Btu/lbm-°F; therefore,

$$\dot{m} = \frac{C_{He}}{c_p} = \frac{\left(400 \dfrac{\text{Btu}}{\text{min-°F}}\right)}{1.242 \text{ Btu/lbm-°F}} = 322 \text{ lbm/min}$$

An energy balance gives the helium exit temperature, i.e.

$$[\dot{m}c_p(T_i - T_o)]_w = [\dot{m}c_p(T_o - T_i)]_{He}$$
$$(80 \text{ Btu/min-°F})[(180 - 120) \text{ °F}] = (400 \text{ Btu/min-°F})[(T_o - 20) \text{ °F}]$$

$$T_o\Big|_{He} = 12 + 20 = 32 \text{ °F}$$

10.18. Hot oil is used in a crossflow heat exchanger to heat a dye solution in a carpet manufacturing plant. The mixed-flow dye solution ($c_p = 1.12$ Btu/lbm-°F) enters at 60 °F and exits at 130 °F at a flow rate of 3000 lbm/hr. The unmixed-flow oil ($c_p = 0.46$ Btu/lbm-°F) enters at 400 °F, and the flow system produces an overall heat-transfer coefficient of 50 Btu/hr-ft²-°F. For an exchanger surface area of 80 ft², what mass-flow rate of oil is required, and what is its exit temperature?

The given conditions do not permit the direct determination of the minimum fluid. If the minimum fluid is the dye solution, we can solve for ϵ and **NTU** and use Fig. 10-15 to get $C_{mixed}/C_{unmixed}$, which would permit the oil flow rate to be determined. If the minimum fluid is the oil, a solution requires trial-and-error.

Assuming that the minimum fluid is the dye solution, the second equation (*10.18*) gives

$$\epsilon_{dye} = \epsilon_c = \frac{T_{co} - T_{ci}}{T_{hi} - T_{ci}} = \frac{130 - 60}{400 - 60} = 0.21$$

Under the same assumption,

$$C_{mixed} = C_{dye} = (\dot{m}c_p)_{dye} = \left(3000 \frac{\text{lbm}}{\text{hr}}\right)\left(1.12 \frac{\text{Btu}}{\text{lbm-°F}}\right) = 3360 \text{ Btu/hr-°F}$$

and

$$\textbf{NTU} = \frac{UA}{C_{mixed}} = \frac{\left(50 \dfrac{\text{Btu}}{\text{hr-ft}^2\text{-°F}}\right)(80 \text{ ft}^2)}{3360 \text{ Btu/hr-°F}} = 1.19$$

With these parameters, it is apparent from Fig. 10-15 that the ratio $C_{mixed}/C_{unmixed}$ does not exist; therefore, *the minimum fluid is the oil.*

We must now assume a value of $C_{unmixed}$ (which amounts to assuming $\dot{m}_{oil}$), determine the pertinent parameters and compare calculated results with Fig. 10-15 until a solution is found. A first trial in this procedure is outlined below; subsequent assumptions and results are presented in Table 10-4. Assume

$$C_{unmixed} = C_{oil} = C_{min} = 1000 \text{ Btu/hr-°F}$$

Then

$$\frac{C_{mixed}}{C_{unmixed}} = \frac{3360}{1000} = 3.36 \qquad \textbf{NTU} = \frac{UA}{C_{min}} = \frac{50(80)}{1000} = 4.00$$

From Fig. 10-15, $\epsilon \approx 0.84$; an energy balance gives

$$\Delta T_h = \Delta T_{oil} = \frac{(\dot{m}c_p\,\Delta T)_{dye}}{(\dot{m}c_p)_{oil}} = \frac{C_{mixed}\,(\Delta T)_{dye}}{C_{unmixed}} = (3.36)(130-60) = 235.2\ °F$$

$$\epsilon_h = \epsilon_{oil} = \frac{(\Delta T)_{oil}}{T_{hi} - T_{ci}} = \frac{235.2}{400 - 60} = 0.69$$

and the assumed $C_{unmixed}$ is incorrect.

Table 10-4

Trial No.	$C_{unmixed}$ (assumed)	$\dfrac{C_{mixed}}{C_{unmixed}}$	NTU	ΔT_h	ϵ Calculated	ϵ From Fig. 10-15
1	1000 Btu/hr-°F	3.36	4.00	235.2 °F	0.69	0.84
2	2000	1.68	2.00	117.6	0.35	0.67
3	1500	2.24	2.67	156.8	0.46	0.77
4	880	3.82	4.55	267.3	0.79	0.87
5	750	4.48	5.33	313.6	0.92	0.92

Figure 10-17 illustrates how such a trial-and-error solution may be refined from a simple plot of the data from the trial solutions. Curve A is an extrapolation of the data from the first three trials, giving $C_{unmixed} \approx 880$, where $\Delta\epsilon$ is the difference between the value from Fig. 10-15 and the calculated value. Curve B is a refinement using the values determined in trial number 4. No further refinement was necessary.

The required parameters may now be determined.

$$\dot{m}_{oil} = \frac{C_{oil}}{c_{p\,oil}} = \frac{750\ \dfrac{Btu}{hr\text{-}°F}}{0.46\ \dfrac{Btu}{lbm\text{-}°F}} = 1630\ lbm/hr$$

$$(\Delta T)_{oil} = 313.6 = T_{hi} - T_{ho} = 400 - T_{ho}$$

$$T_{ho} = 400 - 313.6 = 86.4\ °F$$

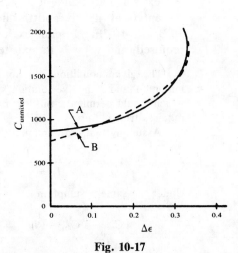

Fig. 10-17

Supplementary Problems

10.19. For a double-pipe heat exchanger in which the overall heat-transfer coefficient varies linearly with the temperature difference, i.e. $U = a + b\,\Delta T$, determine an expression for the heat transfer using subscripts 1 and 2 to represent values at each end of the exchanger.

Ans. $q = A\,\dfrac{U_1\,\Delta T_2 - U_2\,\Delta T_1}{\ln\,(U_1\,\Delta T_2/U_2\,\Delta T_1)}$

10.20. In a one-shell-pass, one-tube-pass (double-pipe) heat exchanger one fluid enters at 120 °F and leaves at 500 °F.   The other fluid enters at 900 °F and leaves at 600 °F.   What is the log-mean temperature difference for (*a*) parallel flow and (*b*) counterflow?   *Ans.*   (*a*) 331 °F; (*b*) 439 °F

10.21. Hot gases enter a parallel-flow heat exchanger at 800 °F and leave at 500 °F, heating 200,000 lbm/hr of water from 90 °F to 180 °F.   For a surface area of 4000 ft^2, what is the overall heat-transfer coefficient?   *Ans.*   9.20 Btu/hr-ft^2-°F

10.22. What is the exit water temperature, in °C, for the heat exchanger of Problem 10.14?   *Ans.*   81.2 °C

10.23. A counterflow heat exchanger has an overall heat-transfer coefficient of 40 Btu/hr-ft^2-°F and a surface area of 900 ft^2.   The hot fluid ($c_p = 0.85$ Btu/lbm-°F) enters at 200 °F and flows at the rate of 20,000 lbm/hr.   The cold fluid ($c_p = 0.40$ Btu/lbm-°F) enters at 60 °F and flows at the rate of 18,000 lbm/hr.   What is the rate of heat transfer?   *Ans.*   9.68 × 10^5 Btu/hr

10.24. How much heat would a crossflow heat exchanger having one fluid mixed and one fluid unmixed transfer under the conditions given in Problem 10.15?   *Ans.*   5.81 × 10^6 Btu/hr

Chapter 11

Radiation

11.1 INTRODUCTION

Radiation is a term applied to many processes which involve energy transfer by electromagnetic wave phenomena. The radiative mode of heat transfer differs in two important respects from the conductive and convective modes: (1) no medium is required and (2) the energy transfer is proportional to the fourth or fifth power of the temperatures of the bodies involved.

The Electromagnetic Spectrum

A major part of the electromagnetic spectrum is illustrated in Fig. 11-1. *Thermal radiation* is defined as the portion of the spectrum between the wavelengths 1×10^{-7} m and 1×10^{-4} m. Of interest also is the very narrow *visible spectrum*, which runs from 3.9×10^{-7} m to 7.8×10^{-7} m.

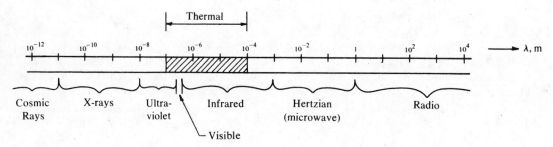

Fig. 11-1

A convenient wavelength unit is the *micrometer*: $1 \mu m = 10^{-6}$ m. In these units, thermal radiation has the range 0.1 to $100 \mu m$, and the visible portion of the spectrum is from 0.39 to $0.78 \mu m$. Another common unit of wavelength is the *angstrom*: $1 \text{ Å} = 10^{-10}$ m.

The propagation velocity for all types of electromagnetic radiation in a vacuum is

$$c = \lambda \nu = 3 \times 10^8 \text{ m/s} \tag{11.1}$$

where λ is the wavelength and ν is the frequency of the radiation.

11.2 PROPERTIES AND DEFINITIONS

The word *spectral* is used to denote dependence upon wavelength for any radiation quantity. The value of the quantity at a given wavelength is called a *monochromatic* value.

Absorptivity, Reflectivity and Transmissivity

Whenever radiant energy is incident upon any surface, part may be absorbed, part may be reflected, and part may be transmitted through the receiving body. Defining

   $\alpha \equiv$ fraction of incident radiation absorbed $\equiv$ absorptivity

   $\rho \equiv$ fraction of incident radiation reflected $\equiv$ reflectivity

   $\tau \equiv$ fraction of incident radiation transmitted $\equiv$ transmissivity it is clear that

$$\alpha + \rho + \tau = 1 \qquad (11.2)$$

Most solids, other than those which are visibly transparent or translucent, do not transmit radiation, and (11.2) reduces to

$$\alpha + \rho = 1 \qquad (11.3)$$

Frequently, (11.3) is applied to liquids, although the transmissivity of a liquid is strongly dependent upon thickness.

Gases generally reflect very little radiant thermal energy, and (11.2) simplifies to

$$\alpha + \tau = 1 \qquad (11.4)$$

Emissive Power and Radiosity

The *total emissive power*, denoted by E, is the total (over all wavelengths and all directions) emitted radiant thermal energy leaving a surface per unit time and unit area of the emitting surface. Note in particular that this is the energy leaving due to original emission only; it does not include any energy reflected from the surface (and originating elsewhere). Other names are "total hemispherical emissive power," "radiant flux density," or simply "emissive power." The total emissive power of a surface is dependent upon (1) the material or substance, (2) the surface condition (including roughness), and (3) the temperature.

Radiosity, J, denotes the total radiant thermal energy leaving a surface per unit time and unit area of the surface. Thus the radiosity is the sum of the emitted and the reflected radiant energy fluxes from a surface. Like total emissive power, total radiosity represents an integration over the spectral and directional distribution.

Specular and Diffuse Surfaces

Reflection of radiant thermal energy from a surface can be described with the help of two ideal models. The perfect *specular reflector* is shown in Fig. 11-2(a); in this case the angle of incidence, ϕ_i, is equal to the angle made by the reflected ray, ϕ_r. A *diffuse reflector* is shown in Fig. 11-2(b); in this case the magnitude of the reflected energy in a specific direction ϕ_r is proportional to the cosine of ϕ_r, ϕ_r being measured from the normal N.

If the roughness dimension (height) for a real surface is considerably smaller than the wavelength of incident irradiation, the surface behaves as a specular one; if the roughness dimension is large with respect to wavelength, the surface reflects diffusely.

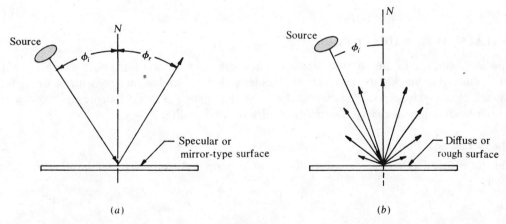

Fig. 11-2

Intensity of Radiation

We shall define the *radiation intensity*, *I*, as the radiant energy per unit time per unit solid angle per unit area of the emitter projected normal to the line of view of the receiver from the radiating element. For the geometry depicted in Fig. 11-3, the energy radiated from element dA_1 and intercepted by element dA_2 is

$$dq_{1\to2} = I(\cos\phi \, dA_1)d\omega \tag{11.5}$$

Here

$$d\omega \equiv \frac{dA_2}{r^2} = \sin\phi \, d\theta \, d\phi \tag{11.6}$$

is the solid angle subtended by dA_2, and $\cos\phi \, dA_1$ is the area of the emitting surface projected normal to the line of view to the receiving surface. Substituting (*11.6*) into (*11.5*) and integrating over the hemispherical surface results in

$$\frac{q_{1\to2}}{dA_1} = \int_0^{2\pi}\int_0^{\pi/2} I \cos\phi \sin\phi \, d\phi \, d\theta \tag{11.7}$$

which is the general relationship between the total emissive power of a body (in this case, the element dA_1) and the intensity of radiation.

If the emitting surface is diffuse, $I = $ constant, and (*11.7*) integrates to

$$\frac{q_{1\to2}}{dA_1} = E = \pi I \tag{11.8}$$

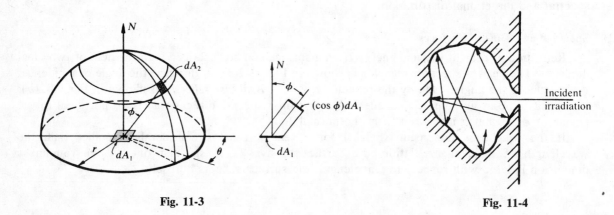

Fig. 11-3   Fig. 11-4

11.3 BLACKBODY RADIATION

The ideal surface in the study of radiative heat transfer is the *blackbody*, which is defined by $\alpha_b = 1$. Thus the blackbody absorbs all incident thermal radiation, regardless of spectral or directional characteristics. As shown in Problem 11.4, such a body can be approximated by a small hole leading into a cavity (German, *Hohlraum*). See Fig. 11-4.

Blackbody Emissive Power

The total (hemispherical) emissive power of a blackbody is given by the *Stefan-Boltzmann equation*:

$$E_b = \sigma T^4 \tag{11.9}$$

where σ, the Stefan-Boltzmann constant, is 0.1714×10^{-8} Btu/hr-ft²-°R⁴ or 5.6697×10^{-8} W/m²-K⁴.

Blackbody Spectral Distribution

In general, a surface emits different amounts of energy at different wavelengths. The total emissive power can be expressed as

$$E = \int_0^\infty E_\lambda \, d\lambda \tag{11.10}$$

where E_λ is the monochromatic emissive power at wavelength λ. For a blackbody,

$$E_b = \int_0^\infty E_{b\lambda} \, d\lambda = \sigma T^4 \tag{11.11}$$

The first accurate expression for $E_{b\lambda}$ was determined by Max Planck; it is

$$E_{b\lambda} = \frac{C_1 \lambda^{-5}}{\exp(C_2/\lambda T) - 1} \tag{11.12}$$

in which

$$C_1 = 1.187 \times 10^8 \, \frac{\text{Btu-}\mu\text{m}^4}{\text{hr-ft}^2} = 3.742 \times 10^8 \, \frac{\text{W-}\mu\text{m}^4}{\text{m}^2}$$

$$C_2 = 2.5896 \times 10^4 \, \mu\text{m-}^\circ\text{R} = 1.4387 \times 10^4 \, \mu\text{m-K}$$

Plots of $E_{b\lambda}$ versus λ for several different temperatures are given in Fig. 11-5. The shift in location of the maximum value of the monochromatic emissive power to shorter wavelengths with increasing temperature is evident. This wavelength shift is described by *Wien's displacement law*,

$$\lambda_{\max} T = 5215.6 \, \mu\text{m-}^\circ\text{R} = 2897.6 \, \mu\text{m-K} \tag{11.13}$$

which plots as the dashed curve through the peak values of emissive power in Fig. 11-5.

It is frequently necessary to determine the amount of energy radiated by a blackbody over a specified portion of the thermal radiation waveband. The energy emitted within the range 0 to λ

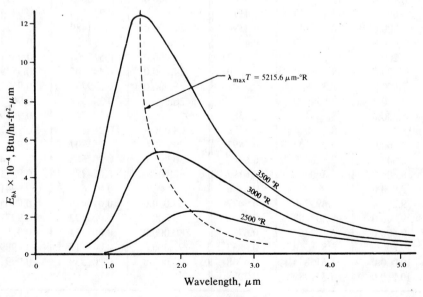

Fig. 11-5

Table 11-1. (From R. V. Dunkle, "Thermoradiation Tables and Applications," *Trans. ASME*, **76**, 1954, pp. 549–552.)

λT, μm-°R	$\dfrac{E_{b\lambda} \times 10^5}{\sigma T^5}$, $(\mu\text{m-°R})^{-1}$	$\dfrac{E_{b(0-\lambda T)}}{\sigma T^4}$	λT, μm-°R	$\dfrac{E_{b\lambda} \times 10^5}{\sigma T^5}$, $(\mu\text{m-°R})^{-1}$	$\dfrac{E_{b(0-\lambda T)}}{\sigma T^4}$
1,000.0	0.000039	0.0000	10,400.0	5.142725	0.7183
1,200.0	0.001191	0.0000	10,600.0	4.921745	0.7284
1,400.0	0.012008	0.0000	10,800.0	4.710716	0.7380
1,600.0	0.062118	0.0000	11,000.0	4.509291	0.7472
1,800.0	0.208018	0.0003	11,200.0	4.317109	0.7561
2,000.0	0.517405	0.0010	11,400.0	4.133804	0.7645
2,200.0	1.041926	0.0025	11,600.0	3.959010	0.7726
2,400.0	1.797651	0.0053	11,800.0	3.792363	0.7803
2,600.0	2.761875	0.0098	12,000.0	3.633505	0.7878
2,800.0	3.882650	0.0164	12,200.0	3.482084	0.7949
3,000.0	5.093279	0.0254	12,400.0	3.337758	0.8017
3,200.0	6.325614	0.0368	12,600.0	3.200195	0.8082
3,400.0	7.519353	0.0507	12,800.0	3.069073	0.8145
3,600.0	8.626936	0.0668	13,000.0	2.944084	0.8205
3,800.0	9.614973	0.0851	13,200.0	2.824930	0.8263
4,000.0	10.463377	0.1052	13,400.0	2.711325	0.8318
4,200.0	11.163315	0.1269	13,600.0	2.602997	0.8371
4,400.0	11.714711	0.1498	13,800.0	2.499685	0.8422
4,600.0	12.123821	0.1736	14,000.0	2.401139	0.8471
4,800.0	12.401105	0.1982	14,200.0	2.307123	0.8518
5,000.0	12.559492	0.2232	14,400.0	2.217411	0.8564
5,200.0	12.613057	0.2483	14,600.0	2.131788	0.8607
5,400.0	12.576066	0.2735	14,800.0	2.050049	0.8649
5,600.0	12.462308	0.2986	15,000.0	1.972000	0.8689
5,800.0	12.284687	0.3234	16,000.0	1.630989	0.8869
6,000.0	12.054971	0.3477	17,000.0	1.358304	0.9018
6,200.0	11.783688	0.3715	18,000.0	1.138794	0.9142
6,400.0	11.480102	0.3948	19,000.0	0.960883	0.9247
6,600.0	11.152254	0.4174	20,000.0	0.815714	0.9335
6,800.0	10.807041	0.4394	21,000.0	0.696480	0.9411
7,000.0	10.450309	0.4607	22,000.0	0.597925	0.9475
7,200.0	10.086964	0.4812	23,000.0	0.515964	0.9531
7,400.0	9.721078	0.5010	24,000.0	0.447405	0.9579
7,600.0	9.355994	0.5201	25,000.0	0.389739	0.9621
7,800.0	8.994419	0.5384	26,000.0	0.340978	0.9657
8,000.0	8.638524	0.5561	27,000.0	0.299540	0.9689
8,200.0	8.290014	0.5730	28,000.0	0.264157	0.9717
8,400.0	7.950202	0.5892	29,000.0	0.233807	0.9742
8,600.0	7.620072	0.6048	30,000.0	0.207663	0.9764
8,800.0	7.300336	0.6197	40,000.0	0.074178	0.9891
9,000.0	6.991475	0.6340	50,000.0	0.032617	0.9941
9,200.0	6.693786	0.6477	60,000.0	0.016479	0.9965
9,400.0	6.407408	0.6608	70,000.0	0.009192	0.9977
9,600.0	6.132361	0.6733	80,000.0	0.005521	0.9984
9,800.0	5.868560	0.6853	90,000.0	0.003512	0.9989
10,000.0	5.615844	0.6968	100,000.0	0.002339	0.9991
10,200.0	5.373989	0.7078			

at a specified temperature T can be expressed as

$$E_{b(0-\lambda T)} \equiv \int_0^{\lambda T} \frac{1}{T} E_{b\lambda} \, d(\lambda T) \tag{11.14}$$

The fraction of the total energy within this range is then

$$\frac{E_{b(0-\lambda T)}}{E_b} = \frac{E_{b(0-\lambda T)}}{\sigma T^4} = \int_0^{\lambda T} \frac{E_{b\lambda}}{\sigma T^5} \, d(\lambda T) \tag{11.15}$$

Values of the integrand and of the integral in (11.15) are presented in Table 11-1.

Blackbody Intensity

Emission from a blackbody is independent of direction, so that (11.8) gives

$$E_b = \sigma T^4 = \pi I_b \tag{11.16}$$

11.4 REAL SURFACES AND THE GRAY BODY

A real surface has a total emissive power E less than that of a blackbody. The ratio of the total emissive power of a body to that of a blackbody at the same temperature is the *total emissivity* (or *total hemispherical emissivity*), ϵ:

$$\epsilon \equiv \frac{E}{E_b} \tag{11.17}$$

Some numerical values of total emissivity are presented in Table B-6 of Appendix B; general trends are shown in Fig. 11-6.

The *monochromatic (hemispherical) emissivity*, ϵ_λ, will be useful in dealing with real surfaces which exhibit spectrally selective emittance values. This is

$$\epsilon_\lambda \equiv \frac{E_\lambda}{E_{b\lambda}} \tag{11.18}$$

where E_λ is the emissive power of the real surface at wavelength λ, and $E_{b\lambda}$ is that of a blackbody, both being at the same temperature.

Kirchhoff's Law

Consider a black enclosure as shown in Fig. 11-7. Suppose that this contains a small body, say body 1, which is also a blackbody. Under equilibrium conditions, energy is absorbed by the small

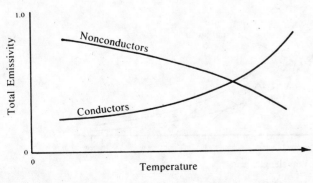

Fig. 11-6

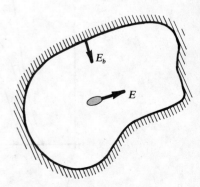

Fig. 11-7

body 1 at the same rate as it is emitted, $A_1 E_b$. Now suppose that the small body is replaced by another body having the same size, shape, geometric position, and orientation. Denote this as body 2. Clearly the energy impinging upon this second body is the same as before, $A_1 E_b = A_2 E_b$. An energy balance on the second body under steady-state conditions is

$$\alpha_2 A_2 E_b = A_2 E$$

and consequently, for any arbitrary body,

$$\alpha = \frac{E}{E_b} = \epsilon \tag{11.19}$$

since E/E_b is by definition the total hemispherical emissivity. Equation (11.19) is known as *Kirchhoff's law.*

Note that (11.19) was obtained under conditions of thermal equilibrium in an isothermal black enclosure. It is applicable to a surface receiving blackbody irradiation from surroundings at the same temperature as itself; serious errors may result from extending this result to other situations.

For monochromatic values, however, it can be shown that Kirchhoff's law is

$$\alpha_\lambda(T) = \epsilon_\lambda(T) \tag{11.20}$$

that is, for specified wavelength, the monochromatic absorptivity and the monochromatic emissivity of a surface at a given surface temperature are equal, regardless of the temperature of the source of the incoming irradiation.

Real Surface Emission and the Gray Body Approximation

As seen in Fig. 11-8, the monochromatic emissive power of a real surface is not a constant fraction of that of a black surface. A very useful idealization is that of a *gray body*, defined by

$$(\epsilon_\lambda)_{\text{gray}} \equiv \text{constant}$$

The computational advantages of this are apparent from a consideration of the expression for the total emissive power of a body:

$$E = \int_0^\infty E_\lambda \, d\lambda = \int_0^\infty \epsilon_\lambda E_{b\lambda} \, d\lambda \tag{11.21}$$

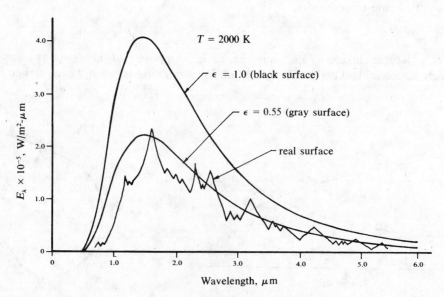

Fig. 11-8

which for constant ϵ_λ simplifies to

$$E = \epsilon \int_0^\infty E_{b\lambda} \, d\lambda = \epsilon\sigma T^4 \qquad (11.22)$$

Directional Dependence of Properties

In addition to the previously discussed variables which influence surface properties, the emissivity of a smooth surface depends strongly upon the polar angle ϕ between the direction of the incoming radiation and a normal to the surface. In general, nonconductors emit more strongly in the direction normal to the surface (or at small polar angles), whereas conductors emit more strongly at large polar angles.

11.5   RADIANT EXCHANGE: BLACK SURFACES

In this section, and in those that follow, we will limit our discussion to *diffuse surface behavior*, and further, we will consider only *total hemispherical properties* unless otherwise specifically stated. The following subscript notation will be used in the remainder of this chapter:

q_i = net rate of energy given up by surface i

$q_{i \to j}$ = rate of energy leaving surface i and striking surface j

$q'_{i\text{-}j}$ = rate of energy emitted by surface i and absorbed by surface j

$q_{i\text{-}j}$ = net rate of energy exchange between surface i and surface j

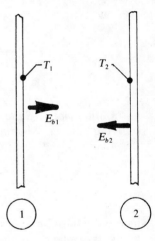

Fig. 11-9

The "net rate of energy given up" is that above the rate of absorption of energy.

Consider the simplest physical configuration, that of two infinite, black, parallel planes maintained at different (but constant) temperatures, T_1 and T_2, as shown in Fig. 11-9. The net exchange of energy between surfaces 1 and 2 is

$$q_{1\text{-}2} = q'_{1\text{-}2} - q'_{2\text{-}1} = \alpha_2 E_{b1} A_1 - \alpha_1 E_{b2} A_2 = E_{b1} A_1 - E_{b2} A_2$$

since $\alpha_1 = \alpha_2 = 1$. Then, per unit area,

$$q_{1\text{-}2}/A = E_{b1} - E_{b2} = \sigma(T_1^4 - T_2^4) \qquad (11.23)$$

Configuration Factors

Engineering problems of practical interest invariably involve one or more surfaces of finite size, and the radiant exchange is strongly dependent upon the geometry. Hence, we must determine the configuration effect upon the radiant heat-transfer rate. This is accounted for by introduction of the *configuration factor*, which is defined as the fraction of radiant energy leaving one surface which strikes a second surface directly, both surfaces assumed to be emitting energy diffusely. The term "directly" means that none of the energy is transferred by reflection or reradiation from other surfaces. Other names in the literature for this factor include the "radiation shape factor," the "shape factor," the "view factor" and the "angle factor."

Infinitesimal areas. To develop a general expression for the configuration factor, consider the surface elements shown in Fig. 11-10. The energy leaving dA_1 and incident upon dA_2 is, by (11.5),

$$dq_{1\to 2} = I_1 \cos \phi_1 \, dA_1 \, d\omega_{1\text{-}2} \qquad (11.24)$$

where $d\omega_{1\text{-}2}$ is the solid angle subtended by dA_2 at dA_1, i.e.

$$dw_{1\text{-}2} = \frac{\cos\phi_2\, dA_2}{r^2} \tag{11.25}$$

The total energy radiated from dA_1 is [compare (11.8)]

$$dq_1 = I_1\pi\, dA_1 \tag{11.26}$$

The configuration factor, $F_{dA_1\to dA_2}$, is by definition the ratio of $dq_{1\to2}$ to dq_1:

$$F_{dA_1\to dA_2} = \frac{\cos\phi_1\cos\phi_2\, dA_2}{\pi r^2} \tag{11.27}$$

Note that the configuration factor involves geometrical quantities only.

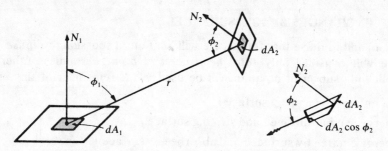

Fig. 11-10

Infinitesimal-to-finite area. Consider the case of a very small emitter and a finite-sized receiving surface. Forming the ratio of $q_{1\to2}$ from (11.24) to dq_1 from (11.26),

$$F_{dA_1\to A_2} = \frac{\displaystyle\int_{A_2} I_1\cos\phi_1\, dA_1\cos\phi_2\, dA_2/r^2}{\pi I_1\, dA_1} = \int_{A_2}\frac{\cos\phi_1\cos\phi_2\, dA_2}{\pi r^2} \tag{11.28}$$

since both I_1 and dA_1 are independent of the integration. It should be noted that this equation is simply the integral over A_2 of (11.27), the configuration factor for infinitesimal-to-infinitesimal areas. The integral is evaluated for one configuration of interest in Fig. 11-11.

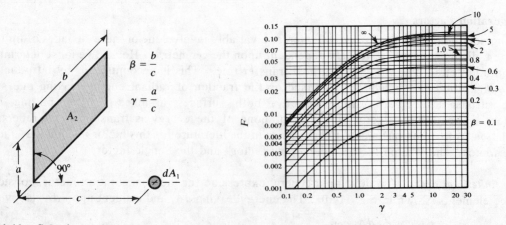

Fig. 11-11. Spherical point source to a plane rectangle. (Adapted from D.C. Hamilton and W.R. Morgan, *NACA Tech. Note TN-2836*, 1952.)

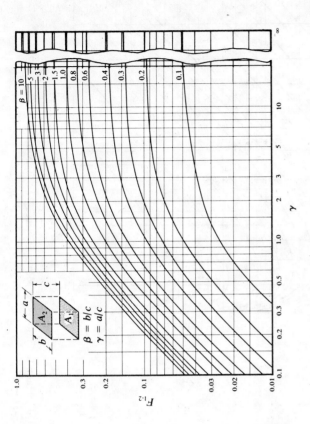

Fig. 11-12. Two identical, parallel, directly opposed flat plates. (Adapted from D.C. Hamilton and W.R. Morgan, *NACA Tech. Note TN-2836,* 1952.)

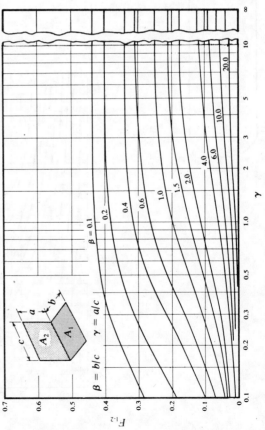

Fig. 11-14. Parallel, concentric discs. (Adapted from D.C. Hamilton and W.R. Morgan, *NACA Tech. Note TN-2836,* 1952.)

Fig. 11-13. Two perpendicular flat plates with a common edge. (Adapted from D.C. Hamilton and W.R. Morgan, *NACA Tech. Note TN-2836,* 1952.)

269

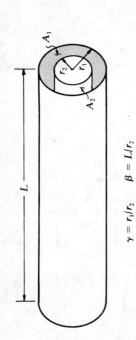

$\gamma = r_1/r_2 \qquad \beta = L/r_2$

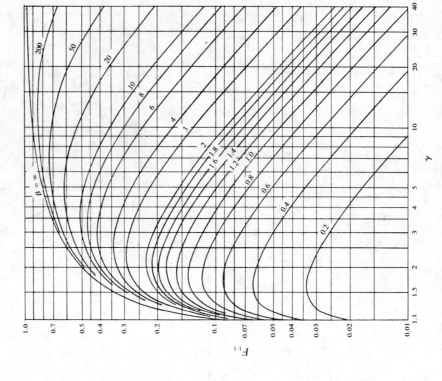

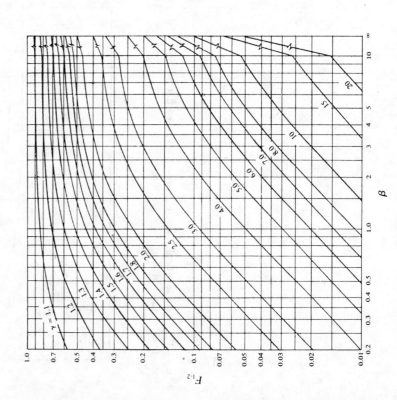

Fig. 11-15. **Concentric cylinders of finite length.** (Adapted from D.C. Hamilton and W.R. Morgan, *NACA Tech. Note TN-2836*, 1952.)

Finite-to-finite area.   From the definition of configuration factor,

$$F_{A_1 \to A_2} = \frac{\int_{A_2} \int_{A_1} I_1 \cos \phi_1 \, dA_1 \cos \phi_2 \, dA_2 / r^2}{\int_{A_1} \pi I_1 \, dA_1}$$

which for constant I_1 (valid for diffuse surfaces) becomes

$$F_{A_1 \to A_2} = \frac{1}{\pi A_1} \int_{A_2} \int_{A_1} \frac{\cos \phi_1 \cos \phi_2}{r^2} \, dA_1 \, dA_2 \qquad (11.29)$$

Some graphical evaluations of (*11.29*) are presented in Figs. 11-12 through 11-15, in which $F_{A_1 \to A_2}$ is abbreviated to $F_{1\text{-}2}$.   In the last of these configurations, area A_1 "sees itself," and consequently $F_{1\text{-}1}$ is of practical importance.

Configuration Factor Properties

There are four useful properties that should be understood prior to attempting to calculate radiant heat-transfer rates between finite areas.

1. Subdivision of receiving surface.   By examination of (*11.27*) and (*11.28*), we see that

$$F_{dA_1 \to A_2} = \int_{A_2} \frac{\cos \phi_1 \cos \phi_2 \, dA_2}{\pi r^2} = \int_{A_2} F_{dA_1 \to dA_2} \qquad (11.30)$$

which reveals that the fractions of energy radiated from dA_1 to each of the subareas dA_2 are simply summed to obtain the total fraction radiated to A_2.

2. Subdivision of emitting surface.   The expression for the configuration factor from a finite-sized area to a finite-sized area is, by (*11.29*),

$$F_{A_1 \to A_2} = \frac{1}{A_1} \int_{A_1} \int_{A_2} \frac{\cos \phi_1 \cos \phi_2 \, dA_2}{\pi r^2} \, dA_1 = \frac{1}{A_1} \int_{A_1} F_{dA_1 \to A_2} \, dA_1 \qquad (11.31)$$

i.e. the average value of $F_{dA_1 \to A_2}$ over the area A_1.   This can be approximated as

$$F_{A_1 \to A_2} \approx \frac{1}{A_1} \sum_i \Delta A_i F_{\Delta A_i \to A_2} \qquad (11.32)$$

where the ΔA_i are the subdivisions of A_1.   As an example, consider the system shown in Fig. 11-16, where A_1 (irradiating surface) is divided into subareas A_1' and A_1''.   In this case, (*11.32*) gives

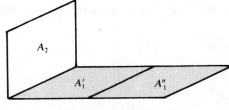

Fig. 11-16

$$F_{(A_1' + A_1'') \to A_2} \approx \frac{1}{A_1' + A_1''} (A_1' F_{A_1' \to A_2} + A_1'' F_{A_1'' \to A_2}) \qquad (11.33)$$

3. Enclosure property.   The third important property is that the summation of configuration factors for any surface has the value unity.   For an emitter A_i completely enclosed by n receiving surfaces A_j (which possibly include A_i itself),

$$\sum_{j=1}^{n} F_{A_i \to A_j} = 1 \qquad (11.34)$$

4. Reciprocity theorem. Multiplying both sides of (*11.29*) by A_1 yields

$$A_1 F_{A_1 \to A_2} = \int_{A_2} \int_{A_1} \frac{\cos \phi_1 \cos \phi_2}{\pi r^2} \, dA_1 \, dA_2$$

Clearly, we can write the similar expression

$$A_2 F_{A_2 \to A_1} = \int_{A_1} \int_{A_2} \frac{\cos \phi_2 \cos \phi_1}{\pi r^2} \, dA_2 \, dA_1$$

since the assignment of subscripts 1 and 2 is arbitrary.   The integrand of both expressions is the same, and it is a continuous function; hence the order of integration is immaterial, and

$$\blacksquare \qquad A_1 F_{A_1 \to A_2} = A_2 F_{A_2 \to A_1} \qquad\qquad\qquad (11.35)$$

which is known as the *reciprocity theorem*.

Black Enclosures

For an enclosure system consisting of n black surfaces, the net rate of energy exchange between any two of the surfaces is

$$q_{i\text{-}j} = q'_{i\text{-}j} - q'_{j\text{-}i} = A_i E_{bi} F_{i\text{-}j} - A_j E_{bj} F_{j\text{-}i}$$

where we have simplified the notation $F_{A_i \to A_j}$ to $F_{i\text{-}j}$, and have used the fact that all the radiation that strikes a blackbody is absorbed by it.   Application of the reciprocity theorem gives

$$q_{i\text{-}j} = A_i F_{i\text{-}j}(E_{bi} - E_{bj}) = -q_{j\text{-}i} \qquad\qquad\qquad (11.36)$$

The net heat-transfer rate from any one of the surfaces is

$$q_i = \sum_{j=1}^{n} q'_{i\text{-}j} - \sum_{j=1}^{n} q'_{j\text{-}i} = \sum_{j=1}^{n} q_{i\text{-}j} \qquad\qquad\qquad (11.37)$$

Substituting $q_{i\text{-}j}$ from (*11.36*) and using (*11.34*), we obtain

$$q_i = A_i \left[E_{bi} - \sum_{j=1}^{n} F_{i\text{-}j} E_{bj} \right] \qquad\qquad\qquad (11.38)$$

where the summation includes the term for $j = i$. The system of equations (*11.37*) ($i = 1, 2, \ldots, n$), with $q_{i\text{-}j}$ given by (*11.36*), suggests an electrical network analogy. For $n = 3$ the

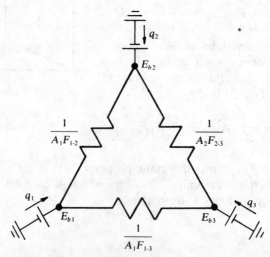

Fig. 11-17

analog is illustrated in Fig. 11-17, where the nodal points are maintained at potentials equal to their respective blackbody emissive powers. The resistances between nodes are spatial resistances, $R_{i-j} = 1/A_i F_{i-j}$.

Reradiating Surfaces

Often encountered in engineering practice are *adiabatic walls* which can have no net gain or loss of thermal energy ($q_i = 0$). Such a surface must reflect and/or reemit all radiant energy incident upon it, and it reaches an equilibrium temperature dependent upon its absorptivity and emissivity and the temperatures of the surrounding surfaces.

If an enclosure system includes reradiating surfaces, the quantities q'_{i-j} cannot be evaluated as before, because of reflection. However, the type of reasoning that led to (*11.37*) gives in this case

$$q_a = \sum_s q_{a \to s} - \sum_s q_{s \to a} = \sum_s A_a F_{a-s}(J_a - J_s) \tag{11.39}$$

where the indices a and s run over all surfaces. The J's are the surface radiosities (Section 11.2); for a black surface, $J = E_b$. Thus, if an enclosure consists of n black (active) surfaces $(1, 2, \ldots, n)$ and m reradiating surfaces $(r1, r2, \ldots, rm)$, we can rewrite (*11.39*) as

$$q_i = \sum_{j=1}^{n} A_i F_{i-j}(E_{bi} - E_{bj}) + \sum_{k=1}^{m} A_i F_{i-rk}(E_{bi} - J_{rk}) \tag{11.40}$$

for each active surface, $i = 1, 2, \ldots, n$; and

$$0 = \sum_{j=1}^{n} A_{rk} F_{rk-j}(J_{rk} - E_{bj}) + \sum_{l=1}^{m} A_{rk} F_{rk-rl}(J_{rk} - J_{rl}) \tag{11.41}$$

for each of the reradiating surfaces, $k = 1, 2, \ldots, m$. Equations (*11.40*) and (*11.41*) constitute a set of $m + n$ equations in the n unknown q_i's and m unknown J_{rk}'s.

Modified Configuration Factors

In order to determine the net heat-exchange rate between two of the active surfaces, A_i and A_j, in the enclosure described by (*11.40*) and (*11.41*), we define the *modified configuration factor* $\bar{F}_{i-j}$ to be the fraction of the radiation from A_i that reaches A_j, directly or via reflection or reradiation. Thus,

$$\bar{F}_{i-j} = \frac{q'_{i-j}}{A_i E_{bi}} \qquad \text{or} \qquad q'_{i-j} = A_i E_{bi} \bar{F}_{i-j} \tag{11.42}$$

Consider now a "reduced problem" in which $E_{bi} = 1$ and all other active surfaces have $E_b = 0$; denote by $q_{j,i}$ the net rate of heat transfer from A_j in this reduced problem. Since all the radiation striking A_j must ultimately have been emitted by A_i, and since A_j itself emits nothing,

$$q'_{i-j} = -q_{j,i} \tag{11.43}$$

But, on physical grounds, it is clear that the *fraction* of the output of A_i that reaches A_j has the same value in the reduced and original problems. Therefore,

$$\bar{F}_{i-j} = \frac{q'_{i-j}}{A_i \cdot 1} = \frac{-q_{j,i}}{A_i} \tag{11.44}$$

The modified configuration factor also obeys the reciprocity theorem:

■ $$A_i \bar{F}_{i-j} = A_j \bar{F}_{j-i} \tag{11.45}$$

From (*11.42*) and (*11.45*),

$$q_{i-j} = q'_{i-j} - q'_{j-i} = A_i \bar{F}_{i-j}(E_{bi} - E_{bj}) = -q_{j-i} \tag{11.46}$$

This being identical in form to (11.36), the electrical network analogy also applies when reradiating surfaces are present, provided the spatial resistances are chosen as $1/A_i\bar{F}_{i-j}$.

While solving the system (11.40) and (11.41) for the appropriate $q_{j,i}$ will always permit determination of the modified configuration factors, expressions for $\bar{F}_{i-j}$ for some often-encountered configurations are available.

One reradiating zone. Denoting the single reradiating zone as A_r, the modified configuration factor is

$$\bar{F}_{i-j} = F_{i-j} + \frac{F_{i-r}F_{r-j}}{1 - F_{r-r}} \tag{11.47}$$

One reradiating zone, only two active surfaces. In this case, (11.47) gives for the two active surfaces:

$$\bar{F}_{1-2} = F_{1-2} + \frac{1}{\dfrac{A_1}{A_2}\left(\dfrac{1}{F_{2-r}}\right) + \dfrac{1}{F_{1-r}}} \tag{11.48}$$

One reradiating zone, two planar or convex active surfaces. Whenever $F_{1-1} = F_{2-2} = 0$,

$$\bar{F}_{1-2} = \frac{(A_2/A_1) - (F_{1-2})^2}{(A_2/A_1) + (1 - 2F_{1-2})} \tag{11.49}$$

Note that the geometry of the reradiating surface is immaterial in the last case, since there is no F_{i-r} in the expression.

11.6 RADIANT EXCHANGE: GRAY SURFACES

To begin a study of radiant heat transfer between gray bodies, consider the system composed of two infinite parallel gray planes depicted in Fig. 11-18. The upper plane is at uniform temperature T_1, the lower plane is at uniform temperature T_2, and both planes are assumed to have uniform radiation characteristics, α and ρ. For simplicity, the diffuse radiant energy per unit area leaving plane 1 is shown as a single ray. The radiant energy transfer rate per unit area from plane 1 to plane 2 is

$$\frac{q'_{1-2}}{A} = \alpha_2 E_1 + \alpha_2 \rho_1 \rho_2 E_1 + \alpha_2 \rho_1^2 \rho_2^2 E_1 + \cdots + \alpha_2 \rho_1^n \rho_2^n E_1 + \cdots = \frac{\alpha_2 E_1}{1 - \rho_1 \rho_2}$$

Likewise, the rate of energy transfer from plane 2 to plane 1 is

$$\frac{q'_{2-1}}{A} = \frac{\alpha_1 E_2}{1 - \rho_2 \rho_1}$$

Therefore the net rate of energy exchange from plane 1 to plane 2 per unit area is

$$\frac{q_{1-2}}{A} = \frac{q'_{1-2}}{A} - \frac{q'_{2-1}}{A} = \frac{\alpha_2 E_1 - \alpha_1 E_2}{1 - \rho_1 \rho_2} \tag{11.50}$$

Since we are considering gray surfaces, E may be expressed as $\epsilon \sigma T^4$. Also, assuming solid surfaces with zero transmissivity, $\rho = 1 - \alpha = 1 - \epsilon$ [this follows from $\epsilon_\lambda = \text{constant} = \epsilon$ and (11.20)]. Then (11.50) can be simplified to

$$\blacksquare \qquad \frac{q_{1-2}}{A} = \frac{\sigma(T_1^4 - T_2^4)}{\dfrac{1}{\epsilon_1} + \dfrac{1}{\epsilon_2} - 1} \tag{11.51}$$

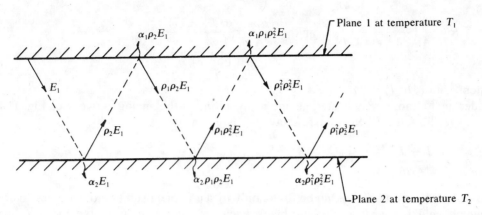

Fig. 11-18

A different approach to the same problem affords insight into the electrical analogy. With reference to Fig. 11-18, an energy balance on either surface in terms of the radiosity and the *irradiation G* (the total radiant thermal energy incident on the surface per unit time and unit area of the surface) is

$$\frac{q}{A} = J - G \tag{11.52}$$

But

$$J = \epsilon E_b + \rho G \tag{11.53}$$

Eliminating G between (*11.52*) and (*11.53*),

$$q = \frac{E_b - J}{(1 - \epsilon)/\epsilon A} \tag{11.54}$$

Equation (*11.54*) provides the basis for the gray body electrical analog. The numerator, $E_b - J$, can be regarded as a potential difference, while the denominator, $(1 - \epsilon)/\epsilon A$, can be considered as a surface resistance. Thus, the unknown potential J can be replaced by the known potential $E_b = \sigma T^4$ by means of the surface resistance. Also, the previously discussed spatial resistance to radiative heat transfer between two surfaces i and j, $1/A_i F_{i-j}$, can be used together with the surface resistances to form a complete electrical analog such as that for the two-gray-body system shown as Fig. 11-19.

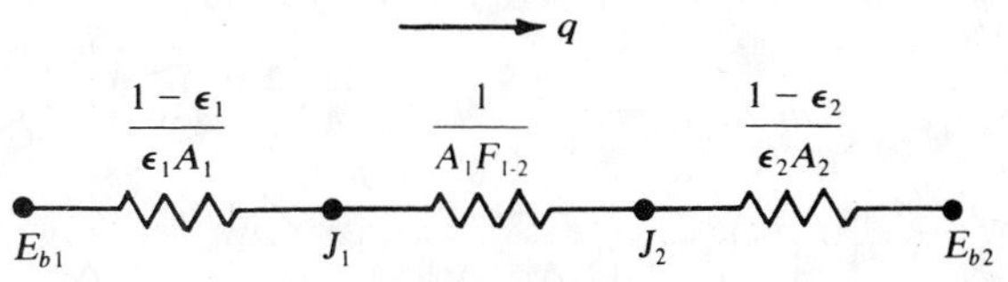

Fig. 11-19

The net radiant heat-transfer rate between two gray surfaces is, by the electrical analogy,

$$q_{1\text{-}2} = \frac{E_{b1} - E_{b2}}{\Sigma R} = \frac{E_{b1} - E_{b2}}{\dfrac{1 - \epsilon_1}{\epsilon_1 A_1} + \dfrac{1}{A_1 F_{1\text{-}2}} + \dfrac{1 - \epsilon_2}{\epsilon_2 A_2}} \tag{11.55}$$

For the case of two infinite gray planes with $A_1 = A_2$ and $F_{1\text{-}2} = 1$,

$$\Sigma R = \frac{1}{A}\left(\frac{1}{\epsilon_1} + \frac{1}{\epsilon_2} - 1\right) \qquad \text{and} \qquad \frac{q_{1\text{-}2}}{A} = \frac{\sigma(T_1^4 - T_2^4)}{\dfrac{1}{\epsilon_1} + \dfrac{1}{\epsilon_2} - 1}$$

in agreement with (11.51).

Consider next a more complicated geometry, resulting in the analog network of Fig. 11-20. In this case, a specific rate of heat transfer, say $q_{1\text{-}2}$, is given by

$$q_{1\text{-}2} = \frac{J_1 - J_2}{1/A_1 F_{1\text{-}2}} \tag{11.56}$$

but J_1 and J_2 are unknown. They can be found only by a complete solution of the general equations (11.39), which hold for gray as well as black enclosures. Again using the index i for active surfaces, k for reradiating surfaces, and s for all surfaces, we have from (11.54) and from the adiabatic condition,

$$q_i = \frac{E_{bi} - J_i}{(1 - \epsilon_i)/\epsilon_i A_i} \qquad i = 1, 2, \ldots, n$$

$$q_{rk} = 0 \qquad\qquad k = 1, 2, \ldots, m$$

Substituting these expressions in the left side of (11.39) gives $m + n$ linear equations in the $m + n$ unknown J's:

$$\sum_s A_i F_{i\text{-}s}(J_i - J_s) + A_i \frac{\epsilon_i}{1 - \epsilon_i}(J_i - E_{bi}) = 0 \tag{11.57}$$

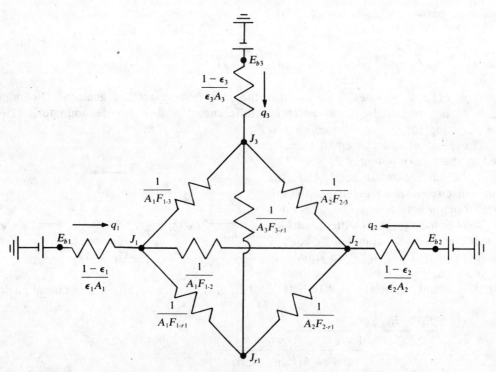

Fig. 11-20. Electrical analog network for an enclosure consisting of three active gray surfaces and one reradiating surface.

for $i = 1, 2, \ldots, n$; and

$$\sum_s A_{rk} F_{rk\text{-}s} (J_{rk} - J_s) = 0 \qquad (11.58)$$

for $k = 1, 2, \ldots, m$.

Gray Body Configuration Factors

A convenient way to determine the rate of radiant heat transfer between two gray surfaces i and j is by use of the *gray body configuration factor* $\mathscr{F}_{i\text{-}j}$. Analogous to (11.44),

$$\mathscr{F}_{i\text{-}j} \equiv \frac{-q_{j,i}}{A_i} = \frac{A_j}{A_i} \frac{\epsilon_j}{1 - \epsilon_j} J_{j,i} \qquad (11.59)$$

the last equality holding for $j \neq i$. The $J_{j,i}$ are obtained by solving the set of equations (11.57) and (11.58) when $E_{bi} = 1$ and all other E_b's are zero. Analogous to (11.46),

$$q_{i\text{-}j} = A_i \mathscr{F}_{i\text{-}j} E_{bi} - A_j \mathscr{F}_{j\text{-}i} E_{bj} = A_i \mathscr{F}_{i\text{-}j} (E_{bi} - E_{bj}) = -q_{j\text{-}i} \qquad (11.60)$$

where the reciprocity property has been used. Some special cases are given below.

Two-gray-surface enclosure. For a complete enclosure consisting of two gray surfaces, A_1 and A_2,

$$\frac{1}{\mathscr{F}_{1\text{-}2}} = \frac{1}{F_{1\text{-}2}} + \frac{A_1}{A_2}\left(\frac{1}{\epsilon_2} - 1\right) + \left(\frac{1}{\epsilon_1} - 1\right) \qquad (11.61)$$

One gray surface enclosing a second. For a gray surface A_1 which does not "see" itself and which is completely enclosed by a second gray surface A_2, $F_{1\text{-}2} = 1$, and (11.61) becomes

$$\frac{1}{\mathscr{F}_{1\text{-}2}} = \frac{1}{\epsilon_1} + \frac{A_1}{A_2}\left(\frac{1}{\epsilon_2} - 1\right) \qquad (11.62)$$

Two planar or convex gray surfaces with no other radiation present. For two surfaces which do not "see" themselves ($F_{1\text{-}1} = F_{2\text{-}2} = 0$) but do not form a complete enclosure and are located in an otherwise radiation-free environment, an enclosure may be considered to exist by defining a third surface having $J = 0$. Then

$$\mathscr{F}_{1\text{-}2} = \frac{F_{1\text{-}2}}{\dfrac{1}{\epsilon_1 \epsilon_2} - \left(\dfrac{1}{\epsilon_1} - 1\right)\left(\dfrac{1}{\epsilon_2} - 1\right)\dfrac{A_1}{A_2}(F_{1\text{-}2})^2} \qquad (11.63)$$

One reradiating zone enclosing two active gray surfaces. In this case,

$$\frac{1}{\mathscr{F}_{1\text{-}2}} = \frac{1}{\bar{F}_{1\text{-}2}} + \frac{A_1}{A_2}\left(\frac{1}{\epsilon_2} - 1\right) + \left(\frac{1}{\epsilon_1} - 1\right) \qquad (11.64)$$

where the modified blackbody configuration factor $\bar{F}_{1\text{-}2}$ is given by (11.48) or (11.49), as appropriate.

11.7 RADIATION SHIELDING

An important application is the use of shielding to reduce the rate of heat transfer by radiant exchange. A specific example is the use of aluminum-foil-backed insulation in building walls, with the foil serving as the shield. Consider the infinite gray walls of Fig. 11-21(a), for which (11.51) gives

$$q_{1\text{-}3} = -q_{3\text{-}1} = \frac{A\sigma(T_1^4 - T_3^4)}{\dfrac{1}{\epsilon_1} + \dfrac{1}{\epsilon_3} - 1} \qquad (11.65)$$

If a thin radiation shield, body 2, is placed between the two walls [Fig. 11-22(a)], the heat transfer rate per unit area for steady state is

$$\frac{q_{1\text{-}2}}{A} = \frac{q_{2\text{-}3}}{A} = \frac{q}{A}$$

or

$$\frac{q}{A} = \frac{\sigma(T_1^4 - T_2^4)}{\dfrac{1}{\epsilon_1} + \dfrac{1}{\epsilon_2} - 1} = \frac{\sigma(T_2^4 - T_3^4)}{\dfrac{1}{\epsilon_2} + \dfrac{1}{\epsilon_3} - 1} \qquad (11.66)$$

For the case where $\epsilon_1 = \epsilon_3$, we obtain from (11.66)

$$T_2^4 = \frac{1}{2}(T_1^4 + T_3^4)$$

so that, if also $\epsilon_1 = \epsilon_2$, the resulting heat transfer rate is

$$\frac{q}{A} = \frac{1}{2}\left[\frac{\sigma(T_1^4 - T_3^4)}{\dfrac{2}{\epsilon} - 1}\right] = \frac{1}{2}\frac{(q_{1\text{-}3})_0}{A} \qquad (11.67)$$

where the subscript 0 denotes no radiation shield. Similarly, for n radiation shields each having the same emissivity as the two active walls,

$$\frac{(q_{1\text{-}3})_n}{A} = \frac{1}{n+1}\frac{(q_{1\text{-}3})_0}{A} \qquad (11.68)$$

A second shielding problem of importance is that of protecting a thermocouple used to measure the temperature of a flowing hot gas from loss by radiation to a cooler duct wall [Fig. 11-23(a)]. Considering the shield and the duct to be very long, the steady-state heat loss from the

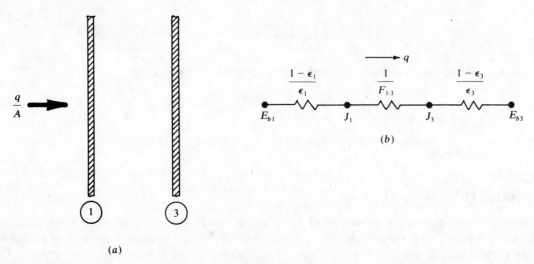

Fig. 11-21

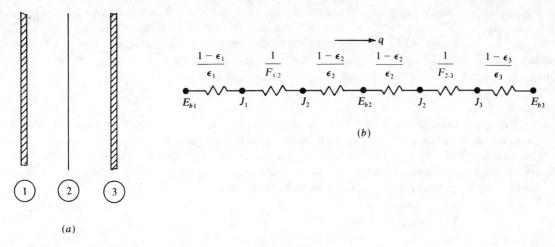

Fig. 11-22

thermocouple can be approximated by

$$q_{1\text{-}3} = q_{1\text{-}2} = \frac{A_1\sigma(T_1^4 - T_2^4)}{\dfrac{1}{\epsilon_1} + \dfrac{A_1}{A_2}\left(\dfrac{1}{\epsilon_2} - 1\right)}$$

$$= q_{2\text{-}3} = \frac{A_2\sigma(T_2^4 - T_3^4)}{\dfrac{1}{\epsilon_2} + \dfrac{A_2}{A_3}\left(\dfrac{1}{\epsilon_3} - 1\right)} \tag{11.69}$$

and clearly the heat loss rate is dependent upon the size and the emissivity of the shielding material. (The shield's inner surface "sees" itself; however, there is no net heat transfer with itself, and so, for clarity, the path joining node J_2 to itself has been omitted from Fig. 11-23(b).)

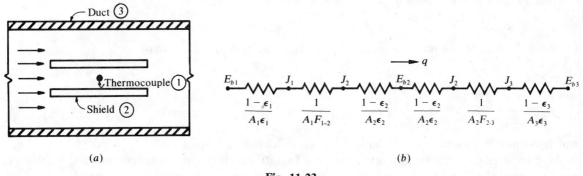

Fig. 11-23

11.8   RADIATION INVOLVING GASES AND VAPORS

In general, gases with nonpolar symmetrical molecules do not emit or absorb thermal energy within the temperature ranges encountered in engineering applications. Typical of this class are O_2, N_2, H_2, and mixtures of these, including dry air.   On the other hand, gases with nonsymmetrical molecules, such as CO_2, H_2O, SO_2 and many of the hydrocarbons, exhibit absorption and emission over certain wavelength ranges or bands. The absorptivity of such a gas is dependent upon molecular structure, wavelength and gas layer thickness.   A typical case is shown in Fig. 11-24, in which (a) refers to a 50-mm layer of CO_2 at 1 atm and (b) refers to a 30-mm layer.

As an aid to understanding the absorption phenomenon, consider a beam of monochromatic radiation having intensity $I_{\lambda 0}$ as it enters a layer of gas shown in Fig. 11-25.   The decrease in intensity is dependent upon wavelength and thickness, and is given by

$$dI_{\lambda x} = -\kappa_\lambda I_{\lambda x}\, dx \tag{11.70}$$

where $I_{\lambda x}$ is the monochromatic intensity at depth x and κ_λ is the *monochromatic absorption coefficient*, which depends upon the state of the gas (i.e. temperature and pressure) as well as upon the wavelength.   To a first approximation, κ_λ increases linearly with pressure at constant temperature; and in a gas mixture involving both absorbing and nonabsorbing gases, κ_λ should be proportional to the partial pressure of the absorbing species.

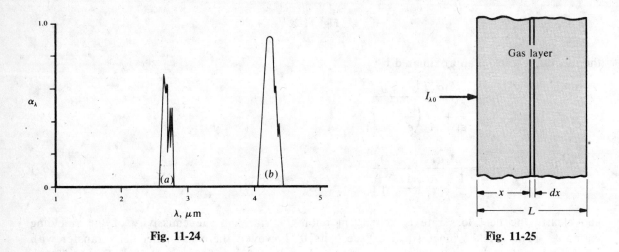

Fig. 11-24   Fig. 11-25

Separating variables and integrating (11.70) for constant κ_λ yields

$$I_{\lambda x} = I_{\lambda 0}e^{-\kappa_\lambda x} \tag{11.71}$$

By the definition of the monochromatic transmissivity, $\tau_\lambda = I_{\lambda x}/I_{\lambda 0}$, so that

$$\tau_\lambda = e^{-\kappa_\lambda x} \tag{11.72}$$

and the absorptivity is

$$\alpha_\lambda = 1 - \tau_\lambda = 1 - e^{-\kappa_\lambda x} \tag{11.73}$$

which is also the monochromatic emissivity, ϵ_λ, if Kirchhoff's law is valid.

A typical engineering computation requires that all radiant heat exchange between a mass of gas and each element of a surrounding solid boundary (container, etc.) be accounted for.   The geometry of most configurations results in a rather complicated integration to yield the absorptivity (or emissivity) of a gas mass with respect to a boundary element.

In the special case of a hemispherical mass of gas, the emissivity for radiant exchange from the gas to the center of the hemispherical base can readily be analytically determined.   Using this approach, H. C. Hottel and R. B. Egbert determined the effective emissivity of a hemispherical gas system of radius L at a partial pressure p_i radiating to a black surface element located at the center of the hemispherical base.   Their results for carbon dioxide and water vapor are given in Figs. 11-26 through 11-29.   These results are also applicable to other shapes of practical interest by use of the equivalent beam lengths given in Table 11-2.

Table 11-2

Configuration	Equivalent Beam Length, L
Space between infinite planes	1.8 × distance between planes
Sphere	2/3 × diameter
Infinitely long cylinder	1 × diameter
Cube	2/3 × side
Arbitrary Surface	≈3.6 × (volume/surface area)

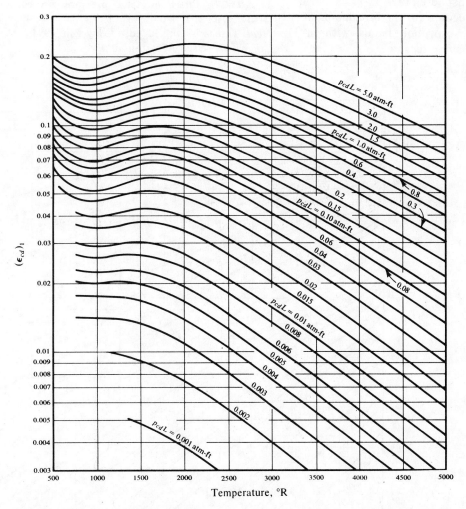

Fig. 11-26. Emissivity of carbon dioxide at 1 atm total pressure. (Adapted from H.C. Hottel, Chap. 4 in W.C. McAdams, *Heat Transmission*, 3d ed. Copyright 1954. McGraw-Hill Book Company. Used by permission.)

To determine the effective emissivity of a mass of carbon dioxide gas, Fig. 11-26 may be used to obtain the emissivity of a hypothetical gas system at one atmosphere total pressure, having radius L, partial CO_2 pressure p_{cd}, and a given uniform system temperature. For other system geometry, the equivalent beam length for use in the parameter $p_{cd}L$ is obtained from Table 11-2. If the total gas system pressure is different than 1 atmosphere, Fig. 11-27 is used to account for the broadening of the absorption bands with increasing pressure.

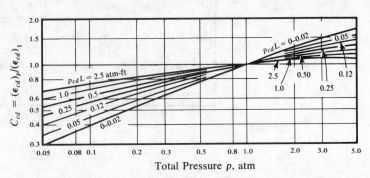

Fig. 11-27. Effect of total pressure on carbon dioxide emissivity. (Adapted from H.C. Hottel, Chap. 4 in W.C. McAdams, *Heat Transmission*, 3d ed. Copyright 1954. McGraw-Hill Book Company. Used by permission.)

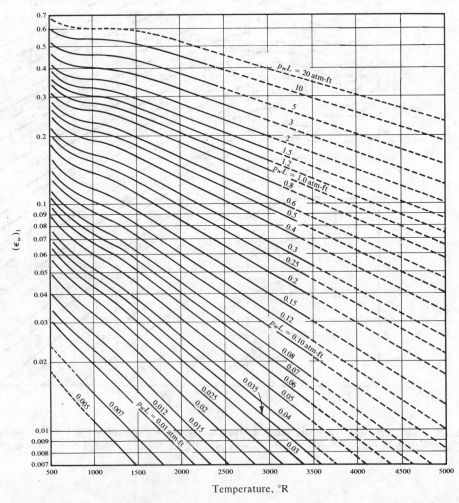

Fig. 11-28. Emissivity of hypothetical water vapor system at 1 atm total pressure. (Adapted from H.C. Hottel, Chap. 4 in W.C. McAdams, *Heat Transmission*, 3d ed. Copyright 1954. McGraw-Hill Book Company. Used by permission.)

Turning to the case of water vapor, Fig. 11-28 gives the emissivity of a hypothetical system having a water-vapor partial pressure of zero (to be corrected later) as a function of the product of the actual partial pressure and the equivalent beam length, $p_w L$, and the uniform gas system temperature. Again, the equivalent beam length is taken from Table 11-2, and the emissivity obtained from Fig. 11-28 is modified by use of Fig. 11-29, which accounts for the fact that the total system pressure may be other than 1 atm and the partial pressure of the H_2O vapor may be other than zero.

For a mixture of gases containing only water vapor, carbon dioxide, and nonactive species having symmetrical molecules, the mixture emissivity may be approximated by simple addition of the individual values for the water vapor and the carbon dioxide.

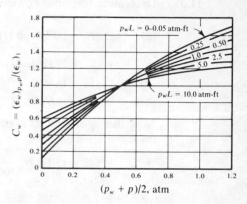

Fig. 11-29. Effect of partial and total pressures on emissivity of water vapor. (Adapted from H.C. Hottel, Chap. 4 in W.C. McAdams, *Heat Transmission*, 3d ed. Copyright 1954. McGraw-Hill Book Company. Used by permission.)

Solved Problems

11.1. The incident solar radiant flux at the earth's mean orbital radius from the sun is $G \approx$ 444.7 Btu/hr-ft^2. Using this fact, determine the solar flux in the vicinity of (*a*) the planet Mercury, which has a mean orbital radius of 36,000,000 miles; (*b*) the planet Pluto, which has a mean orbital radius of 3,671,000,000 miles.

Considering the sun as a point source, the incident energy upon a sphere having radius equal to the earth's orbit is the total $\dot{Q}$ from the sun, i.e.

$$\dot{Q}_{sun} \approx \frac{444.7 \text{ Btu}}{\text{hr-ft}^2} (4\pi) \left[(93 \times 10^6 \text{ mi}) \left(5280 \frac{\text{ft}}{\text{mi}} \right) \right]^2 = 1.347 \times 10^{27} \text{ Btu/hr}$$

(*a*) At Mercury, the incident radiation is

$$G_m = \frac{\dot{Q}_{sun}}{4\pi (\text{mercury orbital radius})^2} \approx \frac{1.347 \times 10^{27}}{(4\pi)(36 \times 10^6 \times 5280)^2} = 2967.76 \text{ Btu/hr-ft}^2$$

(*b*) Similarly, at Pluto,

$$G_p \approx \frac{1.347 \times 10^{27}}{(4\pi)(3.671 \times 10^9 \times 5280)^2} = 0.2854 \text{ Btu/hr-ft}^2$$

11.2. The total incident radiant energy upon a body which partially reflects, absorbs and transmits radiant energy is 2200 W/m^2. Of this amount, 450 W/m^2 is reflected and 900 W/m^2 is absorbed by the body. Find the transmissivity τ.

$$\tau = 1 - \rho - \alpha = 1 - \frac{450}{2200} - \frac{900}{2200} = 0.386$$

11.3. Determine the total emissive power of a blackbody at (a) 1000 °F and (b) 1000 °C.

(a) $\qquad E_b = \sigma T^4 = \left(0.1714 \times 10^{-8} \dfrac{\text{Btu}}{\text{hr-ft}^2\text{-°R}^4}\right)(1460 \text{ °R})^4$

$\qquad\qquad = (0.1714)(14.60)^4 = 7787.93 \text{ Btu/hr-ft}^2$

(b) $\qquad E_b = (5.6697 \times 10^{-8})(1273.16)^4 \text{ W/m}^2$

$\qquad\qquad = (5.6697)(12.7316)^4 = 148{,}967.67 \text{ W/m}^2 \qquad \text{or} \qquad 148.97 \text{ kW/m}^2$

Note that in either unit system it is convenient to divide the absolute temperature by 100, which results in

$$E_b = (0.1714)\left(\frac{T}{100}\right)^4 \frac{\text{Btu}}{\text{hr-ft}^2} = (5.6697)\left(\frac{T}{100}\right)^4 \text{ W/m}^2$$

11.4. Show that a hohlraum approximates a blackbody.

With reference to Fig. 11-4, if the entering irradiation is G_0, the irradiation after the first internal reflection is ρG_0; after the second, $\rho^2 G_0$; ... ; after the nth, $\rho^n G_0$ (which approaches 0 as $n \to \infty$). Since the few rays that emerge from the hole will have suffered many reflections, the emergent flux is essentially zero, i.e. $\alpha_{\text{hole}} = 1$.

11.5. For a blackbody maintained at 240 °F, determine (a) the total emissive power, (b) the wavelength at which the maximum monochromatic emissive power occurs, (c) the maximum monochromatic emissive power.

(a) $\qquad E_b = \sigma T^4 = \left(0.1714 \times 10^{-8} \dfrac{\text{Btu}}{\text{hr-ft}^2\text{-°R}^4}\right)[(240 + 460)^4 \text{ °R}^4] = 411.531 \text{ Btu/hr-ft}^2$

(b) $\qquad \lambda_{\max} T = 5215.6 \ \mu\text{m-°R}$

$$\lambda_{\max} = \frac{5215.6 \ \mu\text{m-°R}}{700 \text{ °R}} = 7.4509 \ \mu\text{m}$$

(c) By Planck's law, (11.12),

$$E_{b\lambda} = \frac{1.187 \times 10^8}{(7.4509)^5[\exp(25{,}896/5215.6) - 1]} = 36.319 \text{ Btu/hr-ft}^2\text{-}\mu\text{m}$$

11.6. Derive Wien's law, (11.13), from Planck's formula, (11.12).

Differentiating (11.12) at constant T,

$$\frac{\partial E_{b\lambda}}{\partial \lambda} = \frac{(e^{C_2/\lambda T} - 1)(-5C_1\lambda^{-6}) - (C_1\lambda^{-5})(e^{C_2/\lambda T})(-C_2/\lambda^2 T)}{(e^{C_2/\lambda T} - 1)^2}$$

For a maximum, the numerator on the right must vanish. This gives, after cancellation of common factors,

$$x - \frac{C_2}{5(1 - e^{-C_2/x})} = 0 \qquad\qquad (1)$$

where $x = \lambda_{\max} T$. For $\lambda \approx 1 \ \mu\text{m}$, $T \approx 10^3$ K, we have

$$\frac{C_2}{\lambda T} \approx 10$$

Thus, as a first approximation, we neglect $e^{-C_2/x}$ in comparison to 1, and (1) gives

$$x = \frac{C_2}{5} = 2877 \ \mu\text{m-K}$$

which is quite close.   If this value is used as the first approximation, x_0, in *Newton's iterative method*, a
single iteration gives

$$x_1 = \frac{1 - 6e^{-5}}{1 - 7e^{-5} + e^{-10}} \left(\frac{C_2}{5}\right) = \frac{0.9595}{0.9528}(2877) = 2897 \ \mu\text{m-K}$$

11.7.   Reflectivity measurements, which are relatively easy to make, are often used to obtain
other surface radiation properties.   A set of reflectivity measurements for a certain solid
surface at 1000 °R is roughly graphed in Fig. 11-30.   Estimate the total emissive power at
this temperature.

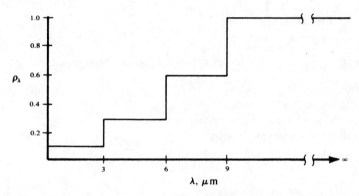

Fig. 11-30

By (*11.21*) and Kirchhoff's law,

$$E = \int_0^\infty \epsilon_\lambda E_{b\lambda} \ d\lambda = \int_0^\infty \alpha_\lambda E_{b\lambda} \ d\lambda$$

But for a solid surface, $\alpha_\lambda = 1 - \rho_\lambda$, so

$$E = \int_0^\infty (1 - \rho_\lambda)E_{b\lambda} \ d\lambda = (1 - \rho_\lambda)_{0-3} \int_0^3 E_{b\lambda} \ d\lambda + (1 - \rho_\lambda)_{3-6} \int_3^6 E_{b\lambda} \ d\lambda$$

$$+ (1 - \rho_\lambda)_{6-9} \int_6^9 E_{b\lambda} \ d\lambda + (1 - \rho_\lambda)_{9-\infty} \int_9^\infty E_{b\lambda} \ d\lambda$$

From Fig. 11-30,

$$(1 - \rho_\lambda)_{0-3} = 1.0 - 0.1 = 0.9 \qquad (1 - \rho_\lambda)_{6-9} = 1.0 - 0.6 = 0.4$$
$$(1 - \rho_\lambda)_{3-6} = 1.0 - 0.3 = 0.7 \qquad (1 - \rho_\lambda)_{9-\infty} = 1.0 - 1.0 = 0$$

Using Table 11-1,

$$\int_0^3 E_{b\lambda} \ d\lambda = \int_0^{3000} \frac{1}{1000} E_{b\lambda} \ d(\lambda T)$$

$$= E_{b(0-3000)} = (0.0254)\sigma(1000)^4 = 43.54 \ \text{Btu/hr-ft}^2$$

$$\int_3^6 E_{b\lambda} \ d\lambda = E_{b(0-6000)} - E_{b(0-3000)}$$

$$= (0.3477 - 0.0254)\sigma(1000)^4 = 552.42 \ \text{Btu/hr-ft}^2$$

$$\int_6^9 E_{b\lambda} \ d\lambda = E_{b(0-9000)} - E_{b(0-6000)}$$

$$= (0.6340 - 0.3477)\sigma(1000)^4 = 490.72 \ \text{Btu/hr-ft}^2$$

Substituting into the expression for E,

$$E = (0.9)(43.54) + (0.7)(552.42) + (0.4)(490.72) = 622.17 \ \text{Btu/hr-ft}^2$$

11.8. Using the definitions of total emissive power and total absorptivity, show that ϵ and α are not necessarily equal for irradiation of a surface at temperature T from a source at a different temperature, T^*.

The total emissivity is, by (11.17), (11.10), (11.11) and (11.18),

$$\epsilon = \frac{E}{E_b} = \frac{\int_0^\infty \epsilon_\lambda(T) E_{b\lambda}(T)\, d\lambda}{\sigma T^4} = \epsilon(T)$$

The total absorptivity is, by definition,

$$\alpha = \frac{\text{energy absorbed}}{\text{energy incident}} = \frac{\int_0^\infty \alpha_\lambda(T) G_\lambda(T^*)\, d\lambda}{G(T^*)} = \alpha(T, T^*)$$

Since α depends on T^*, while ϵ does not, the two will generally be unequal.

11.9. Approximate the radiant energy leaving a 30-mm-dia. sphere at 1200 K and impinging upon a 1-m by 1.5-m wall 1 meter away from the sphere (Fig. 11-31). Assume all surfaces to be blackbodies.

The sphere is small enough to be treated as an infinitesimal disc,

$$dA_1 = \pi R^2$$

From Fig. 11-11 with

$$\beta = 0.75 \qquad \text{and} \qquad \gamma = 0.5$$

the configuration factor to one-fourth the wall is approximately 0.021. Thus, for the entire wall A_2,

$$F_{dA_1 \to A_2} \approx 4(0.021) = 0.084$$

and

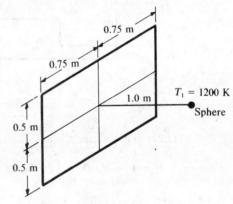

Fig. 11-31

$$q_{dA_1 \to A_2} = F_{dA_1 \to A_2}(\sigma T_1^4)(dA_1)$$
$$= (0.084)(5.6697 \times 10^{-8}\ \text{W/m}^2\text{-K}^4)(1200\ \text{K})^4 \pi (15 \times 10^{-3}\ \text{m})^2$$
$$= 6.98\ \text{W}$$

11.10. Two blackbody rectangles, 6 ft by 12 ft, are parallel and directly opposed, as shown in Fig. 11-12; they are 12 ft apart. If surface 1 is at $T_1 = 200\ °\text{F}$ and surface 2 is at $T_2 = 600\ °\text{F}$, determine (a) the net rate of heat transfer $q_{1\text{-}2}$; (b) the net energy loss rate from the 200 °F surface (side facing surface 2 only) if the surroundings other than surface 2 behave as a blackbody at (i) 0 °R, (ii) 70 °F.

(a) Using Fig. 11-12 with $\beta = 12/12$ and $\gamma = 6/12$, $F_{1\text{-}2} \approx 0.12$. Thus, the net rate of heat transfer is, by (11.36),

$$q_{1\text{-}2} = A_1 F_{1\text{-}2} \sigma(T_1^4 - T_2^4)$$
$$= (6 \times 12)(0.12)(0.1714 \times 10^{-8})(660^4 - 1060^4) = -15{,}886\ \text{Btu/hr}$$

(b) (i) Surface 1 is irradiated only by surface 2; hence the net energy loss rate from surface 1 is

$$q_1 = A_1 E_{b1} - A_2 F_{2\text{-}1} E_{b2} = A_1(E_{b1} - F_{1\text{-}2} E_{b2})$$
$$\doteq 72(0.1714 \times 10^{-8})[(660)^4 - (0.12)(1060)^4] = 4720\ \text{Btu/hr}$$

(ii)   The summation of configuration factors from any surface to its total surroundings is 1, hence

$$F_{1\text{-space}} = 1 - F_{1\text{-}1} - F_{1\text{-}2} = 1 - 0 - 0.12 = 0.88$$

Thus, the net energy loss rate from surface 1 is

$$q_1 = A_1 E_{b1}' - A_2 F_{2\text{-}1} E_{b2} - A_{\text{space}} F_{\text{space-}1} E_{b\,\text{space}}$$
$$= A_1 (E_{b1} - F_{1\text{-}2} E_{b2} - F_{1\text{-space}} E_{b\,\text{space}})$$
$$= 72(0.1714 \times 10^{-8})[(660)^4 - (0.12)(1060)^4 - (0.88)(530)^4] = -3849 \text{ Btu/hr}$$

11.11.   A large black enclosure consists of a box as shown in Fig. 11-32.   Surface 1 (bottom) is at 500 °F, surface 2 (top) is at 350 °F, and all vertical surfaces (including the back wall, 3) are at 400 °F.   Find (a) the net heat-transfer rate $q_{1\text{-}2}$, (b) the net heat-transfer rate $q_{1\text{-}3}$.

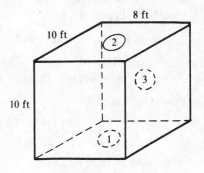

Fig. 11-32

(a)   From Fig. 11-12 with $\beta = 1.0$ and $\gamma = 0.8$, $F_{1\text{-}2} \approx 0.168$.

$$q_{1\text{-}2} = A_1 F_{1\text{-}2}(E_{b1} - E_{b2})$$
$$\approx 80(0.168)(0.1714 \times 10^{-8})(960^4 - 810^4)$$
$$= 9649 \text{ Btu/hr}$$

(b)   From Fig. 11-13 with $\beta = 10/8$ and $\gamma = 10/8$, $F_{1\text{-}3} \approx 0.185$.

$$q_{1\text{-}3} \approx 80(0.185)(0.1714 \times 10^{-8})(960^4 - 860^4)$$
$$= 7669 \text{ Btu/hr}$$

11.12.   For the configuration illustrated in Fig. 11-15, with the flat annular area between the two cylinders at one end designated as A_3, obtain expressions for $F_{1\text{-}3}$, $F_{3\text{-}1}$, $F_{3\text{-}2}$ and $F_{3\text{-}3}$ in terms of $F_{1\text{-}1}$, $F_{1\text{-}2}$ and the three areas A_1, A_2 and A_3. $F_{3\text{-}3}$ is the configuration factor between the two annular areas at opposite ends.

By (11.34),

$$2F_{1\text{-}3} + F_{1\text{-}2} + F_{1\text{-}1} = 1 \quad \text{or} \quad F_{1\text{-}3} = \frac{1}{2}(1 - F_{1\text{-}2} - F_{1\text{-}1})$$

and by reciprocity

$$F_{3\text{-}1} = \frac{A_1}{2A_3}(1 - F_{1\text{-}2} - F_{1\text{-}1})$$

Also, since $F_{2\text{-}2} = 0$,

$$2F_{2\text{-}3} + F_{2\text{-}1} = 1 \quad \text{or} \quad F_{2\text{-}3} = \frac{1}{2}(1 - F_{2\text{-}1})$$

and reciprocity gives

$$F_{2\text{-}3} = \frac{1}{2}\left(1 - \frac{A_1}{A_2}F_{1\text{-}2}\right) \quad \text{and} \quad F_{3\text{-}2} = \frac{A_2}{2A_3}\left(1 - \frac{A_1}{A_2}F_{1\text{-}2}\right)$$

Again by (11.34),

$$F_{3\text{-}3} + F_{3\text{-}1} + F_{3\text{-}2} = 1$$

or

$$F_{3\text{-}3} = 1 - F_{3\text{-}1} - F_{3\text{-}2}$$

$$= 1 - \frac{A_1}{2A_3}(1 - F_{1\text{-}2} - F_{1\text{-}1}) - \frac{A_2}{2A_3}\left(1 - \frac{A_1}{A_2}F_{1\text{-}2}\right)$$

$$= 1 - \frac{A_1 + A_2}{2A_3} + \frac{A_1}{2A_3}(2F_{1\text{-}2} + F_{1\text{-}1})$$

11.13. For two concentric cylinders as shown in Fig. 11-15, having $r_1 = 2.0$ ft, $r_2 = 1.0$ ft, and $L = 2.0$ ft, determine $F_{1\text{-}1}$, $F_{1\text{-}2}$ and $F_{3\text{-}3}$, where the end annular plane area is A_3. Use the results of Problem 11.12.

From Fig. 11-15, with $\gamma = 2.0$ and $\beta = 2.0$, $F_{1\text{-}1} = 0.23$; $F_{1\text{-}2} = 0.335$. The areas are:

$$A_1 = 2\pi(2.0)(2.0) = 25.133 \text{ ft}^2 \qquad A_2 = 2\pi(1.0)(2.0) = 12.566 \text{ ft}^2$$
$$A_3 = \pi(2.0^2 - 1.0^2) = 9.425 \text{ ft}^2$$

From Problem 11.12,

$$F_{3\text{-}3} = 1 - \frac{25.133 + 12.566}{2(9.425)} + \frac{25.133}{2(9.425)}[2(0.335) + 0.23] = 0.200$$

(See Problem 11.29 also.)

11.14. Determine the configuration factor $F_{1\text{-}2}$ for Fig. 11-33.

By reciprocity, $F_{1\text{-}2} = (A_2/A_1)F_{2\text{-}1}$. By subdivision of the receiving surface,

$$F_{2\text{-}(1,3)} = F_{2\text{-}1} + F_{2\text{-}3} \qquad \text{or} \qquad F_{2\text{-}1} = F_{2\text{-}(1,3)} - F_{2\text{-}3}$$

Both $F_{2\text{-}3}$ and $F_{2\text{-}(1,3)}$ are obtainable from Fig. 11-13.

$$F_{2\text{-}3}: \quad \beta = \frac{3}{5} = 0.6, \ \gamma = \frac{1}{5} = 0.2, \ F_{2\text{-}3} \approx 0.125$$

$$F_{2\text{-}(1,3)}: \quad \beta = \frac{3}{5} = 0.6, \ \gamma = \frac{2}{5} = 0.4, \ F_{2\text{-}(1,3)} \approx 0.190$$

Hence,

$$F_{2\text{-}1} \approx 0.190 - 0.125 = 0.065 \qquad \text{and} \qquad F_{1\text{-}2} \approx \left(\frac{15}{5}\right)(0.065) = 0.195$$

11.15. Determine the configuration factor $F_{1\text{-}2}$ for Fig. 11-34.

Using the graphical data of Fig. 11-13,

$$F_{(1,3)\text{-}(2,4)}: \quad \beta = \frac{6}{4} = 1.5, \ \gamma = \frac{4}{4} = 1.0, \ F_{(1,3)\text{-}(2,4)} \approx 0.150$$

$$F_{(1,3)\text{-}4}: \quad \beta = \frac{6}{4} = 1.5, \ \gamma = \frac{2}{4} = 0.5, \ F_{(1,3)\text{-}4} \approx 0.108$$

By subdivision of the receiving surface,

$$F_{(1,3)\text{-}2} = F_{(1,3)\text{-}(2,4)} - F_{(1,3)\text{-}4} \approx 0.150 - 0.108 = 0.042$$

Again from Fig. 11-13,

$$F_{3\text{-}(2,4)}: \quad \beta = \frac{3}{4} = 0.75, \ \gamma = \frac{4}{4} = 1.0, \ F_{3\text{-}(2,4)} \approx 0.250$$

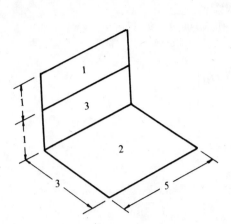

Fig. 11-33

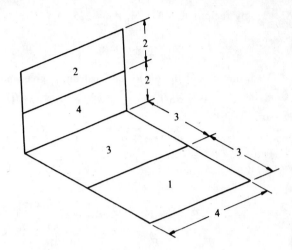

Fig. 11-34

$$F_{3-4}: \quad \beta = \frac{3}{4} = 0.75, \quad \gamma = \frac{2}{4} = 0.5, \quad F_{3-4} \approx 0.195$$

and, by subdivision of the receiving surface,

$$F_{3-2} = F_{3-(2,4)} - F_{3-4} \approx 0.250 - 0.195 = 0.055$$

Now by subdivision of the emitting surface, (*11.33*),

$$F_{(1,3)-2} \approx \frac{1}{A_1 + A_3} (A_1 F_{1-2} + A_3 F_{3-2})$$

$$0.042 \approx \left(\frac{1}{12 + 12}\right) [12 F_{1-2} + 12(0.055)]$$

$$F_{1-2} \approx 0.029$$

11.16. Suppose that in Fig. 11-14 the two concentric parallel discs are 3 ft apart, with disc 1 (1-ft radius) being at 200 °F and disc 2 (1.5-ft radius) being at 400 °F. Calculate q_i and q_{i-j} as appropriate for: (*a*) the two discs being blackbodies, with no other surfaces present (no other radiation present); (*b*) both discs being blackbodies, with a single reradiating surface in the form of a right frustrum of a cone enclosing them; (*c*) both discs being blackbodies, with a single black surface (right frustrum of a cone) at 0 °F enclosing them.

In each case the enclosing surface shall be denoted as surface 3 [this may be chosen as a black surface at 0 °R in (*a*)]. To find the configuration factors we shall use, instead of the graphed results for F_{1-2}, the exact equation

$$F_{1-2} = \tfrac{1}{2}[x - \sqrt{x^2 - 4(\beta\gamma)^2}]$$

where $x \equiv 1 + (1 + \beta^2)\gamma^2$, and β and γ are as defined in Fig. 11-14. Here, $\beta = 1.5/3 = 0.5$ and $\gamma = 3/1 = 3$; hence,

$$x = 1 + (1 + 0.25)9 = 12.25$$

and

$$F_{1-2} = \tfrac{1}{2}[12.25 - \sqrt{(12.25)^2 - 4(1.5)^2}] = 0.1865$$

Then

$$F_{1-3} = 1 - 0.1865 = 0.8135$$

and, by reciprocity ($A_1 = \pi$, $A_2 = 2.25\,\pi$),

$$F_{2\text{-}1} = \frac{\pi}{2.25\,\pi}\,(0.1865) = 0.0829 \qquad \text{and} \qquad F_{2\text{-}3} = 1 - 0.0829 = 0.9171$$

Also, the area of the enclosing frustrum is given by

$$A_3 = \pi S(r_1 + r_2) \qquad \text{where} \qquad S = \sqrt{(r_2 - r_1)^2 + (L)^2}$$

so

$$A_3 = \pi\sqrt{(1.5 - 1)^2 + (3)^2}\,(1.5 + 1) = 23.887 \text{ ft}^2$$

Thus

$$A_1 F_{1\text{-}3} = A_3 F_{3\text{-}1}; \quad F_{3\text{-}1} = \frac{\pi(0.8135)}{23.887} = 0.1070$$

$$A_2 F_{2\text{-}3} = A_3 F_{3\text{-}2}; \quad F_{3\text{-}2} = \frac{\pi(2.25)(0.9171)}{23.887} = 0.2714$$

$$F_{3\text{-}3} = 1 - F_{3\text{-}1} - F_{3\text{-}2} = 1 - 0.1070 - 0.2714 = 0.6216$$

(a) By (11.36),

$$q_{1\text{-}2} = A_1 F_{1\text{-}2}\sigma(T_1^4 - T_2^4) = \pi(0.1865)(0.1714 \times 10^{-8})[(660)^4 - (860)^4]$$
$$= -358.78 \text{ Btu/hr} = -q_{2\text{-}1}$$

By (11.38) with $F_{1\text{-}1} = F_{2\text{-}2} = 0$,

$$q_1 = A_1\sigma[T_1^4 - F_{1\text{-}2}T_2^4]$$
$$= \pi(0.1714 \times 10^{-8})[(660)^4 - (0.1865)(860)^4] = 472.4 \text{ Btu/hr}$$

$$q_2 = A_2\sigma[T_2^4 - F_{2\text{-}1}T_1^4]$$
$$= (2.25\,\pi)(0.1714 \times 10^{-8})[(860)^4 - (0.0829)(660)^4] = 6436.7 \text{ Btu/hr}$$

(b) By (11.49),

$$\bar{F}_{1\text{-}2} = \frac{(A_2/A_1) - (F_{1\text{-}2})^2}{(A_2/A_1) + (1 - 2F_{1\text{-}2})} = \frac{2.25 - (0.1865)^2}{2.25 + [1 - 2(0.1865)]} = 0.770$$

Using (11.46),

$$q_{1\text{-}2} = A_1\bar{F}_{1\text{-}2}(E_{b1} - E_{b2})$$
$$= \pi(0.770)(0.1714 \times 10^{-8})[(660)^4 - (860)^4] = -1481.2 \text{ Btu/hr} = -q_{2\text{-}1}$$

It is physically evident that $q_1 = -q_2 = q_{1\text{-}2}$; this may be verified by solving (11.40) and (11.41) for q_1, q_2 and J_3.

(c) By (11.36),

$$q_{1\text{-}3} = A_1 F_{1\text{-}3}(E_{b1} - E_{b3})$$
$$= \pi(0.8135)(0.1714 \times 10^{-8})[(660)^4 - (460)^4] = 635.04 \text{ Btu/hr} = -q_{3\text{-}1}$$

From part (a),

$$q_{1\text{-}2} = -358.78 \text{ Btu/hr} = -q_{2\text{-}1}$$

Also, by (11.36),

$$q_{2\text{-}3} = A_2 F_{2\text{-}3}(E_{b2} - E_{b3})$$
$$= (2.25\,\pi)(0.9171)(0.1714 \times 10^{-8})[(860)^4 - (460)^4] = 5580.46 \text{ Btu/hr} = -q_{3\text{-}2}$$

The q_i's are given by (11.38). Thus, since $F_{1\text{-}1} = F_{2\text{-}2} = 0$,

$$q_1 = A_1[E_{b1} - F_{1\text{-}2}E_{b2} - F_{1\text{-}3}E_{b3}]$$
$$= \pi(0.1714 \times 10^{-8})[(660)^4 - (0.1865)(860)^4 - (0.8135)(460)^4] = 276.27 \text{ Btu/hr}$$

$$q_2 = A_2[E_{b2} - F_{2\text{-}1}E_{b1} - F_{2\text{-}3}E_{b3}]$$
$$= (2.25\,\pi)(0.1714 \times 10^{-8})[(860)^4 - (0.0829)(660)^4 - (0.9171)(460)^4] = 5939.24 \text{ Btu/hr}$$

$$q_3 = A_3[E_{b3} - F_{3\text{-}1}E_{b1} - F_{3\text{-}2}E_{b2} - F_{3\text{-}3}E_{b3}]$$
$$= (23.887)(0.1714 \times 10^{-8})[(460)^4 - (0.1070)(660)^4 - (0.2714)(860)^4 - (0.6216)(460)^4]$$
$$= -6215.79 \text{ Btu/hr}$$

11.17. Two parallel infinite black planes are maintained at 200 °C and 300 °C. (*a*) Determine the net rate of heat transfer per unit area (SI units). (*b*) Repeat for the case where both temperatures are lowered by 100 °C and determine the ratio of the reduced heat transfer to the original value.

(*a*) Denoting as plane 1 the hotter plane,

$$\frac{q_{1\text{-}2}}{A} = \sigma(T_1^4 - T_2^4) = (5.6697 \times 10^{-8})[(573.15)^4 - (473.15)^4] = 3276.78 \text{ W/m}^2$$

(*b*)
$$\frac{q_{1\text{-}2}}{A} = (5.6697 \times 10^{-8})[(473.15)^4 - (373.15)^4] = 1742.31 \text{ W/m}^2$$

$$\frac{(q_{1\text{-}2}/A)_{200\text{-}100}}{(q_{1\text{-}2}/A)_{300\text{-}200}} = \frac{1742.31}{3276.78} = 0.5317$$

A reduction in temperature of 100 °C reduces the net heat-transfer rate approximately 47%!

11.18. Two blackbody rectangles, 0.6 m by 1.2 m, are parallel and directly opposed. The bottom rectangle is at $T_1 = 500$ K and the top rectangle is at $T_2 = 900$ K. The two rectangles are 1.2 m apart. Determine (*a*) the rate of radiant heat transfer between the two surfaces, (*b*) the rate at which the bottom rectangle is losing energy if the surroundings (other than the top rectangle) are considered a blackbody at 0 K, (*c*) the rate at which the bottom rectangle is losing energy if the surroundings (other than the top rectangle) are considered to be a blackbody at 300 K.

(*a*) Using Fig. 11-12 with $\beta = 1.2/1.2 = 1.0$ and $\gamma = 0.6/1.2 = 0.5$, $F_{1\text{-}2} = F_{2\text{-}1} = 0.12$; therefore,

$$q_{1\text{-}2} = A_1 F_{1\text{-}2}\sigma(T_1^4 - T_2^4)$$
$$= (0.72)(0.12)(5.6697 \times 10^{-8})[(500)^4 - (900)^4] = -2907.82 \text{ W}$$

(*b*) By (*11.38*), with $F_{1\text{-}1} = 0$ and $E_{b\text{ space}} = 0$,

$$q_1 = A_1[E_{b1} - F_{1\text{-}2}E_{b2}] = (0.72)(5.6697 \times 10^{-8})[(500)^4 - (0.12)(900)^4] = -662.62 \text{ W}$$

i.e. a net gain for surface 1.

(*c*)
$$F_{1\text{-space}} = 1 - F_{1\text{-}1} - F_{1\text{-}2} = 1 - 0 - 0.12 = 0.88$$

Again by (*11.38*),

$$q_1 = A_1[E_{b1} - F_{1\text{-}2}E_{b2} - F_{1\text{-space}}E_{b\text{ space}}]$$
$$= (0.72)(5.6697 \times 10^{-8})[(500)^4 - (0.12)(900)^4 - (0.88)(300)^4] = -953.60 \text{ W}$$

i.e. an even larger gain for surface 1 than in part (*b*).

11.19. Repeat Problem 11.16(*a*) for both discs being gray with $\epsilon_1 = \epsilon_2 = 0.7$ and no other radiation present.

By (*11.60*),

$$q_{1\text{-}2} = A_1 \mathscr{F}_{1\text{-}2}\sigma(T_1^4 - T_2^4) = -q_{2\text{-}1}$$

where $\mathscr{F}_{1\text{-}2}$ is given by (*11.63*) with $F_{1\text{-}2} = 0.1865$, $A_1 = \pi$, and $A_2 = 2.25\pi$ (from Problem 11.16). So

$$\mathscr{F}_{1\text{-}2} = \frac{0.1865}{\dfrac{1}{(0.7)^2} - \left(\dfrac{1}{0.7} - 1\right)\left(\dfrac{1}{0.7} - 1\right)\left(\dfrac{\pi}{2.25\,\pi}\right)(0.1865)^2} = 0.092$$

$$q_{1\text{-}2} = \pi(0.092)(0.1714 \times 10^{-8})[(660)^4 - (860)^4] = -176.05 \text{ Btu/hr} = -q_{2\text{-}1}$$

The net heat transfer from either surface is given by (11.38) or by physical reasoning:

$$q_1 = (\text{energy emitted by } A_1) - q_{2\text{-}1}$$
$$= A_1 \epsilon_1 \sigma T_1^4 - q_{2\text{-}1} = \pi(0.7)(0.1714 \times 10^{-8})(660)^4 - 176.05 = 539.16 \text{ Btu/hr}$$

Also,

$$q_2 = A_2 \epsilon_2 \sigma T_2^4 - q_{1\text{-}2}$$
$$= (2.25\,\pi)(0.7)(0.1714 \times 10^{-8})(860)^4 - (-176.05) = 4815.16 \text{ Btu/hr}$$

11.20. Repeat Problem 11.16(b) for both discs being gray, $\epsilon_1 = \epsilon_2 = 0.7$, and the discs surrounded by a reradiating surface which is a right frustrum of a cone.

By (11.64), and using areas from Problem 11.16 and $\bar{F}_{1\text{-}2}$ from Problem 11.16(b),

$$\frac{1}{\mathscr{F}_{1\text{-}2}} = \frac{1}{0.770} + \left(\frac{\pi}{2.25\,\pi}\right)\left(\frac{1}{0.7} - 1\right) + \left(\frac{1}{0.7} - 1\right) = 1.9177 \qquad \mathscr{F}_{1\text{-}2} = 0.52143$$

By (11.60),

$$q_{1\text{-}2} = \pi(0.52143)(0.1714 \times 10^{-8})[(660)^4 - (860)^4] = -1003.10 \text{ Btu/hr} = -q_{2\text{-}1}$$

As in Problem 11.16(b), $q_{1\text{-}2} = q_1 = -q_2$.

11.21. Repeat Problem 11.19 (both discs gray, $\epsilon_1 = \epsilon_2 = 0.7$, $T_1 = 660\ °\text{R}$, $T_2 = 860\ °\text{R}$) when the surrounding right frustrum of a cone is an active gray body with $\epsilon_3 = 0.4$ and $T_3 = 760\ °\text{R}$.

From Problem 11.16,

$$F_{1\text{-}2} = 0.1865 \qquad F_{1\text{-}3} = 0.8135$$
$$F_{2\text{-}1} = 0.0829 \qquad F_{2\text{-}3} = 0.9171$$
$$F_{3\text{-}1} = 0.1070 \qquad F_{3\text{-}2} = 0.2714 \qquad F_{3\text{-}3} = 0.6216$$

Also,

$$E_{b1} = (0.1714 \times 10^{-8})(660)^4 = 325.23 \text{ Btu/hr-ft}^2$$
$$E_{b2} = (0.1714 \times 10^{-8})(860)^4 = 937.57 \text{ Btu/hr-ft}^2$$
$$E_{b3} = (0.1714 \times 10^{-8})(760)^4 = 571.83 \text{ Btu/hr-ft}^2$$

and

$$\frac{\epsilon_1}{1 - \epsilon_1} = \frac{\epsilon_2}{1 - \epsilon_2} = 2.3333 \qquad \frac{\epsilon_3}{1 - \epsilon_3} = 0.6667$$

Equations (11.57) (there are no reradiating surfaces present) are, after canceling the areas:

$$F_{1\text{-}2}(J_1 - J_2) + F_{1\text{-}3}(J_1 - J_3) + \frac{\epsilon_1}{1 - \epsilon_1}(J_1 - E_{b1}) = 0$$

$$F_{2\text{-}1}(J_2 - J_1) + F_{2\text{-}3}(J_2 - J_3) + \frac{\epsilon_2}{1 - \epsilon_2}(J_2 - E_{b2}) = 0$$

$$F_{3\text{-}1}(J_3 - J_1) + F_{3\text{-}2}(J_3 - J_2) + \frac{\epsilon_3}{1 - \epsilon_3}(J_3 - E_{b3}) = 0$$

Inserting known numerical values and collecting terms,

$$3.3333 J_1 - 0.1865 J_2 - 0.8135 J_3 = 758.859$$
$$-0.0829 J_1 + 3.3333 J_2 - 0.9171 J_3 = 2187.632$$
$$-0.1070 J_1 - 0.2714 J_2 + 1.0451 J_3 = 381.239$$

Gaussian elimination (hand or computer solution) yields

$$J_1 = 427.5 \qquad J_2 = 839.3 \qquad J_3 = 626.5$$

Then, from (11.54) and $A_3 = 23.887$ ft^2,

$$q_1 = \frac{325.23 - 427.5}{(1 - 0.7)/(0.7)\pi} = -749.7 \text{ Btu/hr}$$

$$q_2 = \frac{937.57 - 839.3}{(1 - 0.7)/(0.7)(2.25\ \pi)} = 1620.8 \text{ Btu/hr}$$

$$q_3 = \frac{571.83 - 626.5}{(1 - 0.4)/(0.4)(23.887)} = -870.6 \text{ Btu/hr}$$

To determine the $q_{i\text{-}j}$'s, we solve equations (11.57) for the appropriate $J_{j,i}$'s and then calculate the $\mathscr{F}_{i\text{-}j}$'s.

$i = 1$: $E_{b1} = 1$, $E_{b2} = 0$, $E_{b3} = 0$. Using the previously stated values of F's, ϵ's, etc., we have

$$3.3333 J_{1,1} - 0.1865 J_{2,1} - 0.8135 J_{3,1} = 2.3333$$
$$-0.0829 J_{1,1} + 3.3333 J_{2,1} - 0.9171 J_{3,1} = 0$$
$$-0.1070 J_{1,1} - 0.2714 J_{2,1} + 1.0451 J_{3,1} = 0$$

Solving by Gaussian elimination,

$$J_{1,1} = 0.7230 \qquad J_{2,1} = 0.04129 \qquad J_{3,1} = 0.08474$$

and by (11.59),

$$\mathscr{F}_{1\text{-}2} = \frac{A_2}{A_1}\left(\frac{\epsilon_2}{1 - \epsilon_2}\right) J_{2,1} = \frac{2.25\ \pi}{\pi}(2.3333)(0.04129) = 0.2168$$

$$\mathscr{F}_{1\text{-}3} = \frac{A_3}{A_1}\left(\frac{\epsilon_3}{1 - \epsilon_3}\right) J_{3,1} = \frac{23.887}{\pi}(0.6667)(0.08474) = 0.4296$$

Repeating the procedure for $i = 2$, with $E_{b1} = 0$, $E_{b2} = 1$ and $E_{b3} = 0$,

$$3.3333 J_{1,2} - 0.1865 J_{2,2} - 0.8135 J_{3,2} = 0$$
$$-0.0829 J_{1,2} + 3.3333 J_{2,2} - 0.9171 J_{3,2} = 2.3333$$
$$-0.1070 J_{1,2} - 0.2714 J_{2,2} + 1.0451 J_{3,2} = 0$$

Solving by Gaussian elimination,

$$J_{1,2} = 0.09291 \qquad J_{2,2} = 0.7592 \qquad J_{3,2} = 0.2067$$

Then

$$\mathscr{F}_{2\text{-}1} = \frac{A_1}{A_2}\left(\frac{\epsilon_1}{1 - \epsilon_1}\right) J_{1,2} = \frac{\pi}{2.25\ \pi}(2.3333)(0.09291) = 0.09635$$

$$\mathscr{F}_{2\text{-}3} = \frac{A_3}{A_2}\left(\frac{\epsilon_3}{1 - \epsilon_3}\right) J_{3,2} = \frac{23.887}{2.25\ \pi}(0.6667)(0.2067) = 0.4657$$

A check on the solution is:

$$A_1 \mathscr{F}_{1\text{-}2} = A_2 \mathscr{F}_{2\text{-}1}$$
$$\pi(0.2168) \stackrel{?}{=} (2.25\ \pi)(0.09635)$$
$$0.6811 = 0.6811$$

By reciprocity,

$$\mathscr{F}_{3\text{-}1} = \frac{A_1}{A_3}\mathscr{F}_{1\text{-}3} = \frac{\pi}{23.887}(0.4296) = 0.05650$$

$$\mathscr{F}_{3\text{-}2} = \frac{A_2}{A_3}\mathscr{F}_{2\text{-}3} = \frac{2.25\,\pi}{23.887}(0.4657) = 0.1378$$

Next, using (11.60),

$$q_{1\text{-}2} = A_1\mathscr{F}_{1\text{-}2}\sigma(T_1^4 - T_2^4)$$
$$= \pi(0.2168)(0.1714 \times 10^{-8})[(660)^4 - (860)^4] = -417.07 \text{ Btu/hr} = -q_{2\text{-}1}$$

$$q_{1\text{-}3} = A_1\mathscr{F}_{1\text{-}3}\sigma(T_1^4 - T_3^4)$$
$$= \pi(0.4296)(0.1714 \times 10^{-8})[(660)^4 - (760)^4] = -332.82 \text{ Btu/hr} = -q_{3\text{-}1}$$

$$q_{2\text{-}3} = A_2\mathscr{F}_{2\text{-}3}\sigma(T_2^4 - T_3^4)$$
$$= (2.25\,\pi)(0.4657)(0.1714 \times 10^{-8})[(860)^4 - (760)^4] = 1203.97 \text{ Btu/hr} = -q_{3\text{-}2}$$

11.22. A gray body having a surface area of 4 ft^2 has $\epsilon_1 = 0.35$ and $T_1 = 760$ °F. This is completely enclosed by a gray surface having an area of 36 ft^2, $\epsilon_2 = 0.75$, and $T_2 = 100$ °F. Find the net rate of heat transfer $q_{1\text{-}2}$ between the two surfaces if $F_{1\text{-}1} = 0$.

By (11.62),

$$\frac{1}{\mathscr{F}_{1\text{-}2}} = \frac{1}{\epsilon_1} + \frac{A_1}{A_2}\left(\frac{1}{\epsilon_2} - 1\right) = \frac{1}{0.35} + \frac{4}{36}\left(\frac{1}{0.75} - 1\right) = 2.89$$

$$\mathscr{F}_{1\text{-}2} = 0.3455$$

Thus

$$q_{1\text{-}2} = A_1\mathscr{F}_{1\text{-}2}\sigma(T_1^4 - T_2^4)$$
$$= 4(0.3455)(0.1714 \times 10^{-8})[(1220)^4 - (560)^4] = 5014.61 \text{ Btu/hr}$$

11.23. Two parallel metal walls of a kitchen oven have temperatures $T_1 = 450$ °F and $T_3 = 80$ °F, and emissivities $\epsilon_1 = \epsilon_3 = 0.30$, where subscripts 1 and 3 denote the inner and outer walls, respectively. The space between the walls is filled with a rock-wool-type insulation. Assuming this insulation material to be transparent to thermal radiation, calculate the radiant heat-transfer rate per unit area between the two walls (a) with no radiation shield and (b) for one radiation shield of aluminum foil having $\epsilon_1 = 0.09$.

(a) By (11.51),

$$\frac{q_{1\text{-}3}}{A} = \frac{\sigma(T_1^4 - T_3^4)}{\dfrac{1}{\epsilon_1} + \dfrac{1}{\epsilon_3} - 1} = \frac{(0.1714 \times 10^{-8})[(910)^4 - (540)^4]}{\dfrac{1}{0.3} + \dfrac{1}{0.3} - 1} = 181.70 \text{ Btu/hr-ft}^2$$

(b)

$$\frac{q}{A} = \frac{\sigma(T_1^4 - T_2^4)}{\dfrac{1}{\epsilon_1} + \dfrac{1}{\epsilon_2} - 1} = \frac{\sigma(T_2^4 - T_3^4)}{\dfrac{1}{\epsilon_2} + \dfrac{1}{\epsilon_3} - 1}$$

but since $\epsilon_1 = \epsilon_3$, the two denominators are equal; hence

$$T_1^4 - T_2^4 = T_2^4 - T_3^4$$

$$T_2^4 = \frac{1}{2}(T_1^4 + T_3^4) = \frac{1}{2}[(910)^4 + (540)^4]$$

$$T_2 = 787.91 \text{ °R}$$

Then

$$\frac{q}{A} = \frac{q_{1\text{-}2}}{A} = \frac{(0.1714 \times 10^{-8})[(910)^4 - (787.91)^4]}{\dfrac{1}{0.3} + \dfrac{1}{0.09} - 1} = 38.29 \text{ Btu/hr-ft}^2$$

Checking:

$$\frac{q}{A} = \frac{q_{2\text{-}3}}{A} = \frac{(0.1714 \times 10^{-8})[(787.91)^4 - (540)^4]}{\dfrac{1}{0.09} + \dfrac{1}{0.3} - 1} = 38.29 \text{ Btu/hr-ft}^2$$

11.24. Determine the effective emissivity (for radiation from the gas to the surface) of CO_2 gas at 2500 °R in a very long cylinder which is 2 ft in diameter. The partial pressure of the CO_2 is 0.2 atm and the gas system total pressure is 0.3 atm.

From Table 11-2, $L = 1 \times D = 2$ ft. From Fig. 11-26 at $p_{cd}L = 0.2 \times 2 = 0.4$ atm-ft and $T = 2500$ °R, $(\epsilon_{cd})_1 \approx 0.103$. From Fig. 11-27 at $p = 0.3$ atm and $p_{cd}L = 0.4$ atm-ft, $C_{cd} \approx 0.78$. Thus

$$(\epsilon_{cd})_p \approx (0.78)(0.103) = 0.08$$

11.25. A combustion exhaust gas at 2500 °R has a CO_2 partial pressure of 0.08 atm, a water vapor partial pressure of 0.16 atm, and a total gas system pressure of 2.0 atm. Estimate the effective gas mixture emissivity in a long cylindrical flue 3 ft in diameter. The other major gas constituents are O_2 and N_2.

The O_2 and N_2 constituents do not absorb or emit radiant energy in the temperature range of this problem. We may approximate the gas system emissivity by linear addition of the individual emissivities of the CO_2 and the water vapor; thus,

CO_2: $L = 1 \times D = 3$ ft
      $p_{cd}L = 0.08 \times 3 = 0.24$ atm-ft
      From Fig. 11-26, $(\epsilon_{cd})_1 \approx 0.085$
      From Fig. 11-27, $C_{cd} \approx 1.2$
        and $(\epsilon_{cd})_p \approx (0.085)(1.2) = 0.102$

H_2O: $L = 1 \times D = 3$ ft
      $p_w L = 0.16 \times 3 = 0.48$ atm-ft
      From Fig. 11-28, $(\epsilon_w)_1 \approx 0.115$
      From Fig. 11-29 at $(p_w + p)/2 = 2.16/2 = 1.08$ atm, $C_w \approx 1.5$
        and $(\epsilon_w)_{p_w \cdot p} \approx (0.115)(1.5) = 0.173$

It follows that $\epsilon_{\text{total}} \approx 0.102 + 0.173 = 0.275$.

Supplementary Problems

11.26. Determine the monochromatic emissive power at 2.30 μm of a blackbody at a temperature of 2500 °F.

    Ans.   42,038 Btu/hr-ft^2-μm

11.27. Determine λ_{max} and the maximum value of the monochromatic emissive power of a blackbody at (*a*) 3500 °R, (*b*) 3000 °R, (*c*) 2500 °R, and (*d*) 2000 °R.

    Ans.   (*a*)   1.4902 μm, 1.13499 × 10^5 Btu/hr-ft^2-μm;

              (*b*)   1.7385 μm, 5.2512 × 10^4 Btu/hr-ft^2-μm;

              (*c*)   2.0862 μm, 2.1103 × 10^4 Btu/hr-ft^2-μm;

              (*d*)   2.6078 μm, 6.9152 × 10^3 Btu/hr-ft^2-μm

11.28. Obtain (*11.35*) by thermodynamic reasoning.   (*Hint*: $q_{1\text{-}2}$ must vanish when $T_1 = T_2$.)

11.29. For the situation of Problem 11.13, determine $F_{3\text{-}1}$ and $F_{3\text{-}2}$ where area 3 is the annular area at one end of the set of concentric cylinders.   Use these to confirm the answer to Problem 11.13.

    Ans. $F_{3\text{-}1} = 0.580$, $F_{3\text{-}2} = 0.220$, and $F_{3\text{-}1} + F_{3\text{-}2} + F_{3\text{-}3} = 1.000$

11.30. Two blackbody rectangles, each 6 ft by 12 ft, are parallel, directly opposed, and 12 ft apart.   One rectangle is held at $T_1 = 200$ °F; the other is at $T_2 = 600$ °F.   Find: (*a*) the rate of radiant heat transfer $q_{1\text{-}2}$, (*b*) the rate at which the 200 °F rectangle is losing energy if the surroundings are at 0 °R, and (*c*) the rate at which the 200 °F rectangle is losing energy if the surroundings are considered as a single blackbody at 70 °F.

    Ans.   For $F_{1\text{-}2} \approx 0.115$: (*a*) −15,224 Btu/hr; (*b*) 5499.4 Btu/hr; (*c*) −3118.3 Btu/hr

Appendixes

Table A-1. Conversion Factors for Single Terms

To convert from	to	multiply by
Energy		
Btu (thermochemical)	joule	1054.35026448
calorie (thermochemical)	joule	4.184
foot lbf	joule	1.3558179
foot poundal	joule	0.042140110
kilowatt hour	joule	3.60×10^6
watt hour	joule	3600
Force		
dyne	newton	1.00×10^{-5}
kilogram force (kgf)	newton	9.80665
ounce force (avoirdupois)	newton	0.27801385
pound force, lbf (avoirdupois)	newton	4.44822161526
poundal	newton	0.1382549543
Length		
angstrom	meter	1.00×10^{-10}
foot	meter	0.3048
inch	meter	0.0254
micron	meter	1.00×10^{-6}
mil	meter	2.54×10^{-5}
mile (U.S. statute)	meter	1609.344
yard	meter	0.9144
Mass		
gram	kilogram	1.00×10^{-3}
kgf second^2 meter	kilogram	9.80665
lbm (avoirdupois)	kilogram	0.45359237
ounce mass (avoirdupois)	kilogram	0.028349523
ton (long)	kilogram	1016.0469
ton (metric)	kilogram	1000
ton (short, 2000 pound)	kilogram	907.18474
Temperature		
Celsius	Kelvin	$K = C + 273.15$
Fahrenheit	Celsius	$C = \frac{5}{9}(F - 32)$
Fahrenheit	Kelvin	$K = \frac{5}{9}(F + 459.67)$
Kelvin	Celsius	$C = K - 273.15$
Rankine	Kelvin	$K = \frac{5}{9}R$

Table A-2. Conversion Factors for Compound Terms

To convert from	to	multiply by
Acceleration		
foot/second^2	meter/second^2	0.3048
inch/second^2	meter/second^2	0.0254
Density		
gram/centimeter^3	kilogram/meter^3	1000
lbm/foot^3	kilogram/meter^3	16.018463
slug/foot^3	kilogram/meter^3	515.379
Energy/Area-Time		
*Btu/foot^2-hour	watt/meter^2	3.1524808
*calorie/cm^2-minute	watt/meter^2	697.33333
watt/centimeter^2	watt/meter^2	10,000
Power		
Btu/second	watt	1054.3502644
calorie/second	watt	4.184
foot lbf/second	watt	1.3558179
horsepower (550 ft lbf/second)	watt	745.69987
horsepower (electric)	watt	746.00000
horsepower (metric)	watt	735.499
Pressure		
atmosphere	newton/meter^2	1.01325×10^5
bar	newton/meter^2	1.00×10^5
millimeter of mercury (0 °C)	newton/meter^2	133.322
centimeter of water (4 °C)	newton/meter^2	98.0638
dyne/centimeter^2	newton/meter^2	0.100
kgf/centimeter^2	newton/meter^2	98,066.5
lbf/inch^2 (psi)	newton/meter^2	6894.7572
pascal	newton/meter^2	1.00
torr (0 °C)	newton/meter^2	133.322
Speed		
foot/second	meter/second	0.3048
kilometer/hour	meter/second	0.27777778
knot (international)	meter/second	0.51444444
mile/hour (U.S. statute)	meter/second	0.44704
Thermal Conductivity		
Btu inch/foot^2-second-°F	joule/meter-second-K	518.87315
Btu/foot-hour-°F	joule/meter-second-K	1.7295771

*All Btu and calorie terms in Table A-2 are thermochemical values.

Table A-2 (continued)

To convert from	to	multiply by
Viscosity		
centipoise	newton second/meter^2	1.00×10^{-3}
centistoke	meter^2/second	1.00×10^{-6}
foot^2/second	meter^2/second	0.09290304
lbm/foot-second	newton second/meter^2	1.4881639
lbf second/foot^2	newton second/meter^2	47.880258
poise	newton second/meter^2	0.10
poundal second/ft^2	newton second/meter^2	1.4881639
slug/foot-second	newton second/meter^2	47.880258
stoke	meter^2/second	1.00×10^{-4}
Volume		
fluid ounce (U.S.)	meter^3	$2.95735295 \times 10^{-5}$
foot^3	meter^3	0.0283168465
gallon (British)	meter^3	4.546087×10^{-3}
gallon (U.S. dry)	meter^3	$4.40488377 \times 10^{-3}$
gallon (U.S. liquid)	meter^3	$3.78541178 \times 10^{-3}$
liter (H$_2$O at 4 °C)	meter^3	1.000028×10^{-3}
liter (SI)	meter^3	1.00×10^{-3}
pint (U.S. liquid)	meter^3	$4.73176473 \times 10^{-4}$
quart (U.S. liquid)	meter^3	9.4635295×10^{-4}
yard^3	meter^3	0.764554857

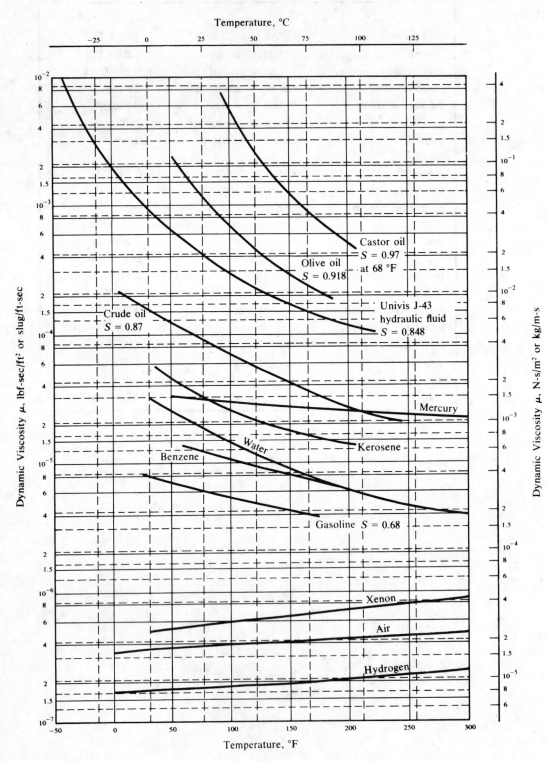

Appendix B

Figure B-1. Dynamic (Absolute) Viscosity of Fluids
Specific gravity (S) values apply at 70 °F

Temperature, °C

Castor oil
S = 0.97
at 68 °F

Olive oil
S = 0.918

Univis J-43
hydraulic fluid
S = 0.848

Crude oil
S = 0.87

Mercury

Water

Kerosene

Benzene

Gasoline S = 0.68

Xenon

Air

Hydrogen

Dynamic Viscosity μ, lbf-sec/ft^2 or slug/ft-sec

Dynamic Viscosity μ, N-s/m^2 or kg/m-s

Temperature, °F

Figure B-2. Kinematic Viscosity of Fluids
Specific gravity (S) values apply at 70 °F

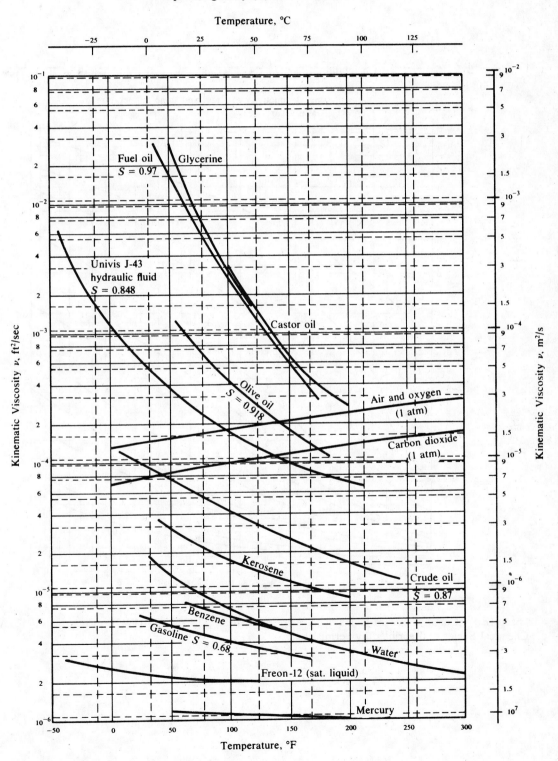

Figure B-3. Generalized Correlation Chart of the Thermal Conductivity of Gases at High Pressures*

k_1 = thermal conductivity at 1 atm and same temperature
p_c = critical pressure.

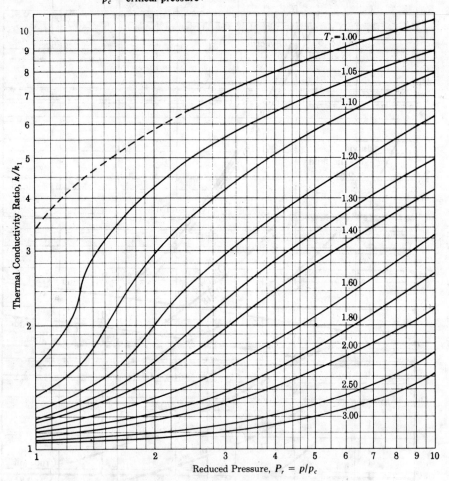

Reduced Pressure, $P_r = p/p_c$

*From E.W. Comings and M.F. Nathan, *Ind. Eng. Chem.*, **39**: 964–970 (1947). Copyright 1947, American Chemical Society. Reprinted by permission.

Figure B-4.   Generalized Correlation Chart of the Dynamic Viscosity of Gases at High Pressures*

μ_1 = dynamic viscosity at 1 atm and same temperature
p_c = critical pressure

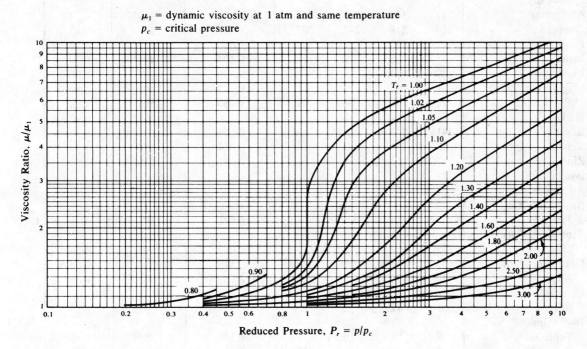

*From E.W. Comings, B.J. Mayland and R.S. Egly, Univ. of Illinois Engineering Experiment Station Bulletin No. 354 (1944).

Table B-1. Properties of Metals

Material	k, Btu/hr-ft-°F				c_p, Btu/lbm-°F	ρ, lbm/ft³	α, ft²/hr
	32 °F 0 °C	212 °F 100 °C	572 °F 300 °C	932 °F 500 °C	32 °F 0 °C	32 °F 0 °C	32 °F 0 °C
Metals—Pure							
Aluminum	117	119	133	156	0.208	169	3.33
Copper	224	218	212	207	0.091	558	4.42
Gold	169	170	...	...	0.030	1203	4.68
Iron	35.8	36.6	...	...	0.104	491	0.70
Lead	20.1	19	18	...	0.030	705	0.95
Magnesium	91	92	...	...	0.232	109	3.60
Molybdenum	72	68	64	62	0.060	638	1.88
Nickel	54	48	37	...	0.106	556	0.92
Silver	241	240	...	...	0.056	655	6.57
Tin	38	34	...	...	0.054	456	1.54
Zinc	65.1	63	58	...	0.091	446	1.60
Alloys							
Admiralty metal	65	64					
Brass, 70% Cu, 30% Zn	61.5	74	85	...	0.092	532	1.26
Bronze, 75% Cu, 25% Sn	15	...	...	...	0.082	541	0.34
Cast iron, Plain	33	31.8	27.7	24.8	0.11	474	0.63
Alloy	30	28.3	27	...	0.10	455	0.66
Constantan, 60% Cu, 40% Ni	12.4	12.8	...	...	0.10	557	0.22
18–8 stainless steel, Type 304	8.0	9.4	10.9	12.4	0.11	488	0.15
Type 347	8.0	9.3	11.0	12.8	0.11	488	0.15
Steel, mild, 1% C	26.5	26	25	22	0.11	490	0.49
SI Units	W/m-K				J/kg-K	kg/m³	m²/s
To convert to SI units multiply tabulated values by	1.729577				4.184 × 10³	1.601846 × 10¹	2.580640 × 10⁻⁵

Substance	T, °F	T, °C	c_p, Btu/lbm-°F	ρ, lbm/ft³	k, Btu/hr-ft-°F	α, ft²/hr
Structural						
Asphalt	68	20			0.43	
Bakelite	68	20	0.38	79.5	0.134	0.0044
Bricks						
Common	68	20	0.20	100	0.40	0.02
Face	68	20		128	0.76	
Carborundum brick	1110	600			10.7	
	2550	1400			6.4	
Chrome brick	392	200			1.34	0.036
	1022	550	0.20	188	1.43	0.038
	1652	900			1.15	0.031
Diatomaceous earth (fired)	400	205			0.14	
	1600	870			0.18	
Fireclay brick (burnt 2426 °F, 1330 °C)	932	500			0.60	0.020
	1472	800	0.23	128	0.62	0.021
	2012	1100			0.63	0.021
Fireclay brick (burnt 2642 °F, 1450 °C)	932	500			0.74	0.022
	1472	800	0.23	145	0.79	0.024
	2012	1100			0.81	0.024
Fireclay brick (Missouri)	392	200			0.58	0.015
	1112	600	0.23	165	0.85	0.022
	2552	1400			1.02	0.027
Magnesite	400	205			2.2	
	1200	650	0.27		1.6	
	2200	1205			1.1	
Cement, Portland				94	0.17	
Cement, mortar	75	24			0.67	
Concrete	68	20	0.21	119–144	0.47–0.81	0.019–0.027
Concrete, cinder	75	24			0.44	
Glass, plate	68	20	0.2	169	0.44	0.013
Glass, borosilicate	86	30		139	0.63	
Plaster, gypsum	70	21	0.2	90	0.28	0.016
Plaster, metal lath	70	21			0.27	
Plaster, wood lath	70	21			0.16	
Stone						
Granite			0.195	165	1.0–2.3	0.031–0.071
Limestone	210–570	100–300	0.217	155	0.73–0.77	0.022–0.023
Marble	68	20	0.193	156–169	1.6	0.054
Sandstone	68	20	0.17	135–144	0.94–1.2	0.041–0.049
SI Units			J/kg-K	kg/m³	W/m-K	m²/s
To convert to SI units multiply tabulated values by			4.184 × 10³	1.601846 × 10¹	1.729577	2.580640 × 10⁻⁵

*Adapted by permission from A. Chapman, *Heat Transfer*, 2nd ed., Macmillan Company, London, 1967. Copyright, 1967.

Table B-2 (continued)

Substance	T, °F	T, °C	c_p, Btu/lbm-°F	ρ, lbm/ft³	k, Btu/hr-ft-°F	α, ft²/hr
Structural (cont.)						
Wood, cross grain:						
Balsa	86	30		8.8	0.032	
Cypress	86	30		29	0.056	
Fir	75	24	0.65	26.0	0.063	0.0037
Oak	86	30	0.57	38–30	0.096	0.0049
Yellow pine	75	24	0.67	40	0.085	0.0032
White pine	86	30		27	0.065	
Wood, radial:						
Oak	68	20	0.57	38–30	0.10–0.12	0.0043–0.0047
Fir	68	20	0.65	26.0–26.3	0.08	0.0048
Insulating						
Asbestos	−328	−200		29.3	0.043	
	32	0			0.090	
Asbestos	32	0			0.087	
	212	100		36.0	0.111	
	392	200			0.120	
	752	400			0.129	
Asbestos	−328	−200		43.5	0.09	
	32	0			0.135	
Asbestos cement					1.2	
Asbestos sheet	124	51			0.096	
Asbestos felt	100	38			0.033	
(40 laminations	300	149			0.040	
per inch)	500	260			0.048	
Asbestos felt	100	38			0.045	
(20 laminations	300	149			0.055	
per inch)	500	260			0.065	
Balsam wool	90	32		2.2	0.023	
Cardboard, corrugated					0.037	
Celotex	90	32			0.028	
Corkboard	86	30		10	0.025	
Cork, ground	86	30		9.4	0.025	
SI Units			J/kg-K	kg/m³	W/m-K	m²/s
To convert to SI units multiply tabulated values by			4.184 × 10³	1.601846 × 10¹	1.729577	2.580640 × 10⁻⁵

Table B-2 (continued)

Substance	T, °F	T, °C	c_p, Btu/lbm-°F	ρ, lbm/ft³	k, Btu/hr-ft-°F	α, ft²/hr
Insulating (cont.)						
Diatomaceous earth (powdered)	200	93			0.033	
	400	204		14	0.039	
	600	316			0.046	
Felt, hair	20	−7			0.0212	
	100	38		11.4	0.0254	
	200	93			0.0299	
Fiber insulating board	70	21		14.8	0.028	
Glass wool	20	−7			0.0217	
	100	38		1.5	0.0313	
	200	93			0.0435	
Glass wool	20	−7			0.0179	
	100	38		4.0	0.0239	
	200	93			0.0317	
Glass wool	20	−7			0.0163	
	100	38		6.0	0.0218	
	200	93			0.0288	
Kapok	86	30			0.020	
Magnesia, 85%	100	38			0.039	
	200	93			0.041	
	300	149		16.9	0.043	
	400	204			0.046	
Rock wool	20	−7			0.0150	
	100	38		4.0	0.0224	
	200	93			0.0317	
Rock wool	20	−7			0.0171	
	100	38		8.0	0.0228	
	200	93			0.0299	
Rock wool	20	−7			0.0183	
	100	38		12.0	0.0226	
	200	93			0.0281	
Miscellaneous						
Aerogel, silica	248	120		8.5	0.013	
Clay	68	20	0.21	91.0	0.739	0.039
Coal, anthracite	68	20	0.30	75–94	0.15	0.005–0.006
Coal, powdered	86	30	0.31	46	0.067	0.005
Cotton	68	20	0.31	5	0.034	0.075
Earth, coarse	68	20	0.44	128	0.30	0.0054
Ice	32	0	0.46	57	1.28	0.048
Rubber, hard	32	0		74.8	0.087	
Sawdust	75	24			0.034	
Silk	68	20	0.33	3.6	0.021	0.017
SI Units			J/kg-K	kg/m³	W/m-K	m²/s
To convert to SI units multiply tabulated values by			4.184 × 10³	1.601846 × 10¹	1.729577	2.580640 × 10⁻⁵

Table B-3. Properties of Liquids in Saturated State*

T, °F	T, °C	ρ, $\dfrac{\text{lbm}}{\text{ft}^3}$	c_p, $\dfrac{\text{Btu}}{\text{lbm-°F}}$	ν, $\dfrac{\text{ft}^2}{\text{sec}}$	k, $\dfrac{\text{Btu}}{\text{hr-ft-°F}}$	α, $\dfrac{\text{ft}^2}{\text{hr}}$	Pr	β, $\dfrac{1}{\text{°R}}$
				Water (H_2O)				
32	0	62.57	1.0074	1.925×10^{-5}	0.319	5.07×10^{-3}	13.6	
68	20	62.46	0.9988	1.083	0.345	5.54	7.02	0.10×10^{-3}
104	40	62.09	0.9980	0.708	0.363	5.86	4.34	
140	60	61.52	0.9994	0.514	0.376	6.02	3.02	
176	80	60.81	1.0023	0.392	0.386	6.34	2.22	
212	100	59.97	1.0070	0.316	0.393	6.51	1.74	
248	120	59.01	1.015	0.266	0.396	6.62	1.446	
284	140	57.95	1.023	0.230	0.395	6.68	1.241	
320	160	56.79	1.037	0.204	0.393	6.70	1.099	
356	180	55.50	1.055	0.186	0.390	6.68	1.004	
392	200	54.11	1.076	0.172	0.384	6.61	0.937	
428	220	52.59	1.101	0.161	0.377	6.51	0.891	
464	240	50.92	1.136	0.154	0.367	6.35	0.871	
500	260	49.06	1.182	0.148	0.353	6.11	0.874	
537	280	46.98	1.244	0.145	0.335	5.74	0.910	
572	300	44.59	1.368	0.145	0.312	5.13	1.019	
				Ammonia (NH_3)				
−58	−50	43.93	1.066	0.468×10^{-5}	0.316	6.75×10^{-3}	2.60	
−40	−40	43.18	1.067	0.437	0.316	6.88	2.28	
−22	−30	42.41	1.069	0.417	0.317	6.98	2.15	
−4	−20	41.62	1.077	0.410	0.316	7.05	2.09	
14	−10	40.80	1.090	0.407	0.314	7.07	2.07	
32	0	39.96	1.107	0.402	0.312	7.05	2.05	
50	10	39.09	1.126	0.396	0.307	6.98	2.04	
68	20	38.19	1.146	0.386	0.301	6.88	2.02	1.36×10^{-3}
86	30	37.23	1.168	0.376	0.293	6.75	2.01	
104	40	36.27	1.194	0.366	0.285	6.59	2.00	
122	50	35.23	1.222	0.355	0.275	6.41	1.99	
				Carbon dioxide (CO_2)				
−58	−50	72.19	0.44	0.128×10^{-5}	0.0494	1.558×10^{-3}	2.96	
−40	−40	69.78	0.45	0.127	0.0584	1.864	2.46	
−22	−30	67.22	0.47	0.126	0.0645	2.043	2.22	
−4	−20	64.45	0.49	0.124	0.0665	2.110	2.12	
14	−10	61.39	0.52	0.122	0.0635	1.989	2.20	
SI Units		$\dfrac{\text{kg}}{\text{m}^3}$	$\dfrac{\text{J}}{\text{kg-K}}$	$\dfrac{\text{m}^2}{\text{s}}$	$\dfrac{\text{W}}{\text{m-K}}$	$\dfrac{\text{m}^2}{\text{s}}$	—	$\dfrac{1}{\text{K}}$
To convert to SI units multiply tabulated values by		1.601846×10^1	4.184×10^3	9.290304×10^{-2}	1.729577	2.580640×10^{-5}	—	1.80

*Adapted by permission from E.R.G. Eckert and R.M. Drake, Jr., *Heat and Mass Transfer*, 2nd. ed., McGraw-Hill Book Company, New York, 1959.

Table B-3 (continued)

T, °F	°C	ρ, $\dfrac{lbm}{ft^3}$	c_p, $\dfrac{Btu}{lbm\text{-}°F}$	ν, $\dfrac{ft^2}{sec}$	k, $\dfrac{Btu}{hr\text{-}ft\text{-}°F}$	α, $\dfrac{ft^2}{hr}$	Pr	β, $\dfrac{1}{°R}$
\multicolumn{9}{c}{Carbon dioxide (CO₂) (cont.)}

T, °F	°C	ρ	c_p	ν	k	α	Pr	β
\multicolumn{9}{l}{**Carbon dioxide (CO_2) (cont.)**}
32	0	57.87	0.59	0.117	0.0604	1.774	2.38	
50	10	53.69	0.75	0.109	0.0561	1.398	2.80	
68	20	48.23	1.2	0.098	0.0504	0.860	4.10	7.78×10^{-3}
86	30	37.32	8.7	0.086	0.0406	0.108	28.7	
\multicolumn{9}{l}{**Sulfur dioxide (SO_2)**}
−58	−50	97.44	0.3247	0.521×10^{-5}	0.140	4.42×10^{-3}	4.24	
−40	−40	95.94	0.3250	0.456	0.136	4.38	3.74	
−22	−30	94.43	0.3252	0.399	0.133	4.33	3.31	
−4	−20	92.93	0.3254	0.349	0.130	4.29	2.93	
14	−10	91.37	0.3255	0.310	0.126	4.25	2.62	
32	0	89.80	0.3257	0.277	0.122	4.19	2.38	
50	10	88.18	0.3259	0.250	0.118	4.13	2.18	
68	20	86.55	0.3261	0.226	0.115	4.07	2.00	1.08×10^{-3}
86	30	84.86	0.3263	0.204	0.111	4.01	1.83	
104	40	82.98	0.3266	0.186	0.107	3.95	1.70	
122	50	81.10	0.3268	0.174	0.102	3.87	1.61	
\multicolumn{9}{l}{**Methylchloride (CH_3Cl)**}
−58	−50	65.71	0.3525	0.344×10^{-5}	0.124	5.38×10^{-3}	2.31	
−40	−40	64.51	0.3541	0.342	0.121	5.30	2.32	
−22	−30	63.46	0.3564	0.338	0.117	5.18	2.35	
−4	−20	62.39	0.3593	0.333	0.113	5.04	2.38	
14	−10	61.27	0.3629	0.329	0.108	4.87	2.43	
32	0	60.08	0.3673	0.325	0.103	4.70	2.49	
50	10	58.83	0.3726	0.320	0.099	4.52	2.55	
68	20	57.64	0.3788	0.315	0.094	4.31	2.63	
86	30	56.38	0.3860	0.310	0.089	4.10	2.72	
104	40	55.13	0.3942	0.303	0.083	3.86	2.83	
122	50	53.76	0.4034	0.295	0.077	3.57	2.97	
\multicolumn{9}{l}{**Dichlorodifluoromethane (Freon = 12)(CCl_2F_2)**}
−58	−50	96.56	0.2090	0.334×10^{-5}	0.039	1.94×10^{-3}	6.2	1.46×10^{-3}
−40	−40	94.81	0.2113	0.300	0.040	1.99	5.4	
−22	−30	92.99	0.2139	0.272	0.040	2.04	4.8	
−4	−20	91.18	0.2167	0.253	0.041	2.09	4.4	
14	−10	89.24	0.2198	0.238	0.042	2.13	4.0	
SI Units		$\dfrac{kg}{m^3}$	$\dfrac{J}{kg\text{-}K}$	$\dfrac{m^2}{s}$	$\dfrac{W}{m\text{-}K}$	$\dfrac{m^2}{s}$	—	$\dfrac{1}{K}$
To convert to SI units multiply tabulated values by		1.601846×10^1	4.184×10^3	9.290304×10^{-2}	1.729577	2.580640×10^{-5}	—	1.80

Table B-3 (continued)

T, °F	°C	ρ, $\dfrac{lbm}{ft^3}$	c_p, $\dfrac{Btu}{lbm\text{-}°F}$	ν, $\dfrac{ft^2}{sec}$	k, $\dfrac{Btu}{hr\text{-}ft\text{-}°F}$	α, $\dfrac{ft^2}{hr}$	Pr	β, $\dfrac{1}{°R}$
\multicolumn{9}{c}{Dichlorodifluoromethane (Freon = 12)(CCl_2F_2) (cont.)}								
32	0	87.24	0.2232	0.230	0.042	2.16	3.8	
50	10	85.17	0.2268	0.219	0.042	2.17	3.6	
68	20	83.04	0.2307	0.213	0.042	2.17	3.5	
86	30	80.85	0.2349	0.209	0.041	2.17	3.5	
104	40	78.48	0.2393	0.206	0.040	2.15	3.5	
122	50	75.91	0.2440	0.204	0.039	2.11	3.5	
\multicolumn{9}{c}{Eutectic calcium chloride solution (29.9% $CaCl_2$)}								
−58	−50	82.39	0.623	39.13×10^{-5}	0.232	4.52×10^{-3}	312	
−40	−40	82.09	0.6295	26.88	0.240	4.65	208	
−22	−30	81.79	0.6356	18.49	0.248	4.78	139	
−4	−20	81.50	0.642	11.88	0.257	4.91	87.1	
14	−10	81.20	0.648	7.49	0.265	5.04	53.6	
32	0	80.91	0.654	4.73	0.273	5.16	33.0	
50	10	80.62	0.660	3.61	0.280	5.28	24.6	
68	20	80.32	0.666	2.93	0.288	5.40	19.6	
86	30	80.03	0.672	2.44	0.295	5.50	16.0	
104	40	79.73	0.678	2.07	0.302	5.60	13.3	
122	50	79.44	0.685	1.78	0.309	5.69	11.3	
\multicolumn{9}{c}{Glycerin [$C_3H_5(OH)_3$]}								
32	0	79.66	0.540	0.0895	0.163	3.81×10^{-3}	84.7×10^3	
50	10	79.29	0.554	0.0323	0.164	3.74	31.0	
68	20	78.91	0.570	0.0127	0.165	3.67	12.5	0.28×10^{-3}
86	30	78.54	0.584	0.0054	0.165	3.60	5.38	
104	40	78.16	0.600	0.0024	0.165	3.54	2.45	
122	50	77.72	0.617	0.0016	0.166	3.46	1.63	
\multicolumn{9}{c}{Ethylene glycol [$C_2H_4(OH_2)$]}								
32	0	70.59	0.548	61.92×10^{-5}	0.140	3.62×10^{-3}	615	
68	20	69.71	0.569	20.64	0.144	3.64	204	0.36×10^{-3}
104	40	68.76	0.591	9.35	0.148	3.64	93	
140	60	67.90	0.612	5.11	0.150	3.61	51	
176	80	67.27	0.633	3.21	0.151	3.57	32.4	
212	100	66.08	0.655	2.18	0.152	3.52	22.4	
SI Units		$\dfrac{kg}{m^3}$	$\dfrac{J}{kg\text{-}K}$	$\dfrac{m^2}{s}$	$\dfrac{W}{m\text{-}K}$	$\dfrac{m^2}{s}$	—	$\dfrac{1}{K}$
To convert to SI units multiply tabulated values by		1.601846×10^1	4.184×10^3	9.290304×10^{-2}	1.729577	2.580640×10^{-5}	—	1.80

Table B-3 (continued)

T, °F	T, °C	$\rho,$ $\dfrac{lbm}{ft^3}$	$c_p,$ $\dfrac{Btu}{lbm\text{-}°F}$	$\nu,$ $\dfrac{ft^2}{sec}$	$k,$ $\dfrac{Btu}{hr\text{-}ft\text{-}°F}$	$\alpha,$ $\dfrac{ft^2}{hr}$	Pr	$\beta,$ $\dfrac{1}{°R}$
Engine oil (unused)								
32	0	56.13	0.429	0.0461	0.085	3.53×10^{-3}	47100	
68	20	55.45	0.449	0.0097	0.084	3.38	10400	0.39×10^{-3}
104	40	54.69	0.469	0.0026	0.083	3.23	2870	
140	60	53.94	0.489	0.903×10^{-3}	0.081	3.10	1050	
176	80	53.19	0.509	0.404	0.080	2.98	490	
212	100	52.44	0.530	0.219	0.079	2.86	276	
248	120	51.75	0.551	0.133	0.078	2.75	175	
284	140	51.00	0.572	0.086	0.077	2.66	116	
320	160	50.31	0.593	0.060	0.076	2.57	84	
Mercury (Hg)								
32	0	850.78	0.0335	0.133×10^{-5}	4.74	166.6×10^{-3}	0.0288	
68	20	847.71	0.0333	0.123	5.02	178.5	0.0249	1.01×10^{-4}
122	50	843.14	0.0331	0.112	5.43	194.6	0.0207	
212	100	835.57	0.0328	0.0999	6.07	221.5	0.0162	
302	150	828.06	0.0326	0.0918	6.64	246.2	0.0134	
392	200	820.61	0.0375	0.0863	7.13	267.7	0.0116	
482	250	813.16	0.0324	0.0823	7.55	287.0	0.0103	
600	316	802	0.032	0.0724	8.10	316	0.0083	
SI Units		$\dfrac{kg}{m^3}$	$\dfrac{J}{kg\text{-}K}$	$\dfrac{m^2}{s}$	$\dfrac{W}{m\text{-}K}$	$\dfrac{m^2}{s}$	—	$\dfrac{1}{K}$
To convert to SI units multiply tabulated values by		1.601846 $\times 10^1$	4.184 $\times 10^3$	9.290304 $\times 10^{-2}$	1.729577	2.580640 $\times 10^{-5}$	—	1.80

°F	T, °C	ρ, $\dfrac{\text{lbm}}{\text{ft}^3}$	c_p, $\dfrac{\text{Btu}}{\text{lbm-}°\text{F}}$	μ_m, $\dfrac{\text{lbm}}{\text{ft-sec}}$	ν, $\dfrac{\text{ft}^2}{\text{sec}}$	k, $\dfrac{\text{Btu}}{\text{hr-ft-}°\text{F}}$	α, $\dfrac{\text{ft}^2}{\text{hr}}$	Pr
Air								
−280	−173	0.2248	0.2452	0.4653×10^{-5}	2.070×10^{-5}	0.005342	0.09691	0.770
−190	−123	0.1478	0.2412	0.6910	4.675	0.007936	0.2226	0.753
−100	−73	0.1104	0.2403	0.8930	8.062	0.01045	0.3939	0.739
−10	−23	0.0882	0.2401	1.074	10.22	0.01287	0.5100	0.722
80	27	0.0735	0.2402	1.241	16.88	0.01516	0.8587	0.708
170	77	0.0623	0.2410	1.394	22.38	0.01735	1.156	0.697
260	127	0.0551	0.2422	1.536	27.88	0.01944	1.457	0.689
350	177	0.0489	0.2438	1.669	31.06	0.02142	1.636	0.683
440	227	0.0440	0.2459	1.795	40.80	0.02333	2.156	0.680
530	277	0.0401	0.2482	1.914	47.73	0.02519	2.531	0.680
620	327	0.0367	0.2520	2.028	55.26	0.02692	2.911	0.680
710	377	0.0339	0.2540	2.135	62.98	0.02862	3.324	0.682
800	427	0.0314	0.2568	2.239	71.31	0.03022	3.748	0.684
890	477	0.0294	0.2593	2.339	79.56	0.03183	4.175	0.686
980	527	0.0275	0.2622	2.436	88.58	0.03339	4.631	0.689
1070	577	0.0259	0.2650	2.530	97.68	0.03483	5.075	0.692
1160	627	0.0245	0.2678	2.620	106.9	0.03628	5.530	0.696
1250	677	0.0232	0.2704	2.703	116.5	0.03770	6.010	0.699
1340	727	0.0220	0.2727	2.790	126.8	0.03901	6.502	0.702
1520	827	0.0200	0.2772	2.955	147.8	0.04178	7.536	0.706
1700	927	0.0184	0.2815	3.109	169.0	0.04410	8.514	0.714
1880	1027	0.0169	0.2860	3.258	192.8	0.04641	9.602	0.722
2060	1127	0.0157	0.2900	3.398	216.4	0.04880	10.72	0.726
2240	1227	0.0147	0.2939	3.533	240.3	0.05098	11.80	0.734
2420	1327	0.0138	0.2982	3.668	265.8	0.05348	12.88	0.741
2600	1427	0.0130	0.3028	3.792	291.7	0.05550	14.00	0.749
2780	1527	0.0123	0.3075	3.915	318.3	0.05750	15.09	0.759
2960	1627	0.0116	0.3128	4.029	347.1	0.0591	16.40	0.767
3140	1727	0.0110	0.3196	4.168	378.8	0.0612	17.41	0.783
3320	1827	0.0105	0.3278	4.301	409.9	0.0632	18.36	0.803
3500	1927	0.0100	0.3390	4.398	439.8	0.0646	19.05	0.831
3680	2027	0.0096	0.3541	4.513	470.1	0.0663	19.61	0.863
3860	2127	0.0091	0.3759	4.611	506.9	0.0681	19.92	0.916
4160	2293	0.0087	0.4031	4.750	546.0	0.0709	20.21	0.972
Helium								
−456	−271		1.242	5.66×10^{-7}		0.0061		
−400	−240	0.0915	1.242	33.7	3.68×10^{-5}	0.0204	0.1792	0.74
−200	−129	0.211	1.242	84.3	39.95	0.0536	2.044	0.70
−100	−73	0.0152	1.242	105.2	69.30	0.0680	3.599	0.694
0	−18	0.0119	1.242	122.1	102.8	0.0784	5.299	0.70
200	93	0.00829	1.242	154.9	186.9	0.0977	9.490	0.71
SI Units		$\dfrac{\text{kg}}{\text{m}^3}$	$\dfrac{\text{J}}{\text{kg-K}}$	$\dfrac{\text{kg}}{\text{m-s}}$	$\dfrac{\text{m}^2}{\text{s}}$	$\dfrac{\text{W}}{\text{m-K}}$	$\dfrac{\text{m}^2}{\text{s}}$	—
To convert to SI units multiply tabulated values by		1.601846×10^1	4.184×10^3	1.488164	9.290304×10^{-2}	1.729577	2.580640×10^{-5}	—

*Adapted by permission from E.R.G. Eckert and R.M. Drake, Jr., *Heat and Mass Transfer*, 2nd. ed., McGraw-Hill Book Company, New York, 1959.

Table B-4 (continued)

$T,$ °F	°C	$\rho,$ $\dfrac{lbm}{ft^3}$	$c_p,$ $\dfrac{Btu}{lbm\text{-}°F}$	$\mu_m,$ $\dfrac{lbm}{ft\text{-}sec}$	$\nu,$ $\dfrac{ft^2}{sec}$	$k,$ $\dfrac{Btu}{hr\text{-}ft\text{-}°F}$	$\alpha,$ $\dfrac{ft^2}{hr}$	Pr
Helium								
400	204	0.00637	1.242	184.8	289.9	0.114	14.40	0.72
600	316	0.00517	1.242	209.2	404.5	0.130	20.21	0.72
800	427	0.00439	1.242	233.5	531.9	0.145	25.81	0.72
1000	538	0.00376	1.242	256.5	682.5	0.159	34.00	0.72
1200	649	0.00330	1.242	277.9	841.0	0.172	41.98	0.72
Hydrogen								
−406	−243	0.05289	2.589	1.079×10^{-6}	2.040×10^{-5}	0.0132	0.0966	0.759
−370	−223	0.03181	2.508	1.691	5.253	0.0209	0.262	0.721
−280	−173	0.01534	2.682	2.830	18.45	0.0384	0.933	0.712
−190	−123	0.01022	3.010	3.760	36.79	0.0567	1.84	0.718
−100	−73	0.00766	3.234	4.578	59.77	0.0741	2.99	0.719
−10	−23	0.00613	3.358	5.321	86.80	0.0902	4.38	0.713
80	27	0.00511	3.419	6.023	117.9	0.105	6.02	0.706
170	77	0.00438	3.448	6.689	152.7	0.119	7.87	0.697
260	127	0.00383	3.461	7.300	190.6	0.132	9.95	0.690
350	177	0.00341	3.463	7.915	232.1	0.145	12.26	0.682
440	227	0.00307	3.465	8.491	276.6	0.157	14.79	0.675
530	277	0.00279	3.471	9.055	324.6	0.169	17.50	0.668
620	327	0.00255	3.472	9.599	376.4	0.182	20.56	0.664
800	427	0.00218	3.481	10.68	489.9	0.203	26.75	0.659
980	527	0.00191	3.505	11.69	612	0.222	33.18	0.664
1160	627	0.00170	3.540	12.62	743	0.238	39.59	0.676
1340	727	0.00153	3.575	13.55	885	0.254	46.49	0.686
1520	827	0.00139	3.622	14.42	1039	0.268	53.19	0.703
1700	927	0.00128	3.670	15.29	1192	0.282	60.00	0.715
1880	1027	0.00118	3.720	16.18	1370	0.296	67.40	0.733
1940	1060	0.00115	3.735	16.42	1429	0.300	69.80	0.736
Oxygen								
−280	−173	0.2492	0.2264	5.220×10^{-6}	2.095×10^{-5}	0.00522	0.09252	0.815
−190	−123	0.1635	0.2192	7.721	4.722	0.00790	0.2204	0.773
−100	−73	0.1221	0.2181	9.979	8.173	0.01054	0.3958	0.745
−10	−23	0.0975	0.2187	12.01	12.32	0.01305	0.6120	0.725
80	27	0.0812	0.2198	13.86	17.07	0.01546	0.8662	0.709
170	77	0.0695	0.2219	15.56	22.39	0.01774	1.150	0.702
260	127	0.0609	0.2250	17.16	28.18	0.02000	1.460	0.695
350	177	0.0542	0.2285	18.66	34.43	0.02212	1.786	0.694
440	227	0.0487	0.2322	20.10	41.27	0.02411	2.132	0.697
530	277	0.0443	0.2360	21.48	48.49	0.02610	2.496	0.700
620	327	0.0406	0.2399	22.79	56.13	0.02792	2.867	0.704
SI Units		$\dfrac{kg}{m^3}$	$\dfrac{J}{kg\text{-}K}$	$\dfrac{kg}{m\text{-}s}$	$\dfrac{m^2}{s}$	$\dfrac{W}{m\text{-}K}$	$\dfrac{m^2}{s}$	—
To convert to SI units multiply tabulated values by		1.601846×10^1	4.184×10^3	1.488164	9.290304×10^{-2}	1.729577	2.580640×10^{-5}	—

Table B-4 (continued)

$T,$ °F	°C	$\rho,$ $\dfrac{lbm}{ft^3}$	$c_p,$ $\dfrac{Btu}{lbm\text{-}°F}$	$\mu_m,$ $\dfrac{lbm}{ft\text{-}sec}$	$\nu,$ $\dfrac{ft^2}{sec}$	$k,$ $\dfrac{Btu}{hr\text{-}ft\text{-}°F}$	$\alpha,$ $\dfrac{ft^2}{hr}$	Pr
				Nitrogen				
−280	−173	0.2173	0.2561	4.611×10^{-6}	2.122×10^{-5}	0.005460	0.09811	0.786
−100	−73	0.1068	0.2491	8.700	8.146	0.01054	0.3962	0.747
80	27	0.0713	0.2486	11.99	16.82	0.01514	0.8542	0.713
260	127	0.0533	0.2498	14.77	27.71	0.01927	1.447	0.691
440	227	0.0426	0.2521	17.27	40.54	0.02302	2.143	0.684
620	327	0.0355	0.2569	19.56	55.10	0.02646	2.901	0.686
800	427	0.0308	0.2620	21.59	70.10	0.02960	3.668	0.691
980	527	0.0267	0.2681	23.41	87.68	0.03241	4.528	0.700
1160	627	0.0237	0.2738	25.19	98.02	0.03507	5.404	0.711
1340	727	0.0213	0.2789	26.88	126.2	0.03741	6.297	0.724
1520	827	0.0194	0.2832	28.41	146.4	0.03958	7.204	0.736
1700	927	0.0178	0.2875	29.90	168.0	0.04151	8.111	0.748
				Carbon dioxide				
−64	−53	0.1544	0.187	7.462×10^{-6}	4.833×10^{-5}	0.006243	0.2294	0.818
−10	−23	0.1352	0.192	8.460	6.257	0.007444	0.2868	0.793
80	27	0.1122	0.208	10.051	8.957	0.009575	0.4103	0.770
170	77	0.0959	0.215	11.561	12.05	0.01183	0.5738	0.755
260	127	0.0838	0.225	12.98	15.49	0.01422	0.7542	0.738
350	177	0.0744	0.234	14.34	19.27	0.01674	0.9615	0.721
440	227	0.0670	0.242	15.63	23.33	0.01937	1.195	0.702
530	277	0.0608	0.250	16.85	27.71	0.02208	1.453	0.685
620	327	0.0558	0.257	18.03	32.31	0.02491	1.737	0.668
				Carbon monoxide				
−64	−53	0.09699	0.2491	9.295×10^{-6}	9.583×10^{-5}	0.01101	0.4557	0.758
−10	−23	0.0525	0.2490	10.35	12.14	0.01239	0.5837	0.750
80	27	0.07109	0.2489	11.990	16.87	0.01459	0.8246	0.737
170	77	0.06082	0.2492	13.50	22.20	0.01666	1.099	0.728
260	127	0.05329	0.2504	14.91	27.98	0.01864	1.397	0.722
350	177	0.04735	0.2520	16.25	34.32	0.0252	1.720	0.718
440	227	0.04259	0.2540	17.51	41.11	0.02232	2.063	0.718
530	277	0.03872	0.2569	18.74	48.40	0.02405	2.418	0.721
620	327	0.03549	0.2598	19.89	56.04	0.02569	2.786	0.724
				Ammonia (NH_3)				
−58	−50	0.0239	0.525	4.875×10^{-6}	2.04×10^{-4}	0.0099	0.796	0.93
32	0	0.0495	0.520	6.285	1.27	0.0127	0.507	0.90
122	50	0.0405	0.520	7.415	1.83	0.0156	0.744	0.88
212	100	0.0349	0.534	8.659	2.48	0.0189	1.015	0.87
302	150	0.0308	0.553	9.859	3.20	0.0226	1.330	0.87
392	200	0.0275	0.572	11.08	4.03	0.0270	1.713	0.84
SI Units		$\dfrac{kg}{m^3}$	$\dfrac{J}{kg\text{-}K}$	$\dfrac{kg}{m\text{-}s}$	$\dfrac{m^2}{s}$	$\dfrac{W}{m\text{-}K}$	$\dfrac{m^2}{s}$	—
To convert to SI units multiply tabulated values by		1.601846 $\times 10^1$	4.184 $\times 10^3$	1.488164	9.290304 $\times 10^{-2}$	1.729577	2.580640 $\times 10^{-5}$	—

Table B-4 (continued)

$T,$ °F	°C	$\rho,$ $\dfrac{lbm}{ft^3}$	$c_p,$ $\dfrac{Btu}{lbm\text{-}°F}$	$\mu_m,$ $\dfrac{lbm}{ft\text{-}sec}$	$\nu,$ $\dfrac{ft^2}{sec}$	$k,$ $\dfrac{Btu}{hr\text{-}ft\text{-}°F}$	$\alpha,$ $\dfrac{ft^2}{hr}$	Pr
colspan over — Steam (H₂O vapor)								
224	107	0.0366	0.492	8.54×10^{-6}	2.33×10^{-4}	0.0142	0.789	1.060
260	127	0.0346	0.481	9.03	2.61	0.0151	0.906	1.040
350	177	0.0306	0.473	10.25	3.35	0.0173	1.19	1.010
440	227	0.0275	0.474	11.45	4.16	0.0196	1.50	0.996
530	277	0.0250	0.477	12.66	5.06	0.0219	1.84	0.991
620	327	0.0228	0.484	13.89	6.09	0.0244	2.22	0.986
710	377	0.0211	0.491	15.10	7.15	0.0268	2.58	0.995
800	427	0.0196	0.498	16.30	8.31	0.0292	2.99	1.000
890	477	0.0183	0.506	17.50	9.56	0.0317	3.42	1.005
980	527	0.0171	0.514	18.72	10.98	0.0342	3.88	1.010
1070	577	0.0161	0.522	19.95	12.40	0.0368	4.38	1.019
SI Units		$\dfrac{kg}{m^3}$	$\dfrac{J}{kg\text{-}K}$	$\dfrac{kg}{m\text{-}s}$	$\dfrac{m^2}{s}$	$\dfrac{W}{m\text{-}K}$	$\dfrac{m^2}{s}$	—
To convert to SI units multiply tabulated values by		1.601846 $\times 10^1$	4.184 $\times 10^3$	1.488164	9.290304 $\times 10^{-2}$	1.729577	2.580640 $\times 10^{-5}$	—

Table B-5. Critical Constants and Molecular Weights of Gases

Gas	Molecular Weight	Critical Constants*	
		p_c, atm	T_c, °R
Air	28.95	37.2	238.4
Oxygen	32	49.7	277.8
Nitrogen	28.02	33.5	226.9
Carbon dioxide	44.01	73.0	547.7
Carbon monoxide	28.01	35.0	241.5
Hydrogen	2.02	12.8	59.9
Ethyl alcohol	46.0	63.1	929.3
Benzene	78.0	47.7	1011
Freon-12	120.92	39.6	692.4
Ammonia	17.03	111.5	730.0
Helium	4.00	2.26	9.47
Mercury vapor	200.61	>200	>3281.7
Methane	16.03	45.8	343.2

*Adapted by permission from N.A. Hall, *Thermodynamics of Fluid Flow*, Prentice-Hall, Englewood Cliffs, N.J., 1951.

Table B-6. Normal Total Emissivity of Various Surfaces*

Surface	T, °F	Emissivity ϵ
Metals and their oxides		
Aluminum:		
Highly polished plate, 98.3% pure	440–1070	0.039–0.057
Commercial sheet	212	0.09
Heavily oxidized	299–940	0.20–0.31
Brass:		
Highly polished:		
73.2% Cu, 26.7% Zn	476–674	0.028–0.031
62.4% Cu, 36.8% Zn, 0.4% Pb, 0.3% Al	494–710	0.033–0.037
82.9% Cu, 17.0% Zn	530	0.030
Hard-rolled, polished, but direction of polishing visible	70	0.038
Dull plate	120–660	0.22
Copper:		
Polished	242	0.023
	212	0.052
Plate, heated long time, covered with thick oxide layer	77	0.78
Gold, pure, highly polished	440–1160	0.018–0.035
Iron and steel (not including stainless):		
Steel, polished	212	0.066
Iron, polished	800–1880	0.14–0.38
Cast iron, newly turned	72	0.44
Cast iron, turned and heated	1620–1810	0.60–0.70
Mild steel	450–1950	0.20–0.32
Oxidized surfaces:		
Iron plate, pickled, then rusted red	68	0.61
Iron, dark-gray surface	212	0.31
Rough ingot iron	1700–2040	0.87–0.95
Sheet steel with strong, rough oxide layer	75	0.80
Lead:		
Unoxidized, 99.96% pure	260–440	0.057–0.075
Gray oxidized	75	0.28
Oxidized at 300 °F	390	0.63
Magnesium, magnesium oxide	530–1520	0.55–0.20
Molybdenum:		
Filament	1340–4700	0.096–0.202
Massive, polished	212	0.071
Monel metal, oxidized at 1110 °F	390–1110	0.41–0.46
Nickel:		
Polished	212	0.072
Nickel oxide	1200–2290	0.59–0.86
Nickel alloys:		
Copper nickel, polished	212	0.059
Nichrome wire, bright	120–1830	0.65–0.79
Nichrome wire, oxidized	120–930	0.95–0.98

*Abstracted by permission from H.C. Hottel in W.H. McAdams (ed.), *Heat Transmission*, 3d ed., pp. 472–478. Copyright 1954, McGraw-Hill Book Company.

Table B-6 (continued)

Surface	T, °F	Emissivity ϵ
Platinum; polished plate, pure	440–1160	0.054–0.104
Silver:		
Polished, pure	440–1160	0.020–0.032
Polished	100–700	0.022–0.031
Stainless steels:		
Polished	212	0.074
Type 301	450–1725	0.54–0.63
Tin, bright tinned iron	76	0.043 and 0.064
Tungsten, filament	6000	0.39
Zinc, galvanized sheet iron, fairly bright	82	0.23

Refractories, building materials, paints, and miscellaneous

Surface	T, °F	Emissivity ϵ
Alumina (85–99.5% Al_2O_3, 0–12% SiO_2, 0–1% Ge_2O_3); effect of mean grain size, μm		
10 μm		0.30–0.18
50 μm		0.39–0.28
100 μm		0.50–0.40
Asbestos, board	74	0.96
Brick:		
Red, rough, but no gross irregularities	70	0.93
Fireclay	1832	0.75
Carbon:		
T-carbon (Gebruder Siemens) 0.9% ash, started with emissivity of 0.72 at 260 °F but on heating changed to values given	260–1160	0.81–0.79
Filament	1900–2560	0.526
Rough plate	212–608	0.77
Lampblack, rough deposit	212–932	0.84–0.78
Concrete tiles	1832	0.63
Enamel, white fused, on iron	66	0.90
Glass:		
Smooth	72	0.94
Pyrex, lead, and soda	500–1000	0.95–0.85
Paints, lacquers, varnishes:		
Snow-white enamel varnish on rough iron plate	73	0.906
Black shiny lacquer, sprayed on iron	76	0.875
Black shiny shellac on tinned iron sheet	70	0.821
Black matte shellac	170–295	0.91
Black or white lacquer	100–200	0.80–0.95
Flat black lacquer	100–200	0.96–0.98
Porcelain, glazed	72	0.92
Quartz, rough, fused	70	0.93
Roofing paper	69	0.91
Rubber, hard, glossy plate	74	0.94
Water	32–212	0.95–0.963

Index

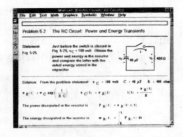

SCHAUM'S SOLVED PROBLEMS SERIES

- ■ Learn the best strategies for solving tough problems in step-by-step detail
- ■ Prepare effectively for exams and save time in doing homework problems
- ■ Use the indexes to quickly locate the types of problems you need the most help solving
- ■ Save these books for reference in other courses and even for your professional library

To order, please check the appropriate box(es) and complete the following coupon.

❑ **3000 SOLVED PROBLEMS IN BIOLOGY**
ORDER CODE 005022-8/**$16.95 406 pp.**

❑ **3000 SOLVED PROBLEMS IN CALCULUS**
ORDER CODE 041523-4/**$19.95 442 pp.**

❑ **3000 SOLVED PROBLEMS IN CHEMISTRY**
ORDER CODE 023684-4/**$20.95 624 pp.**

❑ **2500 SOLVED PROBLEMS IN COLLEGE ALGEBRA & TRIGONOMETRY**
ORDER CODE 055373-4/**$14.95 608 pp.**

❑ **2500 SOLVED PROBLEMS IN DIFFERENTIAL EQUATIONS**
ORDER CODE 007979-x/**$19.95 448 pp.**

❑ **2000 SOLVED PROBLEMS IN DISCRETE MATHEMATICS**
ORDER CODE 038031-7/**$16.95 412 pp.**

❑ **3000 SOLVED PROBLEMS IN ELECTRIC CIRCUITS**
ORDER CODE 045936-3/**$21.95 746 pp.**

❑ **2000 SOLVED PROBLEMS IN ELECTROMAGNETICS**
ORDER CODE 045902-9/**$18.95 480 pp.**

❑ **2000 SOLVED PROBLEMS IN ELECTRONICS**
ORDER CODE 010284-8/**$19.95 640 pp.**

❑ **2500 SOLVED PROBLEMS IN FLUID MECHANICS & HYDRAULICS**
ORDER CODE 019784-9/**$21.95 800 pp.**

❑ **1000 SOLVED PROBLEMS IN HEAT TRANSFER**
ORDER CODE 050204-8/**$19.95 750 pp.**

❑ **3000 SOLVED PROBLEMS IN LINEAR ALGEBRA**
ORDER CODE 038023-6/**$19.95 750 pp.**

❑ **2000 SOLVED PROBLEMS IN Mechanical Engineering THERMODYNAMICS**
ORDER CODE 037863-0/**$19.95 406 pp.**

❑ **2000 SOLVED PROBLEMS IN NUMERICAL ANALYSIS**
ORDER CODE 055233-9/**$20.95 704 pp.**

❑ **3000 SOLVED PROBLEMS IN ORGANIC CHEMISTRY**
ORDER CODE 056424-8/**$22.95 688 pp.**

❑ **2000 SOLVED PROBLEMS IN PHYSICAL CHEMISTRY**
ORDER CODE 041716-4/**$21.95 448 pp.**

❑ **3000 SOLVED PROBLEMS IN PHYSICS**
ORDER CODE 025734-5/**$20.95 752 pp.**

❑ **3000 SOLVED PROBLEMS IN PRECALCULUS**
ORDER CODE 055365-3/**$16.95 385 pp.**

❑ **800 SOLVED PROBLEMS IN VECTOR MECHANICS FOR ENGINEERS**
Vol I: STATICS
ORDER CODE 056582-1/**$20.95 800 pp.**

❑ **700 SOLVED PROBLEMS IN VECTOR MECHANICS FOR ENGINEERS**
Vol II: DYNAMICS
ORDER CODE 056687-9/**$20.95 672 pp.**

ASK FOR THE *S*CHAUM'S *S*OLVED *P*ROBLEMS *S*ERIES AT YOUR LOCAL BOOKSTORE
OR CHECK THE APPROPRIATE BOX(ES) ON THE PRECEDING PAGE
AND MAIL WITH THIS COUPON TO:

M*c*G*RAW*-H*ILL*, I*NC*.
ORDER PROCESSING S-1
PRINCETON ROAD
HIGHTSTOWN, NJ 08520

OR CALL
1-800-338-3987

NAME   (PLEASE PRINT LEGIBLY OR TYPE)

ADDRESS   (NO P.O. BOXES)

CITY   STATE   ZIP

ENCLOSED IS   ☐ A CHECK   ☐ MASTERCARD   ☐ VISA   ☐ AMEX   (✓ ONE)

ACCOUNT # _____   EXP. DATE _____

SIGNATURE _____

MAKE CHECKS PAYABLE TO MCGRAW-HILL, INC. PLEASE INCLUDE LOCAL SALES TAX AND $1.25 SHIPPING/HANDLING
PRICES SUBJECT TO CHANGE WITHOUT NOTICE AND MAY VARY OUTSIDE THE U.S.   FOR THIS
INFORMATION, WRITE TO THE ADDRESS ABOVE OR CALL THE 800 NUMBER.